Communications and Control Engineering

For further volumes:
http://www.springer.com/series/61

Huijun Gao · Xianwei Li

Robust Filtering for Uncertain Systems

A Parameter-Dependent Approach

Springer

Huijun Gao
Xianwei Li
Research Institute of Intelligent Control and Systems
Harbin Institute of Technology
Harbin
China

ISSN 0178-5354 ISSN 2197-7119 (electronic)
ISBN 978-3-319-05902-0 ISBN 978-3-319-05903-7 (eBook)
DOI 10.1007/978-3-319-05903-7
Springer Cham Heidelberg New York Dordrecht London

Library of Congress Control Number: 2014935227

Printed on acid-free paper

Springer is part of Springer Science+Business Media (www.springer.com)

To my parents, Qingxiang Gao
and Yunrong Yan, my wife, Lu Zhao,
and my son, Ying'ao Gao

— Huijun Gao

To my parents, Daiyuan Li
and Helan Chen, and my wife, Yue Gu

— Xianwei Li

Preface

Filtering is one of the basic problems in the fields of systems, control and signal processing. The goal of filtering is to estimate signals that are unmeasurable through processing the measured output signals. Since the development of Kalman's optimal filtering theory for stochastic systems and Luenberger's observer theory for deterministic systems, many scholars have devoted considerable efforts to the problem of filtering. Especially, the well-known Kalman filtering theory plays a significant role in various social and engineering fields such as aerospace, astronautics, industrial processes and economic and financial systems. The Kalman filtering theory is based on the availability of the precisely-known mathematical model of the studied plant and the assumption of strict Gaussian random processes or series. However, it is usually difficult to characterize the dynamics of the studied plant exactly by a mathematical model, inevitably leading to an error between the derived mathematical model and the practical plant; moreover, it is rare for practical external noises to completely satisfy the strong Gaussian assumption. The uncertainties existing in systems and signals would greatly degrade the performance of a traditional Kalman filter and even cause divergence. Hence, it is of practical meaning to research filtering theory for uncertain dynamical systems so as to improve the robustness of a filter against uncertainties.

In systems and control areas, study on uncertain systems has drawn much attention from many researchers for a long time. Robust control theory, originating from 1970s, generally solves analysis and synthesis problems of parametric uncertain systems based on the notion of quadratic stability. Since the late 1980s, this notion has also been gradually employed to solve robust filtering problems of uncertain systems, which results in a great number of quadratic approaches to robust filter design. However, these quadratic approaches have been well recognized to be conservative due to the the utilization of a common quadratic Lyapunov function for the entire uncertainty domain. In recent years, to reduce such conservatism and improve the practical applicability of robust filter design methods, parameter-dependent Lyapunov functions are introduced into robust filtering theory, and parameter-dependent approaches to robust filter design have draw a great deal of attention. The essential idea of the most popular parameter-dependent filtering results is to relax Lyapunov functions to be linearly parameter-dependent and meanwhile to fix some slack matrices, which, though relaxing the quadratic approaches, still has much limitation.

In view of the limitation of the existing parameter-dependent results, some new methods recently have been developed to derive a series of parameter-dependent approaches to robust filter design, including ours, which further release the previous restrictions on Lyapunov functions, showing great potential in conservatism reduction. This book systematically summarizes these recent developments of parameter-dependent filter design methods. Robust H_2 filtering, robust H_∞ filtering and robust energy-to-peak filtering are discussed in the book, where robust H_∞ filtering is employed as the main filtering scheme for various classes of complex uncertain dynamical systems including time-delay systems, two-dimensional systems and networked systems. Moreover, our latest work, some preliminary results on finite frequency H_∞ filtering, will also be involved at the end of the book to show a new filtering theme of future interest. Various examples are provided in the book to illustrate the effectiveness of the presented new parameter-dependent filtering results. All the results are presented in the framework of linear matrix inequality (LMI).

From the book, reader can find an overview of the latest advances in the robust filtering area and grasp the state-of-the-art methods on parameter-dependent filter design. This book can be used as a reference by researchers and engineers working in the areas related to control theory and engineering, and signal processing, and it is especially beneficial for graduate students interested in or focusing on robust filtering theory and its application. Courses like linear systems, modern control theory and robust control theory and basic mathematical background are prerequisite for reading the book. Those familiar with the LMI theory would possibly read the book more efficiently.

Harbin, China, July 2013

Huijun Gao
Xianwei Li

Acknowledgments

The authors would like to dedicate their gratitude to many colleagues and friends who have made direct and indirect contribution to this book. First appreciation is delivered to Prof. Tongwen Chen from University of Alberta, Canada, who offers the first author an opportunity to conduct his postdoctoral research from 2005 to 2007 at University of Alberta, where partial works of the book are completed under the cooperation with Prof. Chen. Special thanks go to Mr. Xiangyu Meng for his efforts and contribution to some pertinent works when he conducted his M.Eng. career under the supervision of the first author. The authors are also grateful to Prof. James Lam from The University of Hong Kong, Hong Kong, Prof. Zidong Wang from Brunel University, UK, Professer Peng Shi from The University of Adelaide, Australia, Prof. Okyay Kaynak from Bogazici University, Turkey, and Prof. Changhong Wang from Harbin Institute of Technology, China, for their long-lasting friendship and cooperation that have provided the authors great support, encouragement and inspiration in study and research. The second author would like to express his sincere thanks to the first author for his guidance and instruction in the second author's study and academic career. Undoubtedly, the authors' deepest gratitude is forever dedicated to their families for their unreservedly understanding, love and encouragement.

The writing of the book is financially supported in part by the National 973 Project under Grant 2009CB320600973, the National Natural Science Foundation of China under Grants 61333012, 61273201, 61329301, 60825303 and 61203035, and the Key Laboratory of Integrated Automation for the Process Industry, Northeast University.

Contents

Acronyms and Symbols

1(2)-D	One(Two)-Dimensional
ARE	Algebraic Riccati Equation
BRL	Bounded Real Lemma
EF	Entire Frequency
FF	Finite Frequency
GKYP	Generalized Kalman–Yakubovich–Popov
HF	High Frequency
LF	Low Frequency
LHS	Left Hand Side
LKF	Lyapunov–Krasovskii Functional
LMI	Linear Matrix Inequality
LTI	Linear Time-Invariant
MF	Middle Frequency
NCS	Networked Control System
NLMI	Nonlinear Matrix Inequality
RHS	Right Hand Side
ZOH	Zero-Order Hold
$\forall$	For all
$\in$	Belong(s) to
$\triangleq$	Be defined as
$\square$	End of proof
$\rightarrow$	Tends to
$\Rightarrow$ ($\Leftarrow$)	Imply or implies (be implied by)
$\Leftrightarrow$	Be equivalent to
max (min)	Maximum (minimum)
sup (inf)	Supremum (infimum)
s.t.	Subject to
$\mathbb{R}$	Set of all real numbers
$\mathbb{C}$	Set of all complex numbers
$\mathbb{N}$, $\mathbb{N}^+$, $\mathbb{N}_0^+$	Set of all integers, positive integers, non-negative integers
$\mathbb{R}^{n\times m}$	Set of all $n \times m$ real matrices
$\mathbb{C}^{n\times m}$	Set of all $n \times m$ complex matrices
$\mathbf{H}_n$	Set of all $n \times n$ Hermitian matrices

I_n	$n \times n$ identity matrix (the subscript is omitted when k can be determined from context)
A^{T}	Transpose of real matrix A
A^*	Conjugate transpose of complex matrix A
A^{-1}	Inverse of square matrix A
$A >, \geq 0$	Symmetric or Hermitian positive definite, semi-definite matrix A
$A <, \leq 0$	Symmetric or Hermitian negative definite, semi-definite matrix A
$\mathrm{diag}\{\cdots\}$	Diagonal-block matrix with $\mathrm{diag}\{A_1, \cdots, A_s\} \triangleq \begin{bmatrix} A_1 & & 0 \\ & \ddots & \\ 0 & & A_s \end{bmatrix}$
$\begin{bmatrix} X & Y \\ * & Z \end{bmatrix}$	$\begin{bmatrix} X & Y \\ * & Z \end{bmatrix} \triangleq \begin{bmatrix} X & Y \\ Y^* & Z \end{bmatrix}$
$\mathrm{Tr}[A]$	Trace of square matrix A
rank (A)	Rank of matrix A
$\mathrm{sym}\{A\}$	$A^{\mathrm{T}}+A$ for real matrix A or A^*+A for complex matrix A
$\sigma_{\max}[A]$	Maximum singular value of matrix A
$A \otimes B$	Kronecker product of matrices A and B
$\mathcal{N}_X$	Matrix with columns being a basis of the nullspace of matrix X
$L_2[0, +\infty)$	Set of square integrable functions on $[0, +\infty)$
$l_2[0, +\infty)$	Set of square summable series on $\{0, 1, 2, \ldots\}$
$\Vert f\Vert_{L_2}$	$\Vert f(t)\Vert_{L_2} \triangleq \left[\int_0^{+\infty} \left(f(t)^{\mathrm{T}} f(t)\right) \mathrm{d}t\right]^{1/2}$
$\Vert f\Vert_{l_2}$	$\Vert f(t)\Vert_{l_2} \triangleq \left[\sum_0^{+\infty} f(t)^{\mathrm{T}} f(t) \mathrm{d}t\right]^{1/2}$ for 1-D signals and $\Vert f(t,s)\Vert_{l_2} \triangleq \left[\sum_{t=0}^{+\infty} \sum_{s=0}^{+\infty} \left(f(t,s)^{\mathrm{T}} f(t,s)\right)\right]^{1/2}$ for 2-D signals
$\Vert f\Vert_{L_\infty}$	$\Vert f(t)\Vert_{L_\infty} \triangleq \sup_{t \in [0,+\infty)} \left(f(t)^{\mathrm{T}} f(t)\right)^{1/2}$
$\Vert f\Vert_{l_\infty}$	$\Vert f(t)\Vert_{l_\infty} \triangleq \sup_{t \in \{0,1,2,\ldots\}} \left(f(t)^{\mathrm{T}} f(t)\right)^{1/2}$
$\Vert\cdot\Vert_2$	H_2 norm of a transfer function
$\Vert\cdot\Vert_\infty$	H_∞ norm of a transfer function
$\mathbf{E}[\cdot]$	Mathematical expectation of a random signal
$\upsilon(G, \Theta)$	For $G \in \mathbb{C}^{n \times m}$, $\Theta \in \mathbf{H}_{n+m}$, a function $\upsilon : \mathbb{C}^{n \times m} \times \mathbf{H}_{n-m} \rightarrow \mathbf{H}_m$ is defined by $\upsilon(G, \Theta) \triangleq \begin{bmatrix} G \\ I_m \end{bmatrix}^* \Theta \begin{bmatrix} G \\ I_m \end{bmatrix}$

Chapter 1
Introduction

1.1 Filtering

Filtering is one of the most fundamental problems in various modern science and engineering fields, such as control systems, signal processing, and communication networks. From the signal point of view, useful signals in the process of transmission and measurement are inevitably contaminated by unwanted signals, namely noise, causing the discrepancy between the obtained signals and the desired signals. It is thus naturally necessary to expect to "minimize the inaccuracies caused by the presence of this noise" and "to make the control close to that desired" [1]. Filtering is rightly aimed for these purposes. By filtering, unwanted signals are "eliminated" while useful signals are retained for other use.

Actually, it has been a long history since humans used the notion of "filtering", from the pristine water purification to the modern ultraviolet filter on a camera [1]. In the relatively earlier modern theory of signal filtering, filters are designed and fixed, in the frequency domain, to amplify and/or attenuate some frequency bands. These frequency-domain results, such as Butterworth filters and Chebyshev filters, have grown up to be classical methods for filter design in signal processing theory [2]. An initiative hypothesis made to ensure the effectiveness of the classical frequency-domain approaches is that useful signals and unwanted noise reside in different frequency bands. Since 1940s, statistical ideas have been gradually applied to filtering problems independently by Wiener [3] and Kolmogorov [4], which greatly promote the development of signal processing theory in the time-domain direction. Instead of focusing on the difference of frequency properties, the statistical approaches, Wiener filtering, are concerned about the differences of statistical properties that are possessed by the useful signals and the unwanted noise.

From the perspective of the time-domain, Wiener–Kolmogorov filtering theory is not so adequate because of the assumption that the involved signal processes (including useful signals and unwanted noise) are stationary, i.e., the statistical properties of processes do not vary with time. For Wiener–Kolmogorov filtering, "there is, thus, a conceptional link with classical filtering", as stated in [1]. To overcome the

H. Gao and X. Li, *Robust Filtering for Uncertain Systems*,
Communications and Control Engineering, DOI: 10.1007/978-3-319-05903-7_1,

drawbacks of Wiener–Kolmogorov filtering theory in coping with the case that the stationarity assumption is not satisfied, Kalman and Bucy, in the early 1960s, developed a filtering theory that did not require the stationarity assumption [5, 6], which now has been thought highly of a milestone in filtering theory with the well-known names, *Kalman filtering theory* for the discrete-time case and *Kalman-Bucy filtering theory* for the continuous-time case. The basic Kalman filter can provide the optimal unbiased estimate of the unknown state of a linear dynamical discrete-time systems in the sense of minimum error variance. More importantly for engineering applications, the Kalman filter is presented in terms of a finite dimensional recursive algorithm that can be conveniently implemented by digital signal processors. Thus, for these two aspects, the Kalman filter up to now has been one of the most widely applied filtering methods, and the related practical applications include radar tracking [7], satellite navigation [8, 9], fault detection [10], automobile industry [11], economics prediction [12, 13], and so on.

1.2 Robust Filtering

1.2.1 Uncertainty in Systems

One of the basic assumptions for the standard Kalman filter is that a system model of the plant that the filter is aimed for is precisely known. However, in many practical engineering applications such as aerospace, astronautics, and industrial processes, it is usually difficult to characterize the dynamics of the considered plant exactly by a mathematical model, which renders an inevitable error to exist between the derived mathematical model and the practical object. The error comes from, for instance, that the original model needs to be simplified due to excessive complexity or high order, although it may exactly describe the practical plant, or that aging, drifting, failure, and breakdown of some components cause the change of the properties of the practical plant.

Since the standard Kalman filter relies greatly on the availability of a precise system model of the plant and does not consider uncertainty, the optimality of the Kalman filter is hardly guaranteed in the sense of error variance when the "nominal" system model is subject to modeling uncertainty. A known, worse fact is that the standard Kalman filter based on the "nominal" system model will usually deteriorate significantly when being applied to the practical plant that has uncertainty compared with the "nominal" system model. In other words, the standard Kalman filter usually does not have enough robustness against parametric uncertainty [14–16]. Hence, it is of practical importance to study robust filtering theory to generalize the standard Kalman filtering method to cope with the allowed modeling uncertainty. It is thus up to now not surprising to find numerous articles addressing robust Kalman filtering (i.e., robust H_2 filtering) [17–34]; and for monographs of related topics, see, e.g., [7, 35].

1.2.2 Uncertainty in Signals

Another basic assumption for the standard Kalman filter is that the exogenous noise inputs have known statistical information and moreover the statistical property must comply with the strict Gaussian white noise processes or series. However, many practical filtering problems cannot completely satisfy this requirement [1, 36]. Moreover, it has been recognized that the Kalman filter is also sensitive to the statistical property of noise [37]. Hence, once the statistical property of the noise is unavailable and/or does not satisfy the assumption of Gaussian white noise process or series, the optimal performance of the Kalman filter cannot be guaranteed anymore, which motivates the development of filtering approaches for non-Gaussian noise inputs or noise with unknown statistical property. Since these approaches do not require the statistical information of the noise, the designed filters are naturally less sensitive to the variation of the statistical property of noise than Kalman filtering. For systems disturbed by non-Gaussian noise or noise with unknown or uncertain statistical property, three schemes for filter design can be generally considered.

1.2.2.1 H_∞ Filtering

The first scheme for non-Gaussian filtering is H_∞ filtering. The objective of H_∞ filtering is to minimize the H_∞ norm of the filtering error system from noise inputs to filtering errors. H_∞ filtering assumes that the noise inputs are energy-bounded signals rather than Gaussian white noise; thus, the design of H_∞ filters do not need any statistical information of the noise signals, and the performance of H_∞ filters is also more robust against uncertainty in the statistical property of noise.

H_∞ filtering was first introduced by Grimble et al. for the scalar system case [38, 39] and later extended to the multivariable system case in [40–43]. Shaked [44] also proposed two approaches to solving the standard H_∞ filtering problem for the multivariable system case. For a brief survey of early results on H_∞ filtering, please refer to [36]. So far, four main approaches to H_∞ filtering can be found:

- Polynomial approach [38–43];
- Interpolation approach [45];
- Riccati equation approach [21, 37, 44, 46–57]; and
- Linear matrix inequality (LMI) approach [58–69].

The early results developed by Grimble et al. in [38–41] were derived by using a polynomial method in control theory [70, 71]. In Reference [45], Fu employed an interpolation approach to H_∞ filtering and exposed the relationship between the optimal H_∞ estimation problem and the open loop transfer recovery problem. The polynomial approach and the interpolation approach both are frequency-domain approaches. One of the merits of these two frequency-domain approaches is that frequency weighting on filtering signals does not increase the system dimension [45, 61].

Due to the dual relationship between H_∞ control theory and H_∞ filtering theory, the use of the Riccati equation approach to H_∞ filtering became popular in the early stage and middle of 1990s, when state-space solving algorithms for the H_∞ control problem were undergoing great progress [72–75]. The forms of these Riccati equation results include algebraic Riccati equation (ARE) [21, 44, 47–51, 53], differential Riccati equation [37, 46, 55, 56], and recursive Riccati equation [52, 57]. It should be noticed that, among the results mentioned above, the differential or discrete-time minimax game theory was employed in [53–57], as well as [37, 42], to solve the corresponding H_∞ filtering problem. In the estimation setting, the game theory views the H_∞-optimization problem as a conflicting game where the objectives of the external noise and the estimator are, respectively, maximizing and minimizing the estimation error, and provides a novel insight into the stochastic interpretation of H_∞ filtering [53, 56].

From the late 1990s up to now, the LMI technique has grown up to the main tool for studying the H_∞ filtering problem, attributed to the emergence of the effective numerical solving algorithms for LMIs [76, 77] and the corresponding LMI Control Toolbox in MATLAB®[1] [78]. LMI approaches to solving the H_∞ filtering problems of both continuous- and discrete-time systems appeared first in 1997 in [58, 59, 61], where [61] is mainly concerned with robust H_∞ filtering with uncertainties described by integral quadratic constraints (IQCs). Later on, by virtue of the LMI technique, Geromel and de Oliveira et al. [62–65] and Jin et al. [69] studied the robust H_∞ filtering problems with polytopic uncertainties; nearly at the same time, Palhares and Peres [66–68] applied the LMI tool to solve the H_∞ filtering problem with pole assignment. In recent years, the LMI technique has been utilized to design H_∞ filters for complex dynamical systems, especially for time-delay systems [79–87].

1.2.2.2 Energy-to-Peak Filtering

The second filtering scheme for non-Gaussian noise is energy-to-peak filtering (i.e., $L_2 - L_\infty$ filtering for the continuous-time case and $l_2 - l_\infty$ filtering for the discrete-time case), which is also called generalized H_2 filtering [88]. The performance index of the energy-to-peak gain of systems was first introduced by Wilson in 1989 [89] and then the control problem in the energy-to-peak sense, referred to as *Generalized H_2 Control Problem*, was investigated by Rotea [90]. Although the noise is also assumed to have bounded energy, similar to the case of H_∞ filtering, the expectation of energy-to-peak filtering is to suppress the peak value of the filtering error, different from the case of H_∞ filtering. Hence, it can be regarded as a deterministic formulation of the Kalman filter [59]. The energy-to-peak filtering problem was first solved by Grigoriadis and Watson in [59, 60], where necessary and sufficient conditions in terms of LMI together with a matrix rank constraint are established for the existence of energy-to-peak filters for both exactly known continuous- and discrete-

[1] MATLAB® is a registered trademark of The MathWorks, Inc.

time systems. Robust energy-to-peak filtering and mixed $L_2 - L_\infty/H_\infty$ filtering for linear systems with polytopic uncertainties were addressed, respectively, in [91, 92], while energy-to-peak filtering for systems with repeated scalar nonlinearities was researched in [88]. All the energy-to-peak filtering results mentioned above are presented in terms of LMI.

1.2.2.3 Peak-to-Peak Filtering

In the above two types of filtering schemes, the noise is both assumed to have bounded energy, although their objectives are different. Another filtering scheme for non-Gaussian noise is peak-to-peak filtering (also termed as L_1 or induced L_∞ filtering for the continuous-time case, and l_1 or induced l_∞ filtering for the discrete-time case). In some situations, the peak value is a very natural measure of the signal size, and the objective of peak-to-peak filtering is to minimize the peak value-induced gain of the filtering error system. Although the optimal L_1 control problem seems to be first formulated in [93] and then the L_1 or l_1 control problem has just been intensively investigated in [94–96], filtering problems in the peak-to-peak sense have been concerned much earlier than the control problems [97, 98]. In [99, 100], Milanese and Tempo addressed the l_1 filtering problem over a finite horizon and presented non-recursive solutions; in [101, 102], Voulgaris studied the l_1 filtering problem over the infinite horizon by an l_1-model matching approach in [94]; since 1996, along with the increasing interest in LMI technique in control theory, Abedor and Nagpal et al. began to apply the LMI tool to solve the peak-to-peak filtering problem [103, 104].

1.2.3 Quadratic Robust Filter Design

Due to the existence of uncertainty in systems and/or signals, robustness has been one of the main concerns in the filtering theory for a long time, and the problem of robust filtering has been considerably studied. The methods for dealing with the uncertainty in signals have been discussed above. With respect to parametric uncertainty in systems, two types of parametric uncertainty are often considered: *norm-bounded uncertainty* [105] and *polytopic uncertainty* [106]. Many examples show that the latter type is a more natural manner for characterizing structured uncertainty than the former [63, 107, 108]. In robust control theory, to handle these two types of uncertainty, the most important technique is the notion of *quadratic stability* [109–111].

Definition 1.1 *(Quadratic Stability)* Consider an uncertain autonomous linear system given by

$$\dot{x}(t) = A(\lambda)x(t); \ \lambda \in \Gamma, \tag{1.1}$$

where Γ represents the parametric uncertainty domain. The system in (1.1) is said to be quadratically stable, if there exists a positive definite matrix P such that

$$A(\lambda)^{\mathrm{T}} P + P A(\lambda) < 0, \forall \lambda \in \Gamma. \tag{1.2}$$

The term "quadratic" comes from the fact that the condition in (1.2) guarantees the existence of a Lyapunov function of the quadratic form as

$$V(t) = x(t)^{\mathrm{T}} P x(t)$$

such that the robustly asymptotically stability of the uncertain system in (1.1) for all $\lambda \in \Gamma$ can be proven by applying the Lyapunov stability theory. Correspondingly, the stability condition in (1.2) is referred to as a quadratic stability condition. Obviously, the notion of quadratic stability implies the robustly asymptotical stability of the uncertain system in (1.1). Hence, for a specific type of uncertainty, the quadratic stability condition in (1.1) can be further developed to solve other robustness analysis and synthesis problems. In particular, due to the application of a simple quadratic Lyapunov function, quadratic stability simplifies analysis and synthesis of uncertain systems significantly; a second merit of this notion is that time-varying uncertainty can also be analyzed in the same framework. Please refer to [74] and [112] for examples of applying this concept to robust controller synthesis with norm-bounded uncertainty and polytopic uncertainty, respectively.

What is more, as is well known in the linear control theory, the quadratic Lyapunov function plays a fundamental role in analyzing input–output performances when external excitations and output signals are considered. Hence, the notion of quadratic stability can be extended to deal with robust input–output performances of uncertain systems, resulting in a series of robust control theory results, see the comprehensive survey and summary book by Boyd et al. [77]. Definition 1.1 is specified for continuous-time systems, and a similar definition can also be made for discrete-time systems. In this chapter, we refer to a robust analysis and synthesis approach that is built on the notion of quadratic stability or its extensions to input–output performances as a *quadratic approach*. Especially, from the late 1980s throughout the entire 1990s, this notion addressing robustness issues also has been extensively adopted to solve robust filtering problems including robust Kalman filtering, robust H_∞ filtering, and robust energy-to-peak filtering, leading to numerous results on quadratic approaches to robust filter design. Table 1.1 gives a brief summary of the literature related to various quadratic approaches to different robust filtering schemes under different types of parametric uncertainty.

Table 1.1 Summary of quadratic robust filtering results

Filtering	Uncertainty	Literature
kalman (H_2)	Bounded norm	[17–27, 29, 30, 34]
		[28, 31, 32, 113]
	Polytope	[62–65, 107]
H_∞	Bounded norm	[37, 48–52, 87]
	Polytope	[62–65, 67–69, 80, 81, 83, 87, 92]
Energy-to-peak	Bounded norm	[114]
	Polytope	[85, 91, 92, 115]

1.3 Parameter-Dependent Robust Filtering

1.3.1 Significance of Parameter Dependence

Although quadratic stability simplifies analysis and synthesis of uncertain systems significantly, this notion requires a *single* quadratic Lyapunov matrix satisfying the stability requirement over the entire parametric uncertainty domain, implying that the aforementioned quadratic approaches to robust filter design inevitably have significant conservatism [116]. To intuitively expose this conservatism, let us consider a simple second-order continuous-time system, as in [117], given by

$$\dot{x}(t) = A(\lambda)x(t) = \begin{bmatrix} 0 & 1 \\ -1+\lambda & -1-\lambda \end{bmatrix} x(t), \tag{1.3}$$

where λ belongs to $\left[-\bar{\lambda}, \bar{\lambda}\right]$. Matrix $A(\lambda)$ with uncertain parameter λ can be denoted by a polytope consisting of two vertices:

$$A(\lambda) = \lambda_1 A_1 + \lambda_2 A_2;\ A_1 = \begin{bmatrix} 0 & 1 \\ -1-\bar{\lambda} & -1+\bar{\lambda} \end{bmatrix},\ A_2 = \begin{bmatrix} 0 & 1 \\ -1+\bar{\lambda} & -1-\bar{\lambda} \end{bmatrix},$$

with $\lambda_1 = \lambda$ and $\lambda_2 = 1 - \lambda$. It is readily known that the maximum stability bound of the time-invariant uncertain scalar λ is $\bar{\lambda} = 1$. The quadratic stability condition in (1.2) can be written as

$$A(\lambda)^{\mathrm{T}} P + PA(\lambda) = (\lambda_1 A_1 + \lambda_2 A_2)^{\mathrm{T}} P + P(\lambda_1 A_1 + \lambda_2 A_2) < 0. \tag{1.4}$$

Note that the inequality in (1.4) is affine with respect to the uncertain but bounded parameters λ_1 and λ_2. Hence, the inequality in (1.4) holds for all $\lambda \in [-\bar{\lambda}, \bar{\lambda}]$ if and only if the inequality holds for λ_1 and λ_2 taking their individual boundary value. Correspondingly, although the inequality in (1.4) has an infinitely dimensional nature, it can be characterized by the requirement that two LMIs

$$A_1^{\mathrm{T}} P + P A_1 < 0, \ A_2^{\mathrm{T}} P + P A_2 < 0 \tag{1.5}$$

hold for a common real-valued positive definite matrix P. By testing the feasibility of the quadratic stability conditions in (1.5), it is found that the maximum stability bound of λ is $\bar{\lambda} = 0.681$ for system (1.3), which is remarkably smaller than the actual exact stability bound $\bar{\lambda} = 1$. Hence, this simple example shows that great conservatism is inherent to the notion of quadratic stability.

As aforementioned, condition (1.5) actually implies that the robust stability of system (1.3) can be guaranteed by a quadratic Lyapunov function $V(t)$ satisfying

$$V(t) = x(t)^{\mathrm{T}} P x(t) > 0$$

and

$$\dot{V}(t) = x(t)^{\mathrm{T}} (A(\lambda)^{\mathrm{T}} P + P A(\lambda)) x(t) < 0$$

for all $\lambda \in [-\bar{\lambda}, \bar{\lambda}]$. More and more scholars have realized that the intrinsic conservatism in condition (1.5) results from the fact that matrix P is fixed over the whole uncertainty domain, and that a key to conservatism reduction is to adopt a *parameter-dependent* Lyapunov function, e.g.,

$$V(t) = x(t)^{\mathrm{T}} P(\lambda) x(t),$$

to study uncertain systems. Obviously, parameter-dependent matrix $P(\lambda)$ is more general than a single matrix P, providing more freedom to the derived results and reducing the conservatism [118–120]. Hence, study of parameter-dependent approaches to robust filtering can not only prosper the development of robust filtering theory, but also reduce the conservatism of quadratic approaches and further improve the applicability of robust filtering theory in the practical industrial fields.

1.3.2 Parameter-Dependent Robust Filter Design

From the middle of 1990s, Haddad, Gahinet and Feron et al. started to apply parameter-dependent Lyapunov functions to stability analysis of uncertain systems [118–120]. Hereafter, many scholars devote themselves to the study of parameter-dependent stability theory of uncertain systems, and a series of creative results have been proposed. In this direction, some main scientific teams and their representative work include

- Geromel and de Oliveira et al. [121–123];
- Peaucelle, Henrion and Arzelier et al. [124–127];
- Apkarian and Tuan et al. [128–131];
- Shaked, Xie and Fridman et al. [117, 132–134]; and
- Daafouz et al. [135, 136].

Most of the above results are based on the idea of decoupling the product terms of Lyapunov matrix and system matrix, developed by de Oliveira et al. in [121]. Exactly speaking, for an uncertain discrete-time autonomous system

$$x(k+1) = A(\lambda)x(k);\ A(\lambda) = \sum_{i=1}^{s} \lambda_i A_i,\ \sum_{i=1}^{s} \lambda_i = 1,\ \lambda_i \geq 0, \tag{1.6}$$

by introducing an extra slack matrix $W(\lambda)$ into the standard Lyapunov stability condition in terms of parameter-dependent LMI

$$A(\lambda)^{\mathrm{T}} P(\lambda) A(\lambda) - P(\lambda) < 0;\ P(\lambda) = P(\lambda)^{\mathrm{T}} > 0,$$

an equivalent one can be obtained as follows:

$$\begin{bmatrix} P(\lambda) - W(\lambda) - W(\lambda)^{\mathrm{T}} & W(\lambda)A(\lambda) \\ A(\lambda)^{\mathrm{T}} W(\lambda)^{\mathrm{T}} & -P(\lambda) \end{bmatrix} < 0, \tag{1.7}$$

which, by considering the polytopic uncertainty of $A(\lambda)$, gives the following parameter-dependent Lyapunov robust stability condition:

$$\begin{bmatrix} P_i - W - W^{\mathrm{T}} & W A_i \\ A_i^{\mathrm{T}} W^{\mathrm{T}} & -P_i \end{bmatrix} < 0,\ (i = 1, 2, \ldots, s) \tag{1.8}$$

with $P_i = P_i^{\mathrm{T}} > 0$ and W to be determined. When using (1.8) to analyze the robust stability of system (1.6), one actually adopts a linearly parameter-dependent Lyapunov function as

$$V(k) = x(k)^{\mathrm{T}} \left(\sum_{i=1}^{s} \lambda_i P_i \right) x(k), \tag{1.9}$$

which is obviously different from the quadratic Lyapunov function

$$V(k) = x(k)^{\mathrm{T}} P x(k). \tag{1.10}$$

Due to the fact that $V(k)$ in (1.9) depends on uncertain parameter λ, condition (1.8) is thus called a *parameter-dependent* stability condition. The reason of introducing matrix $W(\lambda)$ is based on the following considerations. First, it is expected to use a finite number of inequalities, the same number of the vertices of the polytopic domain, to guarantee the robust stability, and second, Lyapunov matrix P should be parameter dependent. The two specifications cannot be directly realized by using the quadratic stability condition for the system in (1.6):

$$A_i^{\mathrm{T}} P A_i - P < 0 \text{ or } \begin{bmatrix} -P & P A_i \\ A_i^{\mathrm{T}} P & -P \end{bmatrix} < 0,\ (i = 1, \ldots, s).$$

Table 1.2 Summary of parameter-dependent robust filtering results

Filtering	Literature
Kalman (H_2)	[116, 130–133, 138–141]
H_∞	[131, 133, 138, 139, 141–148]
Energy-to-peak	[142, 149, 150]

However, by fixing $W(\lambda)$ in (1.7) to be independent of λ, the derived conditions in (1.8) effectively incorporate these two considerations for robust stability analysis.

The crucial technique in the derivation of the parameter-dependent condition (1.8) is to eliminate the product terms between Lyapunov matrix $P(\lambda)$ and system matrix $A(\lambda)$ such that it is ready to choose a linearly parameter-dependent Lyapunov function as in (1.9). Based on this method for robust stability analysis, many researchers also have made efforts to apply the parameter-dependent idea to robust filtering, and many scattered parameter-dependent results upon robust filtering can be found in the literature. Table 1.2 gives a brief summary of relatively early results of parameter-dependent robust filtering in this framework; for parameter-dependent results upon robust H_2 filtering, one can refer to the review paper [137]. From these literature, it can be found that parameter-dependent approaches to robust filtering have significantly improved the corresponding quadratic ones; thus, study of parameter-dependent approaches to robust filtering is a hopeful direction and has been a hotspot in the field of robust filtering.

1.3.3 Motivation

Although, compared with quadratic approaches, the conservatism of the mentioned parameter-dependent results upon robust filtering has been greatly reduced, these results still suffer from some drawbacks with regard to conservatism from the following two aspects:

1. The extra slack matrix is fixed over the entire uncertainty domain. Most of these parameter-dependent filtering results are based on the idea of [121] or its extensions in [117, 124, 129]. For stability conditions (1.8), although the parameter dependence has been realized for Lyapunov matrix $P(\lambda)$, an additional constraint still has been imposed as $W(\lambda) = W$, which leads to conservatism since $W(\lambda)$ also should be parameter dependent.
2. Lyapunov matrix is of the *linear* parameter-dependence form. Although the linear parameter-dependence form of $P(\lambda)$ simplifies the condition (1.7) into a finite number of LMIs (1.8) (equal to the number of the vertices of the convex polytopic uncertainty domain) when $W(\lambda) = W$, this dependence form would be too restrictive. Moreover, when $W(\lambda)$ is also parameter dependent, the linearly parameter-dependent robust stability in (1.8) cannot be obtained anymore.

The above two points actually reveal the sources of conservatism of the condition (1.8) as well as the parameter-dependent filtering results that are based on the stability conditions in [117, 121, 124, 129]. After the well-known work in [121], robust stability conditions based on the *polynomially parameter-dependent* Lyapunov function were proposed in [151–154], which directly extend the quadratic stability results in a more general way such that polynomially parameter-dependent conditions can be directly obtained. The combination of the polynomially parameter-dependent idea and the essential idea in [121] provides a promising direction of the development of parameter-dependent approaches to robust filter design. In recent years, some researchers, including the authors of this book, have made attempt to develop new linearly parameter-dependent approaches [155–164] and moreover to exploit polynomial parameter-dependent approaches to robust filter design [165–169], which further improve the early parameter-dependent results in Table 1.2 that directly use the idea in [121].

Through the foregoing review of the development of robust filtering theory, it can be seen that, although plentiful results upon parameter-dependent robust filtering have been reported, these results, scattering only in numerous conference or journal papers, have not been summarized to form a set of systematic theory, and furthermore there is no preceding academic monograph mainly focusing on the topic of parameter-dependent robust filtering, which are the motivations of this book. In this book, by the LMI technique, we are to systematically discuss the problem of robust filtering for uncertain dynamical systems in the parameter-dependent framework. In addition, we will develop in Chaps. 7 and 8 some results on the robust estimation problem with limited communication capacity and the H_∞ filtering problem with finite frequency (FF) specifications, giving new perspectives of robust filtering.

1.4 Organization of the Book

This book is organized as nine chapters, including three parts of technical results (totally seven chapters). Part I includes Chaps. 2 and 3 that address quadratic and parameter-dependent approaches to robust filter design; Part II includes Chaps. 4–6 concerning uncertain time-delay systems and two-dimensional (2-D) systems; and Part III consists of Chaps. 7 and 8 providing some new problems in robust filtering. An outline of each chapter is as follows:

Chapter 1 briefly introduces the development of filtering theory, especially the quadratic approaches and parameter-dependent approaches to robust filtering, and also states the motivation of this book. Some preliminary facts and results are still provided in this chapter for later use.

Chapter 2 considers the classic quadratic approaches to robust filter design for linear polytopic uncertain systems. A linearization procedure is employed to transform the conditions for filter design into LMIs. First, necessary and sufficient conditions for the existence of H_2, H_∞, and energy-to-peak filters, respectively, are derived for nominal systems, and then directly extended to obtain quadratic robust filter design methods for polytopic uncertain systems.

Chapter 3 focuses on improving the quadratic results presented in Chap. 2 by the parameter-dependent idea. By the Projection Lemma, slack matrices are first introduced to eliminate the product terms between Lyapunov matrices and systems matrices such that Lyapunov matrices can be chosen to be parameter dependent in the design conditions. Then, linearly parameter-dependent approaches and polynomially parameter-dependent approaches to robust H_2 and H_∞ filter design are, respectively, obtained by choosing the corresponding proper parameter-dependence form for the matrix variables. The linearly parameter-dependent approaches are derived by fixing the slack matrices and specifying the Lyapunov matrices to be linearly dependent on uncertain parameters, while polynomially parameter-dependent approaches are attained by choosing the Lyapunov matrices and part of slack matrices to be of homogeneous polynomials of the uncertain parameters.

Chapters 4 and 5 investigate robust H_∞ filtering problem of linear systems with polytopic parametric uncertainty and a time-varying delay. Chapter 4 discusses the continuous-time case while Chap. 5 deals with the discrete-time case. By virtue of the free weighting matrix technique, delay- and parameter-dependent conditions are first obtained for filtering performance analysis. For the discrete-time case, the Projection Lemma is then applied to eliminate the product terms between Lyapunov matrices and systems matrices. Polynomially parameter-dependent conditions in terms of LMIs are finally derived for the existence of robust H_∞ filters for the continuous- and discrete-time cases, respectively. It is shown, by some numerical examples, that the obtained polynomially parameter-dependent results are much less conservative than some existing results.

Chapter 6 is concerned with robust filtering for 2-D systems with polytopic uncertainty. By the polynomial parameter-dependent methods, new approaches to robust filter design for uncertain 2-D systems of the Fornasini–Marchesini (FM) model and the Roesser model are obtained, respectively. The proposed conditions cover the results in the quadratic framework and the linearly parameter-dependent framework.

Chapter 7 is devoted to robust H_∞ estimation for networked control systems (NCSs) with polytopic uncertain parameters and subject to limited communication capacity. First, an NCS model is established by a delay system approach. The model simultaneously considers the three typical phenomena in a network environment, i.e., measurement quantization, signal transmission delay, and data packet dropout. Then, LMI-based conditions are formulated for the existence of admissible filters for the nominal case. In the next, these conditions are extended to the robust filtering case, respectively, by the quadratic stability idea and the linearly parameter-dependent idea, leading to two filter design methods with different degrees of conservativeness and computational complexity. Two illustrative examples are finally provided to show the effectiveness of the proposed H_∞ filter design methods.

Chapter 8 deals with the FF H_∞ filtering problem for continuous time-delay systems. By the Projection Lemma and the generalized Kalman-Yakubovich-Popov (GKYP) Lemma, LMI-based conditions are established for analyzing and designing FF H_∞ filters for continuous-time state-delayed systems. The proposed methods can design filters that guarantee an H_∞ attenuation level over a precise low/middle/high frequency (LF/MF/HF) range, distinct from the existing standard H_∞ filtering methods. Moreover, two methods are presented to prove the analysis condition from a time-domain point and a frequency-domain point of views, respectively. Both delay-dependent and delay-independent conditions are proposed. Finally, two numerical examples are presented to demonstrate the effectiveness and advantages of the design methods.

Chapter 9 provides some concluding remarks on the results of the book and a summary of the algorithms presented in the book.

1.5 Preliminaries

In this section, we present several useful preliminary lemmas for later use.

1.5.1 Schur Complement

The following result is the well-known Schur complement.

Lemma 1.1 ([77]) *Given matrices $S_1 = S_1^{\mathrm{T}} \in \mathbb{R}^{n\times n}$, $S_2 \in \mathbb{R}^{n\times m}$ and $S = S_3^{\mathrm{T}} \in \mathbb{R}^{m\times m}$, the following conditions are equivalent:*

1. $\begin{bmatrix} S_1 & S_2 \\ S_2^{\mathrm{T}} & S_3 \end{bmatrix} < 0$;
2. $S_1 < 0,\ S_3 - S_2^{\mathrm{T}} S_1^{-1} S_2 < 0$;
3. $S_3 < 0,\ S_1 - S_2 S_3^{-1} S_2^{\mathrm{T}} < 0$.

Schur complement plays an essential role in transforming a nonlinear matrix inequality (NLMI) into an LMI. In control or system theory, many conditions are initially expressed in terms of NLMI like the second or third item. By Schur complement, these NLMI conditions can be equivalently written into an LMI or LMIs as in the first item.

1.5.2 Elimination Lemmas

Lemma 1.2 **(Projection Lemma** [77, 170]) *Let appropriately dimensioned matrices X, Z, Φ be given. There exists a matrix Y satisfying*

$$\Phi + X^{\mathrm{T}} Y Z + Z^{\mathrm{T}} Y^{\mathrm{T}} X < 0$$

if and only if the following projection inequalities hold:

$$\mathcal{N}_X^{\mathrm{T}} \Phi \mathcal{N}_X < 0,\ \mathcal{N}_Z^{\mathrm{T}} \Phi \mathcal{N}_Z < 0,$$

where $\mathcal{N}_X$ and $\mathcal{N}_Z$ are any matrices whose columns form the bases of the null spaces of X and Z, respectively.

Lemma 1.3 ([77]) *Let X be a partitioned, real, and symmetric matrix. The LMI*

$$\begin{bmatrix} X_1 & X_2 - Y \\ X_2^{\mathrm{T}} - Y^{\mathrm{T}} & X_3 \end{bmatrix} > 0$$

has a solution with respect to Y if and only if $X_1 = X_1^{\mathrm{T}} > 0$ and $X_3 = X_3^{\mathrm{T}} > 0$. Moreover, if the latter conditions are fulfilled, then a possible solution to the former is $Y = X_2$.

Lemma 1.4 ([171]) *Given appropriately dimensioned matrices $\Sigma_1 = \Sigma_1^{\mathrm{T}}$, Σ_2 and Σ_3, then*

$$\Sigma_1 + \Sigma_3 \Delta(t) \Sigma_2 + \Sigma_2^{\mathrm{T}} \Delta(t)^{\mathrm{T}} \Sigma_3^{\mathrm{T}} < 0 \tag{1.11}$$

holds for all $\Delta(t)$ satisfying $\Delta(t)^{\mathrm{T}} \Delta(t) \leq I$ if and only if for some $\epsilon > 0$

$$\Sigma_1 + \epsilon^{-1} \Sigma_3 \Sigma_3^{\mathrm{T}} + \epsilon \Sigma_2^{\mathrm{T}} \Sigma_2 < 0.$$

Why these three lemmas are called *elimination* lemmas is that there are always some matrices that are eliminated from one side of the equivalence relations in the lemmas. This is important in some situations for reducing the number of variables or simplifying the addressed problem. For example, the first condition in Lemma 1.4 is an infinite-dimensional inequality that often appears in dealing with norm-bounded uncertainty. It is not theoretically and numerically tractable. However, by Lemma 1.4 this infinite-dimensional condition is equivalently characterized by a single inequality with an additional scalar.

Nevertheless, it should be mentioned that the use of Lemma 1.2 in this book is in an opposite way, that is, by Lemma 1.2, we shall not eliminate some matrices but expand the inequalities $\mathcal{N}_X^{\mathrm{T}} \Phi \mathcal{N}_X < 0$ and $\mathcal{N}_Z^{\mathrm{T}} \Phi \mathcal{N}_Z < 0$ with an additional matrix Y, called *slack matrix*.

1.5.3 GKYP Lemma

Define nonempty set $\mathbf{S}$ of complex numbers as

$$\mathbf{S}(\Phi, \Psi) \triangleq \{s \in \mathbb{C} | \upsilon(s, \Phi) = 0, \ \upsilon(s, \Psi) \geq 0\}, \tag{1.12}$$

where $\Phi, \Psi \in \mathbf{H}_2$ are two given matrices. $\mathbf{S}$ is characterized by two matrices Φ and Ψ. Define

$$\Phi_c \triangleq \begin{bmatrix} 0 & 1 \\ 1 & 0 \end{bmatrix}.$$

Table 1.3 The values of Ψ for LF/MF/HF

Ω	LF ($\lvert\omega\rvert \le \omega_l$)	MF ($\omega_1 \le \omega \le \omega_2$)	HF ($\lvert\omega\rvert \ge \omega_h$)
Ψ	$\begin{bmatrix} -1 & 0 \\ 0 & \omega_l^2 \end{bmatrix}$	$\begin{bmatrix} -1 & j\omega_c \\ -j\omega_c & -\omega_1\omega_2 \end{bmatrix}$	$\begin{bmatrix} 1 & 0 \\ 0 & -\omega_h^2 \end{bmatrix}$

According to [172], if letting $\Phi = \Phi_c$, $\Psi = 0$, for any $s \in \mathbf{S}(\Phi_c, 0)$, we have $s = j\omega$ with $\omega \in \mathbb{R}$. Thus, by assigning $\Phi = \Phi_c$, $\mathbf{S}$ denotes a complex set $\{j\omega | \omega \in \mathbb{R}\}$, which corresponds rightly to the frequency variable ω in the continuous-time setting. Furthermore, by appropriately selecting Ψ, $\mathbf{S}(\Phi_c, \Psi)$ can characterize the continuous-time frequency ω in a finite or semi-finite range. According to [172], for the general LF/MF/HF range, the values of Ψ are shown in Table 1.3, where Ω denotes the FF set and $\omega_c = (\omega_1 + \omega_2)/2$. For technical reasons, we assume $\infty \in \mathbf{S}$ if Ω is unbounded.

Lemma 1.5 (GKYP Lemma[172]) *Let* $\Theta \in \mathbf{H}_{n_1+n_2}$, $F \in \mathbb{C}^{2n_1\times(n_1+n_2)}$ *and* Φ, $\Psi \in \mathbf{H}_2$ *be given such that* $\mathbf{S}$ *in* (1.12) *represents curves on the complex plane. Define*

$$\Gamma_s \triangleq \begin{cases} \begin{bmatrix} I_{n_1} & -sI_{n_1} \end{bmatrix} & (s \in \mathbb{C}) \\ \begin{bmatrix} 0 & -I_{n_1} \end{bmatrix} & (s = \infty) \end{cases}. \tag{1.13}$$

The following statements are equivalent.

(i) $\mathcal{N}_{\Gamma_s F}^* \Theta \mathcal{N}_{\Gamma_s F} < 0 \ \forall s \in \mathbf{S}(\Phi, \Psi)$.
(ii) There exist $P, Q \in \mathbf{H}_{n_1}$, *such that* $Q > 0$ *and*

$$F^*(\Phi \otimes P + \Psi \otimes Q) F + \Theta < 0.$$

Moreover, if $\operatorname{rank}\{\Gamma_s F\} = n_1$, *then the following statements are equivalent.*

(i) $\mathcal{N}_{\Gamma_s F}^* \Theta \mathcal{N}_{\Gamma_s F} \le 0 \ \forall s \in \mathbf{S}(\Phi, \Psi)$.
(ii) There exist $P, Q \in \mathbf{H}_{n_1}$, *such that* $Q \ge 0$ *and*

$$F^*(\Phi \otimes P + \Psi \otimes Q) F + \Theta \le 0.$$

1.5.4 Jensen Inequality

Lemma 1.6 ([173]) *For any constant matrix* $M \in \mathbb{C}^{m\times m}$, $M = M^* > 0$, *scalar* $\gamma > 0$, *vector function* $\omega : [0, \gamma] \to \mathbb{C}^m$ *such that the integrations in the following are well defined, then*

$$\gamma \int_0^\gamma \omega(\beta)^* M \omega(\beta) d\beta \ge \left(\int_0^\gamma \omega(\beta) d\beta\right)^* M \left(\int_0^\gamma \omega(\beta) d\beta\right). \tag{1.14}$$

References

1. Anderson, B.D.O., Moore, J.B.: Optimal Filtering. Prentice-Hall, Englewood Cliffs (1979)
2. Oppenheim, A.V., Schafer, R.W., Buck, J.R.: Discrete-Time Signal Processing, 2nd edn. Prentice-Hall, Upper Saddle River (1999)
3. Wiener, N.: Extrapolation, Interpolation, and Smoothing of Stationary Time Series With Engineering Applications. The MIT Press, Cambridge (1949)
4. Kolmogorov, A.N.: Interpolation and extrapolation. Bull. de l'académie des sciences de U.S.S.R. Ser. Math. **5**, 3–14 (1941)
5. Kalman, R.E.: A new approach to linear filtering and prediction problems. Trans. ASME, Ser. D, J. Basic Eng. **82**, 35–45 (1960)
6. Kalman, R.E., Bucy, R.S.: New results in linear filtering and prediction theory. Trans. ASME, Ser. D, J. Basic Eng.**83**, 95–108 (1961)
7. Chui, C.K., Chen, G.: Kalman Filtering with Real-Time Applications. Springer Series in Information Sciences, Springer-Verlag Berlin Heidelberg (2009)
8. Qi, H., Moore, J.B.: Direct Kalman filtering approach for GPS/INS integration. IEEE Trans. Aerosp. Electr. Syst. **38**(2), 687–693 (2002)
9. Grewal, M.S., Andrews, A.P.: Kalman Filtering: Theory and Practice Using MATLAB, 3rd edn. John Wiley & Sons Inc, Hoboken (2008)
10. Hajiyev, C., Caliskan, F.: Fault detection in flight control systems via innovation sequence of kalman filter. In: Tzafestas S., Schmidt G. (eds.) Progress in System and Robot Analysis and Control Design. Lecture Notes in Control and Information Sciences, vol. 243, pp. 63–74. Springer, Berlin / Heidelberg (1999)
11. Yu, F., Zhang, J.W., Crolla, D.A.: A study of a Kalman filter active vehicle suspension system using correlation of front and rear wheel road inputs. Proceedings Institution of Mechanical Engineers, Part D, J. Automob. Eng. **214**, 493–502 (2000)
12. Maybeck, P.S.: Stochastic Models, Estimation, and Control, vol. 2. Academic Press, NY (1982)
13. Liu, X., Zhang, X., Xin, X.: Application of Kalman filter in agricultural economic forecasting. 1996 IEEE International Conference on Systems. Man, and Cybernetics, pp. 2185–2189. Beijing, China (1996)
14. Gelb, A., Gelb, A., et al. (eds.): Applied Optimal Estimation. The MIT Press, Cambridge (1974)
15. D'Appolito, J.A., Hutchinson, C.E.: Low sensitivity filters for state estimation in the presence of large parameter uncertainties. IEEE Trans. Autom. Control AC-**14**(3), 310–312 (1969)
16. Chung, R.C., Belanger, P.R.: Minimum-sensitivity filter for linear time-invariant stochastic systems with uncertain parameter. IEEE Trans. Autom. Control AC-**21**(1) 98–100 (1976)
17. Jain, B.N.: Guaranteed error estimation in uncertain systems. IEEE Trans. Autom. Control AC-**20**, 230–232 (1975)
18. Haddad, W.M., Bernstein, D.S.: Robust, reduced-order, nonstrictly proper state estimation via the optimal projection equations with Petersen-Hollot bounds. Syst. Control Lett. **9**(5), 423–431 (1987)
19. Haddad, W.M., Bernstein, D.S.: Robust, reduced-order, nonstrictly proper state estimation via the optimal projection equations with guaranteed cost bounds. IEEE Trans. Autom. Control **33**(6), 591–595 (1988)
20. Haddad, W.M., Bernstein, D.S.: The optimal projection equations for reduced-order, discrete-time state estimation for linear systems with multiplicativewhite noise. Syst. Control Lett. **8**(4), 381–388 (1987)
21. Bernstein, D.S., Haddad, W.M.: Steady-state Kalman filtering with an H_∞ error bound. Syst. Control Lett. **12**(1), 9–16 (1989)
22. Bernstein, D.S., Haddad, W.M., Mustafa, D.: Mixed-norm H_2/H_∞ regulation and estimation: the discrete-time case. Syst. Control Lett. **16**(4), 235–247 (1991)
23. Petersen, I.R., McFarlane, D.C.: Robust state estimation for uncertain systems. In: Proc. the 30th IEEE Conference on Decision and Control, pp. 2630–2631. Brighton, England (1991)

24. Petersen, I.R., McFarlane, D.C.: Optimal guaranteed cost control and filtering for uncertain linear systems. IEEE Trans. Autom. Control **39**(9), 1971–1977 (1994)
25. Petersen, I.R., McFarlane, D.C.: Optimal guaranteed cost filtering for uncertain discrete-time systems. Int. J. Robust Nonlinear Control **6**, 267–280 (1996)
26. Xie, L., Soh, Y.C., de Souza, C.E.: Robust Kalman filtering for uncertain discrete-time systems. IEEE Trans. Autom. Control **39**(6), 1310–1314 (1994)
27. Xie, L., Soh, Y.C.: Robust Kalman filtering for uncertain systems. Syst. Control Lett. **22**, 123–129 (1994)
28. Bolzern, P., Colaneri, P., Nicolao, G.D.: Optimal robust filtering with time-varying parameter uncertainty. Int. J. Control **63**(3), 557–576 (1996)
29. Bolzern, P., Colaneri, P., Nicolao, G.D.: Optimal design of robust predictors for linear discrete-time systems. Syst. Control Lett. **26**(1), 25–31 (1995)
30. Shaked, U., de Souza, C.E.: Robust minimum variance filtering. IEEE Trans. Signal Process. **43**(11), 2474–2483 (1995)
31. Theodor, Y., Shaked, U.: Robust discrete-time minimum-variance filtering. IEEE Trans. Signal Process. **44**(2), 181–189 (1996)
32. Bolzern, P., Colaneri, P., Nicolao, G.D.: Guaranteed-cost prediction of discrete-time systems: the finite- and infinite-horizon case. In: Proceedings of the 2nd IFAC symposium on robust control design, pp. 471–476. Budapest, Hungary (1997)
33. Fu, M., de Souza, C.E., Luo, Z.Q.: Finite-horizon robust Kalman filter design. IEEE Trans. Signal Process. **49**(9), 2103–2112 (2001)
34. de Souza, C.E., Shaked, U.: Robust H_2 filtering for uncertain systems with measurable inputs. IEEE Trans. Signal Process. **47**(8), 2286–2292 (1999)
35. Petersen, I.R., Savkin, A.V.: Robust Kalman Filtering for Signals and Systems with Large Uncertainties. Birkhauser, Boston (1999)
36. Shaked, U., Theodor, Y.: H_∞-optimal estimation: A tutorial. In: Proceedings of the 31st IEEE Conference on Decision and Control, pp. 2278–2286. Tucson, Arizona (1992)
37. de Souza, C.E., Shaked, U., Fu, M.: Robust H_∞ filtering for continuous time varying uncertain systems with deterministic input signals. IEEE Trans. Signal Process. **43**(3), 709–719 (1995)
38. Grimble, M.J.: H_∞ design of optimal linear filter. In: Byrnes, C.I., Martin, C.F., Saeks, R.E. (eds.) Linear Circuits, Systems, and Signal Processing: Theory and Application, pp. 553–540. North-Holland, Amsterdam, The Netherlands (1988)
39. Elsayed, A., Grimble, M.J.: A new approach to H_∞ design of optimal digital linear filters. IMA J. Math. Control Inf. **6**(6), 233–251 (1989)
40. Elsayed, A., Grimble, M.J., Ho, D.: On H_∞ filtering in multivariable discrete-time systems. In: 1988 IEEE International symposiums on circuits and systems, vol. 2, pp. 1573–1576. Espoo, Finland (1988)
41. Grimble, M.J., Elsayed, A.: Solution of the H_∞ optimal linear filtering problem for discrete-time systems. IEEE Trans. Acoust. Signal Process. **38**(7), 1092–1104 (1990)
42. Grimble, M.J.: Polynomial matrix solution of the H_∞ filtering problem and the relationship to Riccati equation state-space results. IEEE Trans. Signal Process. **41**(1), 67–81 (1993)
43. Grimble, M.J.: H_∞ inferential filtering, prediction and smoothing problems. Signal Process. **60**(3), 289–304 (1997)
44. Shaked, U.: H_∞-minimum error state estimation of linear stationary processes. IEEE Trans. Autom. Control **35**(5), 554–558 (1990)
45. Fu, M.: Interpolation approach to H_∞ optimal estimation and its interconnection to loop transfer recovery. Syst. Control Lett. **17**, 29–36 (1991)
46. Nagpal, K.M., Khargonekar, P.P.: Filtering and smoothing in anH_∞ setting. IEEE Trans. Autom. Control **36**(2), 152–166 (1991)
47. Yaesh, I., Shaked, U.: A transfer function approach to the problems of discrete-time systems: H_∞-optimal linear control and filtering. IEEE Trans. Autom. Control **36**(11), 1264–1271 (1991)
48. Fu, M., de Souza, C.E., Xie, L.: H_∞ estimation for uncertain systems. Int. J. Robust Nonlinear Control **2**(2), 87–105 (1992)

49. Xie, L., de Souza, C.E., Fu, M.: H_∞ estimation for discrete-time linear uncertain systems. Int. J. Robust Nonlinear Control **1**, 111–123 (1991)
50. Xie, L., de Souza, C.E., Fragoso, M.D.: H_∞ filtering for linear periodic systems with parameter uncertainty. Syst. Control Lett. **17**(5), 343–350 (1991)
51. de Souza, C.E., Xie, L., Wang, Y.: H_∞ filtering for a class of uncertain nonlinear systems. Syst. Control Lett. **20**(6), 419–426 (1993)
52. Xie, L., de Souza, C.E., Wang, Y.: Robust filtering for a class of discrete-time uncertain nonlinear systems: An H_∞ approach. Int. J. Robust Nonlinear Control **6**(4), 297–312 (1996)
53. Yaesh, I., Shaked, U.: Game theory approach to optimal linear estimation in the minimum H_∞-norm sense. In: Proceedings of the 28th Conference on Decision and Control, pp. 421–425. Tampa, Florida (1989)
54. Yaesh, I., Shaked, U.: H_∞-optimal estimation-the discrete-time case. In: Kimura, H. (ed.) Recent Advances in Mathematical Theory of Systems, Control, and Signal Proccesing, vol. 1, pp. 261–267. Mita Press, Japan (1991)
55. Banavar, R.N., Speyer, J.L.: A linear-quadratic game approach to estimation and smoothing. In: Proc. of American Control Conference, pp. 2818–2822. Boston, USA (1991)
56. Yaesh, I., Shaked, U.: Game theory approach to optimal linear state estimation and its relation to minimum H_∞-norm estimation. IEEE Trans. Autom. Control **37**(6), 828–831 (1992)
57. Theodor, Y., Shaked, U., de Souza, C.E.: A game theory approach to robust discrete-time H_∞-estimation. IEEE Trans. Signal Process. **42**(6), 1486–1495 (1994)
58. Park, P., Kailath, T.: H_∞ filtering via convex optimization. Int. J. Control **66**(1), 15–22 (1997)
59. Grigoriadis, K.M., Watson, J.T.: Reduced-order H_∞ and L_2-L_∞ filtering via linear matrix inequalities. IEEE Trans. Aerosp. Electr. Syst. **33**(4), 1326–1338 (1997)
60. Watson, J.T., Grigoriadis, K.M.: Optimal unbiased filtering via linear matrix inequalities. Syst. Control Lett. **35**, 111–118 (1998)
61. Li, H., Fu, M.: A linear matrix inequality approach to robust H_∞ filtering. IEEE Trans. Signal Process. **45**(9), 2338–2350 (1997)
62. Geromel, J.C., de Oliveira, M.C.: H_2 and H_∞ robust filtering for convex bounded uncertain systems. In: Proc. of the 37th IEEE Conference on Decision and Control, pp. 146–151. Tampa, Florida USA (1998)
63. Geromel, J.C., Bernussou, J., Garcia, G., de Oliveira, M.C.: H_2 and H_∞ robust filtering for discrete-time linear systems. In: Proc. of the 37th IEEE Conference on Decision and Control, pp. 632–637. Tampa, Florida USA (1998)
64. Geromel, J.C., de Oliveira, M.C.: H_2 and H_∞ robust filtering for convex bounded uncertain systems. IEEE Trans. Autom. Control **46**(1), 100–107 (2001)
65. Geromel, J.C., Bernussou, J., Garcia, G., de Oliveira, M.C.: H_2 and H_∞ robust filtering for discrete-time linear systems. SIAM J. Control Optim. **38**, 1353–1368 (2000)
66. Palhares, R.M., Peres, P.L.D.: Optimal filtering schemes for linear discrete-time systems: a linear matrix inequality approach. Int. J. Syst. Sci. **29**(6), 587–593 (1998)
67. Palhares, R.M., Peres, P.L.D.: Robust H_∞-filtering design with pole placement constraint via linear matrix inequalities. J. Optim. Theory Appl. **102**(2), 239–261 (1999)
68. Palhares, R.M., Peres, P.L.D.: Robust H_∞ filter design with pole constraints for discrete-time systems. J. Frankl. Inst. **337**(6), 713–723 (2000)
69. Jin, S.H., Park, J.B.: Robust H_∞ filter for polytopic uncertain systems via convex optimization. IEE Proc. Control Theory Appl. **148**(1), 55–59 (2001)
70. Kwakernaak, H.: A polynomial approach to minimax frequency domain optimization of multivariable feedback systems. Int. J. Control **44**, 117–156 (1986)
71. Grimble, M.J.: Optimal H_∞ robustness and the relationship to LQG design problems. Int. J. Control **43**(2), 351–372 (1986)
72. Petersen, I.R.: Disturbance attenuation and H_∞ optimization: A design method based on the algebraic Riccati equation. IEEE Trans. Autom. Control AC-**32**(5), 427–429 (1987)
73. Doyle, J.C., Glover, K., Khargonekar, P.P., Francis, B.A.: State-space solutions to standard H_2 and H_∞ control problems. IEEE Trans. Autom. Control **34**(8), 831–847 (1989)

74. Khargonekar, P.P., Petersen, I.R., Zhou, K.: Robust stabilization of uncertain linear systems: quadratic stabilizability and H_∞ control theory. IEEE Trans. Autom. Control **35**(3), 356–361 (1990)
75. Limebeer, D.J.N., Anderson, B.D.O., Khargonekar, P.P., Green, M.: A game theoretic approach to H^∞ control for time-varying systems. SIAM J. Control Optim. **30**(2), 262–283 (1991)
76. Nesterov, Y., Nemirovsky, A.: Interior Point Polynomial Methods in Convex Programming, SIAM Studies in Applied Mathematics, vol. 13. SIAM, Philadelphia (1994)
77. Boyd, S., El Ghaoui, L., Feron, E., Balakrishnan, V.: Linear Matrix Inequalities in Systems and Control Theory. SIAM, Philadelphia (1994)
78. Gahinet, P., Nemirovskii, A., Laub, A.J., Chilali, M.: LMI Control Toolbox User's Guide. The Math. Works Inc., Natick (1995)
79. Pila, A.W., Shaked, U., de. Souza, C.E.: H_∞ filtering for continuous-time linear systems with delay. IEEE Trans. Autom. Control **44**(7), 1412–1417 (1999).
80. de Souza, C.E., Palhares, R.M., Peres, P.L.D.: Robust H_∞ filter design for uncertain linear systems with multiple time-varying state delays. IEEE Trans. Signal Process. **49**(3), 569–576 (2001)
81. Palhares, R.M., de Souza, C.E., Peres, P.L.D.: Robust H_∞ filtering for uncertain discrete-time state-delayed systems. IEEE Trans. Signal Process. **49**(8), 1696–1703 (2001)
82. Fridman, E., Shaked, U.: A new H_∞ filter design for linear time delay systems. IEEE Trans. Signal Process. **49**(11), 2839–2843 (2001)
83. Fridman, E., Shaked, U., Xie, L.: Robust H_∞ filtering of linear systems with time-varying delay. IEEE Trans. Autom. Control **48**(1), 159–165 (2003)
84. Fridman, E., Shaked, U.: An improved delay-dependent H_∞ filtering of linear neutral systems. IEEE Trans. Signal Process. **52**(3), 668–673 (2004)
85. Gao, H., Wang, C.: Delay-dependent robust H_∞ and L_2-L_∞ filtering for a class of uncertain nonlinear time-delay systems. IEEE Trans. Autom. Control **48**(9), 1661–1666 (2003)
86. He, Y., Wang, Q., Lin, C.: An improved H_∞ filter design for systems with time-varying interval delay. IEEE Trans. Circuits Syst. II, Express. Briefs **53**(11), 1235–1239 (2006)
87. Zhang, X.M., Han, Q.L.: Robust H_∞ filtering for a class of uncertain linear systems with time-varying delay. Automatica **44**(1), 157–166 (2008)
88. Gao, H., Lam, J., Wang, C.: Induced l_2 and generalized H_2 filtering for systems with repeated scalar nonlinearities. IEEE Trans. Signal Process. **53**(11), 4215–4226 (2005)
89. Wilson, D.A.: Convolution and Hankel operator norms for linear systems. IEEE Trans. Autom. Control **34**(1), 94–97 (1989)
90. Rotea, M.A.: The generalized H_2 control problem. Automatica **29**(2), 373–385 (1993)
91. Palhares, R.M., Peres, P.L.D.: Robust filtering with guaranteed energy-to-peak performance-an LMI approach. Automatica **36**(6), 851–858 (2000)
92. Palhares, R.M., Peres, P.L.D.: Mixed L_2-L_∞/H_∞ filtering for uncertain linear systems: a linear matrix inequality approach. Int. J. Syst. Sci. **31**(9), 1091–1098 (2000)
93. Vidyasagar, M.: Optimal rejection of persistent bounded disturbances. IEEE Trans. Autom. Control AC-**31**(6), 527–534 (1986)
94. Dahleh, M.A., Pearson, J.B.: l^1-optimal feedback controllers for MIMO discrete-time systems. IEEE Trans. Autom. Control AC-**32**(4), 314–322 (1987)
95. Dahleh, M.A., Pearson, J.B.: L^1-optimal compensators for continuous-time systems. IEEE Trans. Autom. Control AC-**32**(10), 889–895 (1987).
96. Diaz-Bobillo, I.J., Dahleh, M.A.: Minimization of the maximum peak-to-peak gain: The general multiblock problem. IEEE Trans. Autom. Control **38**(10), 1459–1482 (1993)
97. Schweppe, F.C.: Recursive state estimation: Unknown but bounded errors and system inputs. IEEE Trans. Autom. Control AC-**13**(1), 22–28 (1968).
98. Bertsekas, D.P., Rhodes, I.B.: Recursive state estimation for a set-membership description of uncertainty. IEEE Trans. Autom. Control AC-**16**(2), 117–128 (1971).
99. Milanese, M., Tempo, R.: Optimal algorithms theory for robust estimation and prediction. IEEE Trans. Autom. Control AC-30(8), 730–738 (1985).

100. Tempo, R.: Robust estimation and filtering in the presence of bounded noise. IEEE Trans. Autom. Control **33**(9), 864–867 (1988)
101. Voulgaris, P.: Optimal l^{∞} to l^{∞} estimation. Automatica **31**(3), 489–495 (1995)
102. Voulgaris, P.G.: Optimal l_{∞} to l_{∞} estimation for periodic systems. IEEE Trans. Autom. Control **41**(9), 1392–1396 (1996)
103. Abedor, J., Nagpal, K., Poolla, K.: A linear matrix inequality approach to peak-to-peak gain minimization. Int. J. Robust Nonlinear Control **6**(9–10), 899–927 (1996)
104. Vincent, T., Abedor, J., Nagpal, K., Khargonekar, R.P.: Discrete-time estimators with guaranteed peak-to-peak performance. In: Proc. of the 13th IFAC Triennial World Congress, pp. 43–48. San Francisco, USA (1996)
105. Petersen, I.R.: A stabilization algorithm for a class of uncertain linear systems. Syst. Control Lett. **8**, 351–357 (1987)
106. Geromel, J.C., Peres, P.L.D., Bernussou, J.: On a convex parameter space method for linear control design of uncertain systems. SIAM J. Control Optim. **29**(2), 381–402 (1991)
107. Geromel, J.C.: Optimal linear filtering under parameter uncertainty. IEEE Trans. Signal Process. **47**(1), 168–175 (1999)
108. Geromel, J.C., Peres, P.L.D., Souza, S.R.: H_{∞} control of discrete-time uncertain systems. IEEE Trans. Autom. Control **39**(5), 1072–1075 (1994)
109. Barmish, B.R.: Stabilization of uncertain systems via linear control. IEEE Trans. Autom. Control **28**(8), 848–850 (1983)
110. Barmish, B.R., Corless, M., Leitmann, G.: A new class of stabilizing controllers for uncertain dynamical systems. SIAM J. Control Optim. **21**(2), 246–255 (1983)
111. Barmish, B.R.: Necessary and sufficient conditions for quadratic stabilizability of an uncertain linear systems. J. Optim. Theory Appl. **46**(4), 399–408 (1985)
112. Geromel, J.C., Bernussou, J., de Oliveira, M.C.: H_2 norm optimization with constrained dynamic output feedback controllers: Decentralized and reliable control. IEEE Trans. Autom. Control **44**(7), 1449–1454 (1999)
113. Xie, L., Soh, Y.C., Du, C., Zou, Y.: Robust H_2 estimation and control. J. Control Theory Appl. **2**(1), 20–26 (2004)
114. Gao, H., Lam, J., Wang, C.: Robust energy-to-peak filter design for stochastic time-delay systems. Syst. Control Lett. **55**(2), 101–111 (2006)
115. Gao, H., Wang, C.: Robust L_2-L_{∞} filtering for uncertain systems with multiple time-varying state delays. IEEE Trans. Circuits Syst. I, Fundam. Theory Appl. **50**(4), 594–599 (2003)
116. Geromel, J.C., De Oliveira, M.C., Bernussou, J.: Robust filtering of discrete-time linear systems with parameter dependent Lyapunov functions. SIAM J. Control Optim. **41**(3), 700–711 (2002)
117. Shaked, U.: Improved LMI representations for the analysis and the design of continuous-time systems with polytopic type uncertainty. IEEE Trans. Autom. Control **46**(4), 652–656 (2001)
118. Haddad, W.M., Bernstein, D.S.: Parameter-dependent Lyapunov functions and the Popov criterion in robust analysis and synthesis. IEEE Trans. Autom. Control **40**(3), 536–543 (1995)
119. Gahinet, P., Apkarian, P., Chilali, M.: Affine parameter-dependent Lyapunov functions and real parametric uncertainty. IEEE Trans. Autom. Control **41**(3), 436–442 (1996)
120. Feron, E., Apkarian, P., Gahinet, P.: Analysis and synthesis of robust control systems via parameter-dependent Lyapunov functions. IEEE Trans. Autom. Control **41**(7), 1041–1046 (1996)
121. de Oliveira, M.C., Bernussou, J., Geromel, J.C.: A new discrete-time robust stability condition. Syst. Control Lett. **37**, 261–265 (1999)
122. de Oliveira, M.C., Geromel, J.C.: A new absolute stability test for systems with state-dependent perturbations. Int. J. Robust Nonlinear Control **12**(14), 1209–1226 (2002)
123. de Oliveira, M.C., Geromel, J.C., Bernussou, J.: Extended H_2 and H_{∞} norm characterizations and controller parametrizations for discrete-time systems. Int. J. Control **75**(9), 666–679 (2002)
124. Peaucelle, D., Arzelier, D., Bachelier, J., Bernussou, J.: A new robust D-stability condition for real convex polytopic uncertainty. Syst. Control Lett. **40**, 21–30 (2000)

125. Bachelier, O., Peaucelle, D., Arzelier, D.: Computing a robust D-stability bound using a parameter-dependent lyapunov approach. IEE Proc. Control Theory Appl. **149**(6), 505–510 (2002)
126. Peaucelle, D., Arzelier, D.: Robust performance analysis with LMI-based methods for real parametric uncertainty via parameter-dependent Lyapunov functions. IEEE Trans. Autom. Control **46**(4), 624–630 (2001)
127. Arzelier, D., Henrion, D., Peaucelle, D.: Robust D stabilization of a polytope of matrices. Int. J. Control **75**(10), 744–752 (2002)
128. Apkarian, P., Pellanda, P.C., Tuan, H.D.: Mixed H_2/H_∞ multi-channel linear parameter-varying control in discrete time. Syst. Control Lett. **41**(5), 333–346 (2000)
129. Apkarian, P., Tuan, H.D., Bernussou, J.: Continuous-time analysis, eigenstructure assignment, and H_2 synthesis with enhanced linear matrix inequalities (LMI) characterizations. IEEE Trans. Autom. Control **46**(12), 1941–1946 (2001)
130. Tuan, H.D., Apkarian, P., Nguyen, T.Q.: Robust and reduced-order filtering: New LMI-based characterizations and methods. IEEE Trans. Signal Process. **49**(12), 2975–2984 (2001)
131. Tuan, H.D., Apkarian, P., Nguyen, T.Q.: Robust mixed H_2/H_∞ filtering of 2-D systems. IEEE Trans. Signal Process. **50**(7), 1759–1771 (2002)
132. Shaked, U., Xie, L., Soh, Y.C.: New approaches to robust minimum variance filer design. IEEE Trans. Signal Process. **49**(11), 2620–2629 (2001)
133. Xie, L., Lu, L., Zhang, H.: Improved robust H_2 and H_∞ filtering for uncertain discrete-time systerms. Automatica **40**, 873–880 (2004)
134. Fridman, E., Shaked, U.: Parameter dependent stability and stabilization of uncertain time-delay systems. IEEE Trans. Autom. Control **48**(5), 861–866 (2003)
135. Daafouz, J., Bernussou, J.: Parameter dependent Lyapunov functions for discrete time systems with time varying parametric uncertainties. Syst. Control Lett. **43**(5), 355–359 (2001)
136. Daafouz, J., Riedinger, P., Iung, C.: Stability analysis and control synthesis for switched systems: a switched Lyapunov function approach. IEEE Trans. Autom. Control **47**(11), 1883–1887 (2002)
137. Xie, L.: On robust H_2 estimation. Acta Automatica Sinica **31**(1), 1–12 (2005)
138. Tuan, H.D., Apkarian, P., Nguyen, T.Q.: Robust filtering for uncertain nonlinearly parameterized plants. IEEE Trans. Signal Process. **51**(7), 1806–1815 (2003)
139. Hoang, N.T., Tuan, H.D., Apkarian, P., Hosoe, S.: Robust filtering for discrete nonlinear fractional transformation systems. IEEE Trans. Circuits Syst. II, Express. Briefs **51**(11), 587–592 (2004)
140. Barbosa, K.A., de Souza, C.E., Trofino, A.: Robust H_2 filtering for uncertain linear systems: LMI based methods with parametric Lyapunov functions. Syst. Control Lett. **54**(3), 251–262 (2005)
141. Duan, Z., Zhang, J.: C., Z., Mosca, E.: Robust H_2 and H_∞ filtering for uncertain linear systems. Automatica **42**(11), 1919–1926 (2006)
142. Gao, H., Wang, C.: New approaches to robust l_2-l_∞ and H_∞ filtering for uncertain discrete-time systems. Sci. China (Ser. F) **46**(5), 355–370 (2003)
143. Gao, H.J., Wang, C.H.: Robust H_∞ state estimation for systems with uncertain parameters. Control Decis. **19**(2), 147–152 (2004)
144. Wu, A., Duan, G.: Robust H-infinity estimation for continuous-time polytopic uncertain systems. J. Control Theory Appl. **4**, 393–398 (2005)
145. Gao, H., Wang, C.: A delay-dependent approach to robust H_∞ filtering for uncertain discrete-time state-delayed systems. IEEE Trans. Signal Process. **52**(6), 1631–1640 (2004)
146. Zhang, X., Han, Q.: Delay-dependent robust H_∞ filtering for uncertain discrete-time systems with time-varying delay based on a finite sum inequality. IEEE Trans. Circuits Syst. II, Express. Briefs **53**(12), 1466–1470 (2006)
147. He, Y., Liu, G., Rees, D., Wu, M.: H_∞ filtering for discrete-time systems with time-varying delay. Signal Process. **89**(3), 275–282 (2009)
148. He, Y., Liu, G.P., Rees, D., Wu, M.: Improved H_∞ filtering for systems with a time-varying delay. Circuits, Syst. Signal Process. **29**(3), 377–389 (2010)

149. Gao, H., Wang, C.: New approach to robust L_2-L_∞ filter design for uncertain continuous-time systems. Acta Automatica Sinica **29**(6), 809–814 (2003)
150. Gao, H., Wang, C.: Robust energy-to-peak filtering with improved LMI representations. IEE Proc. Vis. Image Signal Process. **150**(2), 82–89 (2003)
151. Chesi, G., Garulli, A., Tesi, A., Vicino, A.: Homogeneous Lyapunov functions for systems with structured uncertainties. Automatica **39**, 1027–1035 (2003)
152. Chesi, G., Garulli, A., Tesi, A., Vicino, A.: Robust stability of time-varying polytopic systems via parameter-dependent homogeneous Lyapunov functions. Automatica **43**, 309–316 (2007)
153. iveira, R.C.L.F., Peres, P.L.D.: LMI conditions for robust stability analysis based on polynomially parameter-dependent Lyapunov functions. Syst. Control Lett. **55**(1), 52–61 (2006)
154. Oliveira, R.C.L.F., Peres, P.L.D.: Parameter-dependent lmis in robust analysis: Characterization of homogeneous polynomially parameter-dependent solutions via LMI relaxations. IEEE Trans. Autom. Control **52**(7), 1334–1340 (2007)
155. Gao, H., Lam, J., Shi, P., Wang, C.: Parameter-dependent filter design with guaranteed H_∞ performance. IEE Proc. Control Theory Appl. **152**(5), 531–537 (2005)
156. Gao, H., Lam, J., Xie, L., Wang, C.: New approach to mixed H_2/H_∞ filtering for polytopic discrete-time systems. IEEE Trans. Signal Process. **53**(8), 3183–3192 (2005)
157. Gao, H., Chen, T.: H_∞ estimation for uncertain systems with limited communication capacity. IEEE Trans. Autom. Control **52**, 2070–2084 (2007)
158. Meng, X., Gao, H., Mou, S.: A parameter-dependent approach to robust energy-to-peak filter design. Circuits, Syst. Signal Process. **26**(4), 451–471 (2007)
159. Zhang, H., Mehr, A.S., Shi, Y.: Improved robust energy-to-peak filtering for uncertain linear systems. Signal Process. **90**(9), 2667–2675 (2010)
160. Qiu, J., Feng, G., Yang, J.: Improved delay-dependent H_∞ filtering design for discrete-time polytopic linear delay systems. IEEE Trans. Circuits Syst. II, Express. Briefs **55**(2), 178–182 (2008)
161. Qiu, J., Feng, G., Yang, J.: A new design of delay-dependent robust H_∞ filtering for continuous-time polytopic systems with time-varying delay. Int. J. Robust Nonlinear Control **20**(3), 346–365 (2010)
162. Li, X., Li, Z., Gao, H.: Further results on H_∞ filtering for discrete-time systems with state delay. Int. J. Robust Nonlinear Control **21**(3), 248–270 (2011)
163. Li, X., Gao, H.: A delay-dependent approach to robust generalized H_2 filtering for uncertain continuous-time systems with interval delay. Signal Process. **91**(10), 2371–2378 (2011)
164. Zhang, W.A., Yu, L., Jiang, X.: Delay-dependent generalized H_2 filtering for uncertain systems with multiple time-varying state delays. Signal Process. **87**(4), 709–724 (2007)
165. Gao, H., Meng, X., Chen, T.: A new design of robust H_2 filters for uncertain systems. Syst. Control Lett. **57**(7), 585–593 (2008)
166. Gao, C.Y., Duan, G.R., Meng, X.Y.: Robust H_∞ filter design for 2D discrete systems in Roesser model. Int. J. Autom. Control **05**(4), 413–418 (2008)
167. Gao, H., Meng, X., Chen, T.: A parameter-dependent approach to robust H_∞ filtering for time-delay systems. IEEE Trans. Autom. Control **53**(10), 2420–2425 (2008)
168. Gao, H., Meng, X., Chen, T.: H_∞ filter design for discrete delay systems: A new parameter-dependent approach. Int. J. Control **82**(6), 993–1005 (2009)
169. Lacerda, M.J., Oliveira, R.C.L.F., Peres, P.L.D.: Robust H_2 and H_∞ filter design for uncertain linear systems via LMIs and polynomial matrices. Signal Process. **91**(5), 1115–1122 (2011)
170. Gahinet, P., Apkarian, P.: A linear matrix inequality approach to H_∞ control. Int. J. Robust Nonlinear Control **4**(4), 421–448 (1994)
171. Xie, L.: Output feedback H_∞ control of systems with parameter uncertainty. Int. J. Control **63**, 741–750 (1996)
172. Iwasaki, T., Hara, S.: Generalized KYP lemma: Unified frequency domain inequalities with design applications. IEEE Trans. Autom. Control **50**, 41–59 (2005)
173. Gu, K.: An integral inequality in the stability problem of ime-delay systems. In: Proceedings of the 39th IEEE Conference on Decision nd Control, pp. 2805–2810. Sydney, Australia (2000)

Part I
Quadratic and Parameter-Dependent Filter Design

Chapter 2
Quadratic Robust Filter Design

Filtering is one of the fundamental issues in the areas of control, circuits and systems, and signal processing. Different filtering methods rely on different assumptions of the noises (including the process noise and the measurement noise) and the corresponding suitable performance index of the filtering error. For H_2 filtering, the noise is assumed to be Gaussian white noise processes with identity power spectrum density matrix, and the objective is to minimize the H_2 norm of the transfer function from the noise to the filtering error. For H_∞ filtering, the noise is assumed to have bounded energy and the objective is to minimize the H_∞ norm of the transfer function from the noise to the filtering error. For energy-to-peak filtering, it is self-explanatory that the noise is still assumed to be energy-bounded but the objective turns to minimizing the energy-to-peak gain of transfer function from the noise to the filtering error.

In this chapter, we present quadratic approaches to robust filtering for polytopic uncertain systems, and some common filtering schemes as mentioned, including H_2 filtering, H_∞ filtering, and energy-to-peak filtering, will be taken into consideration. The so-called *quadratic approach* means that for an uncertain system, design a filter such that the stability and the prescribed disturbance attenuation level of the filtering error system are guaranteed by a *single* quadratic Lyapunov function, the underlying idea of which is the generalization of the notion of quadratic stability.

Filtering for nominal systems will be addressed first and the main attention will be paid to linearizing the NLMI conditions for filtering performance analysis into LMI for the state-space realization of filters. Then, based on the notion of quadratic stability, the filter design methods for nominal systems will be extended to the uncertain case, that is, quadratic approaches to robust filter design. All the filter design methods, for both uncertain continuous- and discrete-time systems, are to be presented in terms of solving the corresponding optimization problem subject to LMI constraints. The effectiveness of the presented filter design methods will be demonstrated by several illustrative examples.

H. Gao and X. Li, *Robust Filtering for Uncertain Systems*,
Communications and Control Engineering, DOI: 10.1007/978-3-319-05903-7_2,

2.1 Quadratic Robust H_2 Filter Design

In this section, we consider the robust H_2 filtering problem in the framework of quadratic approaches. For the nominal systems, the results to be presented are classical and standard. Moreover, it is quite natural to extend the filter design methods in terms of LMI from the nominal case to the polytopic uncertainty by the quadratic stability notion.

2.1.1 Problem Formulation

Consider a stable, uncertain linear time-variant (LTI) system $\mathcal{S}$ described by the following state-space model:

$$\begin{aligned} \mathcal{S}:\ \delta[x(t)] &= A(\lambda)x(t) + B(\lambda)w(t), \\ y(t) &= C(\lambda)x(t) + D(\lambda)w(t), \\ z(t) &= L(\lambda)x(t), \end{aligned} \tag{2.1}$$

where $x(t) \in \mathbb{R}^n$ is the state vector, $y(t) \in \mathbb{R}^m$ is the measured output, $z(t) \in \mathbb{R}^p$ is the signal to be estimated, $w(t) \in \mathbb{R}^l$ is a zero-mean white noise with identity power spectrum density matrix. $\delta[\cdot]$ denotes the shift operator for discrete-time systems and the derivative operator for continuous-time systems, respectively. $A(\lambda)$, $B(\lambda)$, $C(\lambda)$, $D(\lambda)$, and $L(\lambda)$ are appropriately dimensioned real matrices. It is assumed that

$$M(\lambda) \triangleq (A(\lambda), B(\lambda), C(\lambda), D(\lambda), L(\lambda)) \in \mathcal{M} \tag{2.2}$$

where $\mathcal{M}$ is a given convex polyhedral domain bounded by s vertices

$$\mathcal{M} \triangleq \left\{ M(\lambda):\ M(\lambda) = \sum_{i=1}^{s} \lambda_i M_i;\ \lambda \in \Gamma \right\}$$

with $M_i \triangleq (A_i, B_i, C_i, D_i, L_i)$ denoting the vertices of the polytope, and Γ denoting the unit simplex

$$\Gamma \triangleq \left\{ (\lambda_1, \lambda_2, \ldots, \lambda_s) : \sum_{i=1}^{s} \lambda_i = 1,\ \lambda_i \geq 0 \right\}. \tag{2.3}$$

Equation (2.2) indicates that system matrices can be expressed as the convex combination of several vertex matrices.

For system $\mathcal{S}$ in (2.1), the filtering problem can be roughly thought as using the measurable signal $y(t)$ to construct an estimate $z_F(t)$ for the unmeasurable signal $z(t)$. Hence, the filter, denoted by $\mathcal{F}$, actually is an operator from $y(t)$ to $z_F(t)$, that is, $z_F = \mathcal{F}[y]$. In this book, consider that the filter $\mathcal{F}$ is a full-order LTI operator given by the following minimum state-space realization:

$$\begin{aligned}\mathcal{F}: \delta[x_F(t)] &= A_F x_F(t) + B_F y(t),\\ z_F(t) &= C_F x_F(t),\end{aligned} \tag{2.4}$$

where $x_F(t) \in \mathbb{R}^n$ is the filter state vector, and A_F, B_F, and C_F are appropriately dimensioned filter matrices to be determined.

Define the filtering error signal $e(t) \triangleq z(t) - z_F(t)$ and the augmented state vector $\xi(t) = \left[x(t)^{\mathrm{T}},\ x_F(t)^{\mathrm{T}}\right]^{\mathrm{T}}$. By connecting the filter $\mathcal{F}$ to the system $\mathcal{S}$, the filtering error system $\mathcal{E}$ is given by the following augmented state space model:

$$\begin{aligned}\mathcal{E}: \delta[\xi(t)] &= \bar{A}(\lambda)\xi(t) + \bar{B}(\lambda)w(t),\\ e(t) &= \bar{C}(\lambda)\xi(t)\end{aligned} \tag{2.5}$$

with

$$\bar{A}(\lambda) \triangleq \begin{bmatrix} A(\lambda) & 0 \\ B_F C(\lambda) & A_F \end{bmatrix},\ \bar{B}(\lambda) \triangleq \begin{bmatrix} B(\lambda) \\ B_F D(\lambda) \end{bmatrix},\ \bar{C}(\lambda) \triangleq \begin{bmatrix} L(\lambda) & -C_F \end{bmatrix}.$$

For any fixed $\lambda \in \Gamma$, the transfer function of the filtering error system from the noise input $w(t)$ to the filtering error $e(t)$ is defined by

$$T(\delta, \lambda) \triangleq \bar{C}(\lambda)\left[\delta I - \bar{A}(\lambda)\right]^{-1}\bar{B}(\lambda). \tag{2.6}$$

With slight abuse of notations, keep in mind that, in the frequency domain, the operator δ can be seen as, respectively, the Laplacian operator s for the continuous-time case and the shift operator z for the discrete-time case.

It is assumed that the initial condition $x(0)$ of system $\mathcal{S}$ in (2.1) is a zero-mean random signal uncorrelated with the input noise $w(t)$ for all $t \geq 0$, while the filter $\mathcal{F}$ in (2.4) has zero initial condition $x_F(0) = 0$. The robust H_2 filtering problem to be addressed for system $\mathcal{S}$ is formulated as follows.

Robust H_2 Filtering Problem: Given system $\mathcal{S}$ in (2.1), design a filter $\mathcal{F}$ of the form in (2.4) such that the filtering error system $\mathcal{E}$ in (2.5), for all $\lambda \in \Gamma$, is robustly asymptotically stable, and for all zero-mean white noise $w(t)$ with identity power spectrum density matrix, satisfies

$$\sup_{\lambda\in\Gamma} \mathbf{E}\left[e(t)^{\mathrm{T}}e(t)\right] < \gamma \left(\text{i.e.,} \sup_{\lambda\in\Gamma} \|T(\delta, \lambda)\|_2^2 < \gamma\right), \tag{2.7}$$

where $\mathbf{E}[\cdot]$ denotes mathematical expectation of the random signal and γ is a given positive constant. Moreover, a filter satisfying the above conditions is referred to as *a filter with a guaranteed robust H_2 performance bound $\sqrt{\gamma}$*.

2.1.2 H_2 Filtering for Nominal Systems: Continuous-Time

First, we consider the H_2 filtering problem for nominal systems, i.e., the uncertain parameter λ is arbitrary but fixed in the unit simplex Γ.

For nominal systems, necessary and sufficient conditions in terms of LMI have been well-established for the H_2 performance, see [1–3]. In the continuous-time context, the H_2 performance criterion is given in the following lemma, which will be used for H_2 filter design.

Lemma 2.1 ([1, 3]) *Consider the continuous-time system $\mathcal{S}$ in* (2.1) *and assume that $M \in \mathcal{M}$ is fixed but arbitrary. Given filter $\mathcal{F}$ in* (2.4) *and a scalar $\gamma > 0$, the filtering error system $\mathcal{E}$ in* (2.5) *is asymptotically stable and satisfies $\mathbf{E}\left[e(t)^{\mathrm{T}}e(t)\right] < \gamma$ if and only if the following matrix inequalities*

$$\mathrm{Tr}[Z_c] < \gamma, \tag{2.8}$$

$$\begin{bmatrix} P_c & \bar{C}^{\mathrm{T}} \\ \bar{C} & Z_c \end{bmatrix} > 0, \tag{2.9}$$

$$\begin{bmatrix} \bar{A}^{\mathrm{T}} P_c + P_c \bar{A} & P_c \bar{B} \\ \bar{B}^{\mathrm{T}} P_c & -I \end{bmatrix} < 0 \tag{2.10}$$

are feasible in the real matrix variables P_c and Z_c.

It is well-known that as $t \to +\infty$, the filtering error variance

$$\mathbf{E}\left[e(t)^{\mathrm{T}}e(t)\right] = \mathrm{Tr}[\bar{C}P\bar{C}^{\mathrm{T}}],$$

where P is the solution to the Lyapunov equation

$$\bar{A}P + P\bar{A}^{\mathrm{T}} + \bar{B}\bar{B}^{\mathrm{T}} = 0.$$

Note that if system $\mathcal{M}$ is stable, a unique solution exists for this equation. To briefly justify Lemma 2.1, pre- and postmultiplying the both hand sides of (2.10) by $\begin{bmatrix} P_c^{-1} & 0 \\ 0 & I \end{bmatrix}$ result in

$$\begin{bmatrix} P_c^{-1}\bar{A}^{\mathrm{T}} + \bar{A}P_c^{-1} & \bar{B} \\ \bar{B}^{\mathrm{T}} & -I \end{bmatrix} < 0,$$

which, by the Schur complement, is equivalent to

$$\bar{A}P_c^{-1} + P_c^{-1}\bar{A}^{\mathrm{T}} + \bar{B}\bar{B}^{\mathrm{T}} < 0.$$

Due to the monotonicity property of the solution to the Lyapunov equation, it follows that $P < P_c^{-1}$, which together with (2.8) and (2.9) gives

$$\mathbf{E}\left[e(t)^{\mathrm{T}}e(t)\right] = \mathrm{Tr}[\bar{C}P\bar{C}^{\mathrm{T}}] < \mathrm{Tr}[\bar{C}P_c^{-1}\bar{C}^{\mathrm{T}}] < \mathrm{Tr}[Z_c] < \gamma.$$

Remark 2.1 In [4, 5], the LMI constraints corresponding to (2.9) and (2.10) are given, respectively, by

$$\begin{bmatrix} \bar{P} & \bar{P}\bar{C}^{\mathrm{T}} \\ \bar{C}\bar{P} & Z \end{bmatrix} > 0, \quad \begin{bmatrix} \bar{P}\bar{A}^{\mathrm{T}} + \bar{A}\bar{P} & \bar{B} \\ \bar{B}^{\mathrm{T}} & -I \end{bmatrix} < 0.$$

Simply matrix manipulations can verify that these two inequalities are equivalent to (2.9) and (2.10), respectively. Indeed, if letting $P_c \triangleq \bar{P}^{-1}$, it is readily obtained that

$$\begin{bmatrix} P_c & \bar{C}^{\mathrm{T}} \\ \bar{C} & Z \end{bmatrix} = \begin{bmatrix} \bar{P}^{-1} & 0 \\ 0 & I \end{bmatrix} \begin{bmatrix} \bar{P} & \bar{P}\bar{C}^{\mathrm{T}} \\ \bar{C}\bar{P} & Z \end{bmatrix} \begin{bmatrix} \bar{P}^{-1} & 0 \\ 0 & I \end{bmatrix},$$

$$\begin{bmatrix} \bar{A}^{\mathrm{T}}P_c + P_c\bar{A} & P_c\bar{B} \\ \bar{B}^{\mathrm{T}}P_c & -I \end{bmatrix} = \begin{bmatrix} \bar{P}^{-1} & 0 \\ 0 & I \end{bmatrix} \begin{bmatrix} \bar{P}\bar{A}^{\mathrm{T}} + \bar{A}\bar{P} & \bar{B} \\ \bar{B}^{\mathrm{T}} & -I \end{bmatrix} \begin{bmatrix} \bar{P}^{-1} & 0 \\ 0 & I \end{bmatrix}.$$

From these two formulations, it is found that the product terms between Lyapunov matrix P_c and system matrices in the latter formulation appear in both of the LMIs while similar terms in Lemma 2.1 exist only in (2.10). This feature is important for robustness analysis of the filtering error system, which will be further elucidated in the next chapter regarding the parameter-dependent approach to robust H_2 filter design. However, for nominal systems, it does not matter which formulation is employed for H_2 performance analysis. For consistency, this book adopts the one in Lemma 2.1.

When a filter is known, the above matrix inequality constraints are LMIs in the variables P_c, Z_c, and γ. Thus, the optimal H_2 filtering performance level of the given filter can be obtained by minimizing γ subject to (2.8)–(2.10).

When the state-space realization of a filter is unknown, (2.10) is not an LMI with respect to the filter matrices A_F, B_F, and C_F. Hence, Lemma 2.1 does not give an explicit solution to the realization of an H_2 filter. The possibility to handle H_2 filter design with Lemma 2.1 in light of the LMI technique relies on a fact that the inequality in (2.10) can be converted into an LMI by a reversible matrix transformation. In what follows, the main purpose of this subsection is to convert the inequality in (2.10) into an LMI such that the problem of H_2 filter design is cast into a convex optimization problem that can be efficiently solved by mature numerical algorithms.

To this end, partition matrix P_c into four blocks as the following form

$$P_c = \begin{bmatrix} P_{c1} & P_{c2} \\ P_{c2}^{\mathrm{T}} & P_{c3} \end{bmatrix}$$

with each block being an $n \times n$ matrix. Due to the fact that P_c is nonsingular, it can be assumed that P_{c2} is nonsingular by invoking a small perturbation if necessary. Thus, this assumption does not lose generality.

Define multiplier matrix

$$J_{c0} \triangleq \begin{bmatrix} I & 0 \\ 0 & P_{c3}^{-1} P_{c2}^{\mathrm{T}} \end{bmatrix}.$$

Introduce the following matrix transformations

$$\tilde{A}_F \triangleq P_{c2} A_F P_{c3}^{-1} P_{c2}^{\mathrm{T}}, \tag{2.11}$$

$$\tilde{B}_F \triangleq P_{c2} B_F, \tag{2.12}$$

$$\tilde{C}_F \triangleq C_F P_{c3}^{-1} P_{c2}^{\mathrm{T}}, \tag{2.13}$$

$$\begin{bmatrix} \tilde{P}_{c1} & \tilde{P}_{c2} \\ \tilde{P}_{c2} & \tilde{P}_{c2} \end{bmatrix} \triangleq J_{c0}^{\mathrm{T}} P_c J_{c0} = \begin{bmatrix} P_{c1} & P_{c2} P_{c3}^{-1} P_{c2}^{\mathrm{T}} \\ P_{c2} P_{c3}^{-1} P_{c2}^{\mathrm{T}} & P_{c2} P_{c3}^{-1} P_{c2}^{\mathrm{T}} \end{bmatrix}, \tag{2.14}$$

from which, the following relations can be readily obtained:

$$J_{c0}^{\mathrm{T}} P_c \bar{A} J_{c0} = \begin{bmatrix} \tilde{P}_{c1} A + \tilde{B}_F C & \tilde{A}_F \\ \tilde{P}_{c2} A + \tilde{B}_F C & \tilde{A}_F \end{bmatrix} \triangleq \begin{bmatrix} X_{c1} & \tilde{A}_F \\ X_{c2} & \tilde{A}_F \end{bmatrix},$$

$$J_{c0}^{\mathrm{T}} P_c \bar{B} = \begin{bmatrix} \tilde{P}_{c1} B + \tilde{B}_F D \\ \tilde{P}_{c2} B + \tilde{B}_F D \end{bmatrix},$$

$$\bar{C} J_{c0} = \begin{bmatrix} L & -\tilde{C}_F \end{bmatrix}.$$

Now, by virtue of the transformations defined in (2.11)–(2.14) and further noting the following transformations

$$J_{c1}^{\mathrm{T}} \begin{bmatrix} P_c & \bar{C}^{\mathrm{T}} \\ \bar{C} & Z_c \end{bmatrix} J_{c1} = \begin{bmatrix} \tilde{P}_{c1} & \tilde{P}_{c2} & L^{\mathrm{T}} \\ \tilde{P}_{c2} & \tilde{P}_{c2} & -\tilde{C}_F^{\mathrm{T}} \\ L & -\tilde{C}_F & Z_c \end{bmatrix},$$

$$J_{c2}^{\mathrm{T}} \begin{bmatrix} \bar{A}^{\mathrm{T}} P_c + P_c \bar{A} & P_c \bar{B} \\ \bar{B}^{\mathrm{T}} P_c & -I \end{bmatrix} J_{c2} = \begin{bmatrix} X_{c1} + X_{c1}^{\mathrm{T}} & \tilde{A}_F + X_{c2}^{\mathrm{T}} & \tilde{P}_{c1} B + \tilde{B}_F D \\ * & \tilde{A}_F + \tilde{A}_F^{\mathrm{T}} & \tilde{P}_{c2} B + \tilde{B}_F D \\ * & * & -I \end{bmatrix}$$

with $J_{c1} \triangleq \mathrm{diag}\{J_{c0}, I_n\}$ and $J_{c2} \triangleq \mathrm{diag}\{J_{c0}, I_l\}$, it can be shown that the NLMIs in (2.8)–(2.10) with respect to variables

$$P_{c1}, \; P_{c2}, \; P_{c3}, \; Z_c, \; A_F, \; B_F, \; \text{and } C_F$$

are converted into LMIs with respect to new variables

$$\tilde{P}_{c1}, \; \tilde{P}_{c2}, \; Z_c, \; \tilde{A}_F, \; \tilde{B}_F, \; \text{and } \tilde{C}_F.$$

More importantly, since matrix J_{c0} is invertible, the above transformations are reversible; that is, the filter matrices can be obtained from the feasible solution to the transformed LMI constraints.

Now, it is in a position to present the first method for H_2 filter design for nominal systems in the continuous-time case.

Theorem 2.1 *Consider the continuous-time system $\mathcal{S}$ in* (2.1) *and assume that $M \in \mathcal{M}$ is fixed but arbitrary. Given a scalar $\gamma > 0$, a filter $\mathcal{F}$ in* (2.4) *exists such that the filtering error system $\mathcal{E}$ in* (2.5) *is asymptotically stable and satisfies* $\mathbf{E}\left[e(t)^{\mathrm{T}}e(t)\right] < \gamma$ *if and only if the inequalities in* (2.8) *and in the following*

$$\begin{bmatrix} \tilde{P}_{c1} & \tilde{P}_{c2} & L^{\mathrm{T}} \\ \tilde{P}_{c2} & \tilde{P}_{c2} & -\tilde{C}_F^{\mathrm{T}} \\ L & -\tilde{C}_F & Z_c \end{bmatrix} > 0, \tag{2.15}$$

$$\begin{bmatrix} X_{c1} + X_{c1}^{\mathrm{T}} & \tilde{A}_F + X_{c2}^{\mathrm{T}} & \tilde{P}_{c1}B + \tilde{B}_F D \\ * & \tilde{A}_F + \tilde{A}_F^{\mathrm{T}} & \tilde{P}_{c2}B + \tilde{B}_F D \\ * & * & -I \end{bmatrix} < 0 \tag{2.16}$$

are feasible in the real matrix variables $\tilde{P}_{c1}$, $\tilde{P}_{c2}$, Z_c, $\tilde{A}_F$, $\tilde{B}_F$, *and* $\tilde{C}_F$, *where*

$$X_{c1} \triangleq \tilde{P}_{c1}A + \tilde{B}_F C, \; X_{c2} \triangleq \tilde{P}_{c2}A + \tilde{B}_F C.$$

Moreover, if these conditions are feasible, an admissible state-space realization of the filter $\mathcal{F}$ in (2.4) *is given by*

$$\begin{bmatrix} A_F & B_F \\ C_F & 0 \end{bmatrix} = \begin{bmatrix} \tilde{P}_{c2}^{-1} & 0 \\ 0 & I \end{bmatrix} \begin{bmatrix} \tilde{A}_F & \tilde{B}_F \\ \tilde{C}_F & 0 \end{bmatrix} \tag{2.17}$$

or

$$\begin{bmatrix} A_F & B_F \\ C_F & 0 \end{bmatrix} = \begin{bmatrix} \tilde{A}_F & \tilde{B}_F \\ \tilde{C}_F & 0 \end{bmatrix} \begin{bmatrix} \tilde{P}_{c2}^{-1} & 0 \\ 0 & I \end{bmatrix}. \tag{2.18}$$

Proof By combining the above discussion and Lemma 2.1, the necessity of the conditions in (2.15) and (2.16) can be obtained.

The sufficiency of the conditions in (2.15) and (2.16) follows from the fact that the matrix transformations embedded in the relation between Lemma 2.1 and Theorem 2.1 are reversible. To elucidate this more clearly, note that from (2.15), $\tilde{P}_{c2}$ is nonsingular. Thus, it is always possible to find two matrices P_{c2} and P_{c3} such that $\tilde{P}_{c2} = P_{c2}P_{c3}^{-1}P_{c2}^{\mathrm{T}}$ holds with both P_{c2} and P_{c3} invertible. Then, the following definitions are meaningful

$$\begin{bmatrix} A_F & B_F \\ C_F & 0 \end{bmatrix} \triangleq \begin{bmatrix} P_{c2}^{-1} & 0 \\ 0 & I \end{bmatrix} \begin{bmatrix} \tilde{A}_F & \tilde{B}_F \\ \tilde{C}_F & 0 \end{bmatrix} \begin{bmatrix} P_{c2}^{-\mathrm{T}}P_{c3} & 0 \\ 0 & I \end{bmatrix}, \tag{2.19}$$

$$P_c \triangleq J_{c0}^{-\mathrm{T}} \begin{bmatrix} \tilde{P}_{c1} & \tilde{P}_{c2} \\ \tilde{P}_{c2} & \tilde{P}_{c2} \end{bmatrix} J_{c0}^{-1},$$

which are just the reversed ones in (2.11)–(2.14). With these definitions, the conditions in (2.15) and (2.16) can be rewritten as

$$J_{c1}^{\mathrm{T}} \begin{bmatrix} P_c & \bar{C}^{\mathrm{T}} \\ \bar{C} & Z_c \end{bmatrix} J_{c1} < 0,$$

$$J_{c2}^{\mathrm{T}} \begin{bmatrix} \bar{A}^{\mathrm{T}} P_c + P_c \bar{A} & P_c \bar{B} \\ \bar{B}^{\mathrm{T}} P_c & -I \end{bmatrix} J_{c2} < 0.$$

Due to the nonsingularity of J_{c1} and J_{c2}, the above two conditions are equivalent to (2.9) and (2.10), respectively. Hence, (2.15) and (2.16) are sufficient conditions for (2.9) and (2.10), respectively. Consequently, (2.8), (2.15), and (2.15) are also necessary and sufficient conditions for the existence of a filter with a guaranteed H_2 performance level bound $\sqrt{\gamma}$.

From the expressions of A_F, B_F, and C_F in (2.19), the transfer function of the filter satisfies

$$\begin{aligned} T_F(\delta) &= C_F\,(\delta I - A_F)^{-1}\,B_F \\ &= \left(\tilde{C}_F P_{c2}^{-\mathrm{T}} P_{c3}\right)\left(\delta I - P_{c2}^{-1}\tilde{A}_F P_{c2}^{-\mathrm{T}} P_{c3}\right)^{-1}\left(P_{c2}^{-1}\tilde{B}_F\right) \\ &= \tilde{C}_F\left(P_{c2}^{-\mathrm{T}} P_{c3}\right)\left(\delta I - P_{c2}^{-1}\tilde{A}_F P_{c2}^{-\mathrm{T}} P_{c3}\right)^{-1}\left(P_{c2}^{-\mathrm{T}} P_{c3}\right)^{-1}\left(\tilde{P}_{c2}^{-1}\tilde{B}_F\right) \\ &= \tilde{C}_F\left(\delta I - \tilde{P}_{c2}^{-1}\tilde{A}_F\right)^{-1}\left(\tilde{P}_{c2}^{-1}\tilde{B}_F\right) \end{aligned}$$

and

$$\begin{aligned} T_F(\delta) &= \left(\tilde{C}_F P_{c2}^{-\mathrm{T}} P_{c3}\right)\left(\delta I - P_{c2}^{-1}\tilde{A}_F P_{c2}^{-\mathrm{T}} P_{c3}\right)^{-1}\left(P_{c2}^{-1}\tilde{B}_F\right) \\ &= \left(\tilde{C}_F P_{c2}^{-\mathrm{T}} P_{c3}\right)\left(P_{c2}^{-1} P_{c2}\right)\left(\delta I - P_{c2}^{-1}\tilde{A}_F P_{c2}^{-\mathrm{T}} P_{c3}\right)^{-1} P_{c2}^{-1}\tilde{B}_F \\ &= \left(\tilde{C}_F \tilde{P}_{c2}^{-1}\right)\left(\delta I - \tilde{A}_F \tilde{P}_{c2}^{-1}\right)^{-1}\tilde{B}_F, \end{aligned}$$

which show that the transfer function is independent of the choice of P_{c2} and P_{c3} and the matrices in Eqs. (2.17) and (2.18) both are admissible realizations for the filter $\mathcal{F}$. The entire proof is completed. □

Remark 2.2 Given system $\mathcal{S}$ and filter $\mathcal{F}$, Theorem 2.1 can also be applied to filter performance analysis. An interesting fact is that the conditions in Theorem 2.1 involve fewer variables than those in Lemma 2.1. However, as is proven, Theorem 2.1 and Lemma 2.1 actually are equivalent to each other and their equivalence can be established by some reversible matrix transformations in (2.11)–(2.14). This point implies that some variables in Lemma 2.1 are redundant. The procedures in the proof of sufficiency show that the additional variables in Lemma 2.1 are introduced when $\tilde{P}_{c2}$ is decomposed as $\tilde{P}_{c2} = P_{c2} P_{c3}^{-1} P_{c2}^{\mathrm{T}}$. Indeed, the proof of sufficiency can be completed by a simpler setting as $\tilde{P}_{c2} = P_{c2} = P_{c3}$. This observation reveals that matrix P_c in Lemma 2.1 can be set as

$$P_c = \begin{bmatrix} P_{c1} & P_{c2} \\ P_{c2} & P_{c2} \end{bmatrix}$$

without any conservatism introduced. Moreover, based on this interesting observation, the linearization procedure in the proof of Theorem 2.1 can be simplified.

By Theorem 2.1, the H_2 filter design problem for continuous-time nominal systems has been cast into a feasibility problem with LMI constraints. Thus, the H_2 filter design method based on Theorem 2.1 can be readily implemented. Moreover, γ is also a linear variable in (2.8), which can be directly minimized by virtue of the LMI tool. The optimal H_2 filter design based on Theorem 2.1 is summarized in Algorithm 1.

Algorithm 1 H_2 Filter Design I (Continuous-Time)

- Solve the minimization problem

$$\min_{\tilde{P}_{c1},\, \tilde{P}_{c2},\, Z_c,\, \tilde{A}_F,\, \tilde{B}_F,\, \tilde{C}_F,\, \gamma} \gamma \ \text{ s.t. } (2.8),\ (2.15),\ \text{and } (2.16)$$

- Compute filter $\mathcal{F}$ in (2.4) by (2.17) or (2.18).

Revisiting the conditions in Theorem 2.1, it is found that the matrix variable $\tilde{C}_F$ appears only in the LMI (2.15); thus, by utilizing Lemma 1.3, this variable can be eliminated from the condition such that the total number of variables involved in Algorithm 1 can be further reduced.

To verify this point, by using the Schur complement, the LMI in (2.15) can be rewritten in the equivalent form

$$\begin{bmatrix} \tilde{P}_{c2} - \tilde{P}_{c2}\tilde{P}_{c1}^{-1}\tilde{P}_{c2} & -\tilde{C}_F^{\mathrm{T}} - \tilde{P}_{c2}\tilde{P}_{c1}^{-1}L^{\mathrm{T}} \\ -\tilde{C}_F - L\tilde{P}_{c1}^{-1}\tilde{P}_{c2} & Z_c - L\tilde{P}_{c1}^{-1}L^{\mathrm{T}} \end{bmatrix} > 0,$$

which, by applying Lemma 1.3, is further equivalent to the following pair of inequalities

$$\tilde{P}_{c2} - \tilde{P}_{c2}\tilde{P}_{c1}^{-1}\tilde{P}_{c2} > 0,\ Z_c - L\tilde{P}_{c1}^{-1}L^{\mathrm{T}} > 0$$

together with $\tilde{C}_F = -L\tilde{P}_{c1}^{-1}\tilde{P}_{c2}$. Again using the Schur complement, the pair of inequalities can be written in the LMI form

$$\begin{bmatrix} \tilde{P}_{c1} & \tilde{P}_{c2} \\ \tilde{P}_{c2} & \tilde{P}_{c2} \end{bmatrix} > 0,\ \begin{bmatrix} \tilde{P}_{c1} & L^{\mathrm{T}} \\ L & Z_c \end{bmatrix} > 0. \tag{2.20}$$

Furthermore, for any $n \times n$ invertible matrix P_{c2}, $\tilde{P}_{c2}$ can always be decomposed as $\tilde{P}_{c2} = P_{c2}P_{c3}^{-1}P_{c2}^{\mathrm{T}}$ with P_{c3} also being an invertible matrix. Based on the matrix transformations in (2.13) and (2.14), the above variable elimination procedure yields

$$\begin{aligned} C_F &= \tilde{C}_F P_{c2}^{-\mathrm{T}} P_{c3} = -L\tilde{P}_{c1}^{-1}\tilde{P}_{c2}P_{c2}^{-\mathrm{T}}P_{c3} \\ &= -L\tilde{P}_{c1}^{-1}P_{c2}P_{c3}^{-1}P_{c2}^{\mathrm{T}}P_{c2}^{-\mathrm{T}}P_{c3} \\ &= -L\tilde{P}_{c1}^{-1}P_{c2}. \end{aligned}$$

Due to the arbitrariness of P_{c2}, it can be chosen as $P_{c2} = -\tilde{P}_{c1}$ with no conservatism introduced, which results in $C_F = L$. This imposition obviously further reduces the number of free variables in the H_2 filter design method.

According to the above discussion, we present another method for H_2 filter design for continuous-time nominal systems.

Theorem 2.2 *Consider the continuous-time system $\mathcal{S}$ in* (2.1) *and assume that $M \in \mathcal{M}$ is fixed but arbitrary. Given a scalar $\gamma > 0$, a filter $\mathcal{F}$ in* (2.4) *exists such that the filtering error system $\mathcal{E}$ in* (2.5) *is asymptotically stable and satisfies $\mathbf{E}\left[e(t)^{\mathrm{T}}e(t)\right] < \gamma$ if and only if the inequalities in* (2.8), (2.16), *and* (2.20) *are feasible in the real matrix variables $\tilde{P}_{c1}$, $\tilde{P}_{c2}$, Z_c, $\tilde{A}_F$, and $\tilde{B}_F$. Moreover, if these conditions are feasible, an admissible state-space realization of the filter $\mathcal{F}$ in* (2.4) *is given by*

$$A_F = \tilde{P}_{c1}^{-1}\tilde{A}_F\tilde{P}_{c2}^{-1}\tilde{P}_{c1},\ B_F = -\tilde{P}_{c1}^{-1}\tilde{B}_F,\ C_F = L. \qquad (2.21)$$

Proof The sufficiency and necessity of the conditions can be proven according to the above discussion. Now we demonstrate the admissible realization of A_F and B_F in (2.21). From $P_{c2} = -\tilde{P}_{c1}$, $\tilde{P}_{c2} = P_{c2}P_{c3}^{-1}P_{c2}^{\mathrm{T}}$ and the expression of A_F and B_F in (2.19), it follows that

$$\begin{aligned} A_F &= P_{c2}^{-1}\tilde{A}_F(P_{c3}^{-1}P_{c2}^{\mathrm{T}})^{-1} \\ &= -\tilde{P}_{c1}^{-1}\tilde{A}_F(P_{c2}^{-1}P_{c2}P_{c3}^{-1}P_{c2}^{\mathrm{T}})^{-1} \\ &= \tilde{P}_{c1}^{-1}\tilde{A}_F\tilde{P}_{c2}^{-1}\tilde{P}_{c1}, \\ B_F &= P_{c2}^{-1}\tilde{B}_F = -\tilde{P}_{c1}^{-1}\tilde{B}_F. \end{aligned}$$

□

Based on Theorem 2.2, the second method for the optimal H_2 filter design is summarized in Algorithm 2.

Remark 2.3 As shown in Theorem 2.2, to design H_2 filters, the filter matrix C_F can be prescribed as $C_F = L$, which does not bring about conservatism. The basis of this point lies in the relationship between H_2 filtering and Kalman filtering. For nominal systems, it is well known that the optimal H_2 filter and the optimal Kalman filter have the same filtering error variance level bound [6]. Note that $C_F = L$

Algorithm 2 H_2 Filter Design II (Continuous-Time)

- Solve the minimization problem

$$\min_{\tilde{P}_{c1},\, \tilde{P}_{c2},\, Z_c,\, \tilde{A}_F,\, \tilde{B}_F,\, \gamma} \gamma \quad \text{s.t. (2.8), (2.16), and (2.20)}$$

- Compute filter $\mathcal{F}$ in (2.4) by (2.21).

is also the filter gain for the Kalman filter. Thus, the fact that C_F takes L for the optimal H_2 filter is quite natural. Indeed, the result of $C_F = L$ has already been pointed out in [4, 6]. Of course, it should be emphasized that the main advantage of Theorems 2.1 and 2.2 is not to handle precisely known systems or nominal systems, because the optimal H_2 filtering for nominal systems can be designed by the Kalman filtering method, which is much simpler than Theorem 2.1 and 2.2 (see Sect. 2.1.3). The main advantage of Theorem 2.1, by virtue of the quadratic stability notion, is its extendability to polytopic uncertain systems bounded in a convex polyhedral domain $\mathcal{M}$ (see Sect. 2.1.6).

2.1.3 Connection to the Kalman filtering: Continuous-Time

As mentioned in Remark 2.3, the optimal H_2 filter and the Kalman filter achieve the same variance level of the estimation error for nominal systems. In this section, we provide some theoretical analysis to relate the H_2 filter design method in Sect. 2.1.2 and the continuous-time Kalman filtering method.

To design the Kalman filter for the system $\mathcal{S}$ in (2.1) as in [6], we assume

$$BD^{\mathrm{T}} = 0, \; DD^{\mathrm{T}} = I.$$

This assumption implicates that the noise $w(t)$ has two parts of remarkably different meanings: the first one acts as the real process noise through matrix B, the second one as the measurement noise through matrix D, and the two parts are independent of each other. The Kalman filter for the system $\mathcal{S}$ with nominal parameters in the continuous-time case has the form of filter $\mathcal{F}$ in (2.4) with state space matrices given by

$$A_F = A - PC^{\mathrm{T}}C, \; B_F = PC^{\mathrm{T}}, \; C_F = L \tag{2.22}$$

where matrix P is the solution to the following Riccati equation:

$$AP + PA^{\mathrm{T}} - PC^{\mathrm{T}}CP + BB^{\mathrm{T}} = 0. \tag{2.23}$$

To relate the Kalman filter realization in (2.22) with the H_2 filter realization in (2.21), rewrite the LMI in (2.16) into

$$\Phi + X^{\mathrm{T}} \tilde{A}_F Z + Z^{\mathrm{T}} \tilde{A}_F^{\mathrm{T}} X < 0 \tag{2.24}$$

where

$$\Phi = \begin{bmatrix} X_{c1} + X_{c1}^{\mathrm{T}} & X_{c2}^{\mathrm{T}} & \tilde{P}_{c1} B + \tilde{B}_F D \\ * & 0 & \tilde{P}_{c2} B + \tilde{B}_F D \\ * & * & -I \end{bmatrix},$$
$$X = \begin{bmatrix} I & I & 0 \end{bmatrix}, Z = \begin{bmatrix} 0 & I & 0 \end{bmatrix}.$$

Choose $\mathcal{N}_X$ and $\mathcal{N}_Z$ for the nullspaces of X and Z, respectively, as

$$\mathcal{N}_X = \begin{bmatrix} I & 0 \\ -I & 0 \\ 0 & I \end{bmatrix}, \mathcal{N}_Z = \begin{bmatrix} I & 0 \\ 0 & 0 \\ 0 & I \end{bmatrix}.$$

Then by applying the Projection Lemma, the inequality in (2.24) is equivalent to

$$\mathcal{N}_X^{\mathrm{T}} \Phi \mathcal{N}_X = \begin{bmatrix} A^{\mathrm{T}} (\tilde{P}_{c1} - \tilde{P}_{c2}) + (\tilde{P}_{c1} - \tilde{P}_{c2}) A & (\tilde{P}_{c1} - \tilde{P}_{c2}) B \\ * & -I \end{bmatrix} < 0, \tag{2.25}$$

$$\mathcal{N}_Z^{\mathrm{T}} \Phi \mathcal{N}_Z = \begin{bmatrix} X_{c1} + X_{c1}^{\mathrm{T}} & \tilde{P}_{c1} B + \tilde{B}_F D \\ * & -I \end{bmatrix} < 0. \tag{2.26}$$

Since system $\mathcal{S}$ is assumed to be stable, there always exists a symmetric matrix $\hat{P} > 0$ such that

$$A^{\mathrm{T}} \hat{P} + \hat{P} A < 0.$$

Hence, when we choose $\tilde{P}_{c1} - \tilde{P}_{c2} = \epsilon \hat{P}$ with $\epsilon > 0$ being a sufficiently small scalar, the inequality in (2.25) reduces to

$$\mathcal{N}_X^{\mathrm{T}} \Phi \mathcal{N}_X = \begin{bmatrix} \epsilon A^{\mathrm{T}} \hat{P} + \epsilon \hat{P} A & \epsilon \hat{P} B \\ * & -I \end{bmatrix} < 0$$

which automatically holds. What's more, under this choice, the first inequality of (2.20) is also automatically satisfied, because, according to the Schur complement, it is equivalent to

$$\begin{cases} \tilde{P}_{c2} > 0 \\ \tilde{P}_{c1} - \tilde{P}_{c2} = \epsilon \hat{P} > 0 \end{cases}.$$

Note that from the second inequality of (2.20) and the one in (2.8), the H_2 performance bound $\sqrt{\gamma}$ is not directly relevant to $\tilde{P}_{c2}$, which implies that the choice of $\tilde{P}_{c2}$ satisfying $\tilde{P}_{c1} - \tilde{P}_{c2} = \epsilon \hat{P}$ does not introduce conservatism for computing γ. Consequently, γ_{H_2} is larger than but can get arbitrarily close to γ_1, where γ_{H_2} is the minimum of γ in Algorithm 2 and γ_1 is defined as

$$\gamma_1 \triangleq \left(\min_{\tilde{P}_{c1}, \tilde{B}_F, Z_c, \gamma} \quad \mathrm{Tr}[Z_c] \text{ s.t. (2.26) and } \begin{bmatrix} \tilde{P}_{c1} & L^{\mathrm{T}} \\ L & Z_c \end{bmatrix} > 0 \right).$$

By using the Schur complement and the assumption $BD^{\mathrm{T}} = 0$ and $DD^{\mathrm{T}} = I$, the inequality in (2.26) reduces to

$$\begin{aligned} & X_{c1} + X_{c1}^{\mathrm{T}} + \left(\tilde{P}_{c1}B + \tilde{B}_F D\right)\left(\tilde{P}_{c1}B + \tilde{B}_F D\right)^{\mathrm{T}} \\ & = \tilde{P}_{c1}A + A^{\mathrm{T}}\tilde{P}_{c1} + \tilde{P}_{c1}BB^{\mathrm{T}}\tilde{P}_{c1} + \tilde{B}_F C + C^{\mathrm{T}}\tilde{B}_F^{\mathrm{T}} + \tilde{B}_F\tilde{B}_F^{\mathrm{T}} \\ & < 0. \end{aligned} \tag{2.27}$$

Noting the fact that

$$\tilde{B}_F C + C^{\mathrm{T}}\tilde{B}_F^{\mathrm{T}} + \tilde{B}_F\tilde{B}_F^{\mathrm{T}} \geq -C^{\mathrm{T}}C$$

always holds, where the equality holds only for $\tilde{B}_F = -C^{\mathrm{T}}$, one sees that the inequality in (2.27) must hold for $\tilde{B}_F = -C^{\mathrm{T}}$. Hence, the inequality in (2.26) can be replaced by

$$\tilde{P}_{c1}A + A^{\mathrm{T}}\tilde{P}_{c1} + \tilde{P}_{c1}BB^{\mathrm{T}}\tilde{P}_{c1} - C^{\mathrm{T}}C < 0$$

or by performing a congruence transformation with $\tilde{P}_{c1}^{-1}$ and defining $\tilde{P} \triangleq \tilde{P}_{c1}^{-1}$,

$$A\tilde{P} + \tilde{P}A^{\mathrm{T}} + BB^{\mathrm{T}} - \tilde{P}C^{\mathrm{T}}C\tilde{P} < 0. \tag{2.28}$$

Consequently, by the Schur complement, we have

$$\begin{aligned} \gamma_1 &= \left(\min_{\tilde{P}, Z_c} \mathrm{Tr}[Z_c] \text{ s.t. (2.28) and } \begin{bmatrix} \tilde{P}^{-1} & L^{\mathrm{T}} \\ L & Z_c \end{bmatrix} > 0 \right) \\ &> \left(\min_{\tilde{P}} \mathrm{Tr}[L\tilde{P}L^{\mathrm{T}}] \text{ s.t. (2.28)} \right) \triangleq \gamma_2 \end{aligned}$$

Also, γ_1 can get arbitrarily close to γ_2.

In view of (2.28), and according to the monotonicity of the solution to the Riccati equation in (2.23), we have $\tilde{P} > P$, and

$$\gamma_2 = \left(\min_{\tilde{P}} \mathrm{tr}[L\tilde{P}L^{\mathrm{T}}] \text{ s.t. (2.28).}\right) > \mathrm{tr}[LPL^{\mathrm{T}}] \triangleq \gamma_{\mathrm{Kalman}},$$

where γ_{Kalman} is the H_2 performance of the Kalman filter. Moreover, $\gamma_2 \to \gamma_{\mathrm{Kalman}}$ when $\tilde{P} \to P$, so that γ_2 can get arbitrarily close to γ_{Kalman}. Consequently, we have established that γ_{H_2} is larger than but can get arbitrarily close to γ_{Kalman}, that is, γ_{Kalman} is the optimal H_2 filtering performance that can be achieved by general LTI filters. This interprets the optimality of the Kalman filter in the H_2 sense.

From the above derivations, we have recovered the Kalman filter gains

$$B_F = -\tilde{P}_{c1}^{-1}\tilde{B}_F = \tilde{P}C^{\mathrm{T}}, C_F = L.$$

In the following, we show how to obtain the gain A_F for the Kalman filter from the solution of $\tilde{P}$. To this end, performing a congruence transformation to the inequality in (2.16), we have

$$\begin{bmatrix} I & 0 & 0 \\ -I & I & 0 \\ 0 & 0 & I \end{bmatrix} \begin{bmatrix} X_{c1} + X_{c1}^{\mathrm{T}} & \tilde{A}_F + X_{c2}^{\mathrm{T}} & \tilde{P}_{c1}B + \tilde{B}_F D \\ * & \tilde{A}_F + \tilde{A}_F^{\mathrm{T}} & \tilde{P}_{c2}B + \tilde{B}_F D \\ * & * & -I \end{bmatrix} \begin{bmatrix} I & -I & 0 \\ 0 & I & 0 \\ 0 & 0 & I \end{bmatrix}$$
$$= \begin{bmatrix} X_{c1} + X_{c1}^{\mathrm{T}} & \tilde{A}_F - X_{c1} + A^{\mathrm{T}}(\tilde{P}_{c2} - \tilde{P}_{c1})^{\mathrm{T}} & \tilde{P}_{c1}B + \tilde{B}_F D \\ * & (\tilde{P}_{c1} - \tilde{P}_{c2})A + A^{\mathrm{T}}(\tilde{P}_{c2} - \tilde{P}_{c1})^{\mathrm{T}} & (\tilde{P}_{c2} - \tilde{P}_{c1})B \\ * & * & -I \end{bmatrix}$$
$$< 0.$$

Substituting $\tilde{P}_{c1} - \tilde{P}_{c2} = \epsilon\hat{P}$ into the above inequality, we get

$$\begin{bmatrix} X_{c1} + X_{c1}^{\mathrm{T}} & \tilde{A}_F - X_{c1} - \epsilon A^{\mathrm{T}}\hat{P} & \tilde{P}_{c1}B + \tilde{B}_F D \\ * & \epsilon\hat{P}A + \epsilon A^{\mathrm{T}}\hat{P} & -\epsilon\hat{P}B \\ * & * & -I \end{bmatrix} < 0. \tag{2.29}$$

The previous derivations show that the inequality in (2.16) holds for ϵ sufficiently small, so does the one in (2.29). As $\epsilon \to 0^+$, we need to impose $\tilde{A}_F = X_{c1}$, and the inequality in (2.29) finally reduces to the one in (2.26). Moreover, in view of the expression of A_F in (2.21), we have

$$\begin{aligned} A_F &= \tilde{P}_{c1}^{-1}\tilde{A}_F\tilde{P}_{c2}^{-1}\tilde{P}_{c1} \\ &= \tilde{P}_{c1}^{-1}(\tilde{P}_{c1}A + \tilde{B}_F C)\tilde{P}_{c2}^{-1}\tilde{P}_{c1} && (\text{since } \tilde{A}_F = X_{c1} = \tilde{P}_{c1}A + \tilde{B}_F C) \\ &= \left(A - \tilde{P}C^{\mathrm{T}}C\right)\tilde{P}_{c2}^{-1}\tilde{P}^{-1} && (\text{since } \tilde{B}_F = -C^{\mathrm{T}} \text{and} \tilde{P}_{c1}^{-1} = \tilde{P}) \\ &= \left(A - \tilde{P}C^{\mathrm{T}}C\right)(\tilde{P}^{-1} - \epsilon\hat{P})^{-1}\tilde{P}^{-1} && (\text{since } \tilde{P}_{c1} - \tilde{P}_{c2} = \epsilon\hat{P}) \\ &= \left(A - \tilde{P}C^{\mathrm{T}}C\right) && (\text{as } \epsilon \to 0^+) \end{aligned}$$

Table 2.1 Step-by-step procedures relating the optimal H_2 design and the Kalman filter design

Filter	Procedure	Purpose	Matrix transfer
H_2			
$\Updownarrow$	Theorem 2.1 $\leftrightarrow$ Theorem 2.2	Eliminate C_F	$C_F = L$
H_2			
$\Updownarrow$	$\gamma_{H_2} > \gamma_1$ and $\gamma_{H_2} \to \gamma_1$	Eliminate A_F and $\tilde{P}_{c2}$	$\tilde{P}_{c2} \to \tilde{P}_{c1}$
H_2			
$\Updownarrow$	$\gamma_1 > \gamma_2$ and $\gamma_1 \to \gamma_2$	Eliminate B_F	$B_F = -C^{\mathrm{T}}$
H_2			
$\Updownarrow$	$\gamma_2 > \gamma_{\mathrm{Kalman}}$ and $\gamma_2 \to \gamma_{\mathrm{Kalman}}$	Recover Kalman filter	$A_F = A - \tilde{P}C^{\mathrm{T}}C$
Kalman			

which is expressed in a form consistent with the Kalman gain for A_F in (2.22) but with P replaced by $\tilde{P}$. We thus have completed the recovery of all the gain matrices of the Kalman filter.

The above analysis has uncovered the relationship between the H_2 filter design method in Sect. 2.1.2 and the Kalman filtering method. For helping reader's understanding, we summarize these procedures and the main points in Table 2.1.

Remark 2.4 It is shown that the optimal H_2 filters can be designed by the Kalman filtering method. When recovering the Kalman filter gains from Theorem 2.1 or 2.2, we need to specifically choose some matrices that depend on the plant parameters, so as to keep the optimality of each procedure. Once the plant has uncertainty, the optimality of each procedure would be broken, resulting in the loss of optimality for the Kalman filter. Hence, an inevitable drawback of the Kalman filter is that its optimality depends strongly on the plant parameters, such that the Kalman filter inherently is not robust against parametric uncertainty.

2.1.4 H_2 Filtering for Nominal Systems: Discrete-Time

In Sect. 2.1.2, we have considered the analysis and synthesis problems of H_2 filters for continuous-time systems. In this subsection, we deal with the H_2 filtering problem for discrete-time systems. Sufficient and necessary conditions will be presented in terms of LMI for analysis and synthesis of H_2 filters. At first, an H_2 performance criterion for discrete-time nominal systems is given in terms of LMI as follows.

Lemma 2.2 ([1, 3]) *Consider the discrete-time system $\mathcal{S}$ in* (2.1) *and assume that $M \in \mathcal{M}$ is fixed but arbitrary. Given filter $\mathcal{F}$ in* (2.4) *and a scalar $\gamma > 0$, the filtering error system $\mathcal{E}$ in* (2.5) *is asymptotically stable and satisfies* $\mathbf{E}\left[e(t)^{\mathrm{T}}e(t)\right] < \gamma$ *if and only if the following matrix inequalities*

$$Tr[Z_d] < \gamma, \tag{2.30}$$

$$\begin{bmatrix} P_d & \bar{C}^{\mathrm{T}} \\ \bar{C} & Z_d \end{bmatrix} > 0, \tag{2.31}$$

$$\begin{bmatrix} P_d & P_d\bar{A} & P_d\bar{B} \\ \bar{A}^{\mathrm{T}}P_d & P_d & 0 \\ \bar{B}^{\mathrm{T}}P_d & 0 & I \end{bmatrix} > 0 \tag{2.32}$$

are feasible in the real matrix variables P_d and Z_d.

In the context of discrete-time systems, the above result can be briefly justified as follows. First note that $\mathbf{E}\left[e(t)^{\mathrm{T}}e(t)\right] = \mathrm{Tr}[\bar{C}P\bar{C}^{\mathrm{T}}]$, where P is the solution to the discrete-time Lyapunov equation

$$\bar{A}P\bar{A}^{\mathrm{T}} - P + \bar{B}\bar{B}^{\mathrm{T}} = 0.$$

Pre- and post-multiplying the both hand sides of (2.32) by diag $\left\{P_d^{-1},\ P_d^{-1},\ I\right\}$ and using the Schur complement, we have

$$\bar{A}P_d^{-1}\bar{A}^{\mathrm{T}} - P_d^{-1} + \bar{B}\bar{B}^{\mathrm{T}} < 0.$$

The monotonicity property of the solution to the Lyapunov equation ensures $P < P_d^{-1}$, which together with (2.30) and (2.31) further results in

$$\mathbf{E}\left[e(t)^{\mathrm{T}}e(t)\right] = \mathrm{Tr}[\bar{C}P\bar{C}^{\mathrm{T}}] < \mathrm{Tr}[\bar{C}P_d^{-1}\bar{C}^{\mathrm{T}}] < \mathrm{Tr}[Z_d] < \gamma.$$

Remark 2.5 Note that in [7, 8], the LMI constraints corresponding to (2.31) and (2.32) are given, respectively, by

$$\begin{bmatrix} \bar{P} & \bar{P}\bar{C}^{\mathrm{T}} \\ \bar{C}\bar{P} & Z \end{bmatrix} > 0, \begin{bmatrix} \bar{P} & \bar{A}\bar{P} & \bar{B} \\ \bar{P}\bar{A}^{\mathrm{T}} & \bar{P} & 0 \\ \bar{B}^{\mathrm{T}} & 0 & I \end{bmatrix} > 0.$$

By performing some simple matrix manipulations, it can be verified that these two inequalities are equivalent to (2.31) and (2.32), respectively. The difference of these two formulations is very similar to that for the continuous-time case discussed before. For the case of nominal systems, it does not matter which formulation is employed for H_2 performance analysis. Nevertheless, the feature that product terms between Lyapunov matrix P_d and system matrices in Lemma 2.2 exist only in (2.32) is useful for robustness analysis of the filtering error system.

Given a filter, the above lemma gives a tractable LMI-based condition for computing the optimal H_2 filtering performance level. After performing some matrix transformations to the conditions in Lemma 2.2, the first result on H_2 filter design for discrete-time nominal systems is obtained as follows.

Theorem 2.3 *Consider the discrete-time system $\mathcal{S}$ in* (2.1) *and assume that $M \in \mathcal{M}$ is fixed but arbitrary. Given a scalar $\gamma > 0$, a filter $\mathcal{F}$ in* (2.4) *exists such that the filtering error system $\mathcal{E}$ in* (2.5) *is asymptotically stable and satisfies $\mathbf{E}\left[e(t)^{\mathrm{T}}e(t)\right] < \gamma$ if and only if the inequalities in* (2.30) *and in the following*

$$\begin{bmatrix} \tilde{P}_{d1} & \tilde{P}_{d2} & L^{\mathrm{T}} \\ \tilde{P}_{d2} & \tilde{P}_{d2} & -\tilde{C}_F^{\mathrm{T}} \\ L & -\tilde{C}_F & Z_d \end{bmatrix} > 0, \tag{2.33}$$

$$\begin{bmatrix} \tilde{P}_{d1} & \tilde{P}_{d2} & X_{d1} & \tilde{A}_F & \tilde{P}_{d1}B + \tilde{B}_F D \\ * & \tilde{P}_{d2} & X_{d2} & \tilde{A}_F & \tilde{P}_{d2}B + \tilde{B}_F D \\ * & * & \tilde{P}_{d1} & \tilde{P}_{d2} & 0 \\ * & * & * & \tilde{P}_{d2} & 0 \\ * & * & * & * & I \end{bmatrix} > 0 \tag{2.34}$$

are feasible in the real matrix variables $\tilde{P}_{d1}$, $\tilde{P}_{d2}$, Z_d, $\tilde{A}_F$, $\tilde{B}_F$, and $\tilde{C}_F$, where

$$X_{d1} \triangleq \tilde{P}_{d1}A + \tilde{B}_F C, \; X_{d2} \triangleq \tilde{P}_{d2}A + \tilde{B}_F C.$$

Moreover, if these conditions are feasible, an admissible state-space realization of the filter $\mathcal{F}$ in (2.4) *is given by*

$$\begin{bmatrix} A_F & B_F \\ C_F & 0 \end{bmatrix} = \begin{bmatrix} \tilde{P}_{d2}^{-1} & 0 \\ 0 & I \end{bmatrix} \begin{bmatrix} \tilde{A}_F & \tilde{B}_F \\ \tilde{C}_F & 0 \end{bmatrix} \tag{2.35}$$

or

$$\begin{bmatrix} A_F & B_F \\ C_F & 0 \end{bmatrix} = \begin{bmatrix} \tilde{A}_F & \tilde{B}_F \\ \tilde{C}_F & 0 \end{bmatrix} \begin{bmatrix} \tilde{P}_{d2}^{-1} & 0 \\ 0 & I \end{bmatrix}. \tag{2.36}$$

Proof First, we prove that the feasibility of (2.30), (2.33), and (2.34) is a necessary and sufficient condition for a guaranteed H_2 performance of the filtering error system. The proof methodology and the employed techniques are similar to that of the proof of Theorem 2.1 for continuous-time systems.

Necessity: Suppose that there exist P_d and Z_d such that (2.30)–(2.32) are satisfied. Partition matrix P_d into four blocks as the following form

$$P_d = \begin{bmatrix} P_{d1} & P_{d2} \\ P_{d2}^{\mathrm{T}} & P_{d3} \end{bmatrix}$$

with each block being an $n \times n$ matrix. The nonsingularity of P_d enables us to assume that, by invoking a small perturbation if necessary, P_{d2} is nonsingular. Define multiplier matrix

$$J_{d0} \triangleq \begin{bmatrix} I & 0 \\ 0 & P_{d3}^{-1} P_{d2}^{\mathrm{T}} \end{bmatrix}$$

and introduce the following matrix transformations

$$\tilde{A}_F \triangleq P_{d2} A_F P_{d3}^{-1} P_{d2}^{\mathrm{T}}, \tag{2.37}$$

$$\tilde{B}_F \triangleq P_{d2} B_F, \tag{2.38}$$

$$\tilde{C}_F \triangleq C_F P_{d3}^{-1} P_{d2}^{\mathrm{T}}, \tag{2.39}$$

$$\begin{bmatrix} \tilde{P}_{d1} & \tilde{P}_{d2} \\ \tilde{P}_{d2} & \tilde{P}_{d2} \end{bmatrix} \triangleq J_{d0}^{\mathrm{T}} P_d J_{d0} = \begin{bmatrix} P_{d1} & P_{d2} P_{d3}^{-1} P_{d2}^{\mathrm{T}} \\ P_{d2} P_{d3}^{-1} P_{d2}^{\mathrm{T}} & P_{d2} P_{d3}^{-1} P_{d2}^{\mathrm{T}} \end{bmatrix}. \tag{2.40}$$

Then, the following equations can be obtained

$$J_{d0}^{\mathrm{T}} P_d \bar{A} J_{d0} = \begin{bmatrix} X_{d1} & \tilde{A}_F \\ X_{d2} & \tilde{A}_F \end{bmatrix},$$

$$J_0^{\mathrm{T}} P_d \bar{B} = \begin{bmatrix} \tilde{P}_{d1} B + \tilde{B}_F D \\ \tilde{P}_{d2} B + \tilde{B}_F D \end{bmatrix},$$

$$\bar{C} J_{d0} = \begin{bmatrix} L & -\tilde{C}_F \end{bmatrix}.$$

Furthermore, by defining $J_{d1} \triangleq \mathrm{diag}\{J_{d0}, I_n\}$ and $J_{d1} \triangleq \mathrm{diag}\{J_{d0}, J_{d0}, I_l\}$ and noting these given equations, it readily follows that

$$J_{d1}^{\mathrm{T}} \begin{bmatrix} P_d & \bar{C}^{\mathrm{T}} \\ \bar{C} & Z_d \end{bmatrix} J_{d1} = \begin{bmatrix} \tilde{P}_{d1} & \tilde{P}_{d2} & L^{\mathrm{T}} \\ \tilde{P}_{d2} & \tilde{P}_{d2} & -\tilde{C}_F^{\mathrm{T}} \\ L & -\tilde{C}_F & Z_d \end{bmatrix},$$

$$J_{d2}^{\mathrm{T}} \begin{bmatrix} P_d & P_d \bar{A} & P_d \bar{B} \\ \bar{A}^{\mathrm{T}} P_d & P_d & 0 \\ \bar{B}^{\mathrm{T}} P_d & 0 & I \end{bmatrix} J_{d2} = \begin{bmatrix} \tilde{P}_{d1} & \tilde{P}_{d2} & X_{d1} & \tilde{A}_F & \tilde{P}_{d1} B + \tilde{B}_F D \\ * & \tilde{P}_{d2} & X_{d2} & \tilde{A}_F & \tilde{P}_{d2} B + \tilde{B}_F D \\ * & * & \tilde{P}_{d1} & \tilde{P}_{d2} & 0 \\ * & * & * & \tilde{P}_{d2} & 0 \\ * & * & * & * & I \end{bmatrix},$$

which together with (2.30) imply that if there exist variables P_d and Z_d such that (2.30)–(2.32) hold, there always exist variables $\tilde{P}_{d1}$, $\tilde{P}_{d2}$, $\tilde{A}_F$, $\tilde{B}_F$, and $\tilde{C}_F$ defined in (2.37)–(2.40) and variable Z_d such that (2.30), (2.33), and (2.34) hold. Then by virtue of Lemma 2.2, the necessity has been established.

Sufficiency: Suppose that there exist $\tilde{P}_{d1}$, $\tilde{P}_{d2}$, $Z_d \tilde{A}_F$, $\tilde{B}_F$, and $\tilde{C}_F$ such that (2.30), (2.33), and (2.34) are satisfied. Due to (2.33), $\tilde{P}_{d2}$ is nonsingular, which means that it is always possible to find two invertible matrices P_{d2} and P_{d3} such that $\tilde{P}_{d2} = P_{d2} P_{d3}^{-1} P_{d2}^{\mathrm{T}}$. Introduce the following definitions

$$\begin{bmatrix} A_F & B_F \\ C_F & 0 \end{bmatrix} \triangleq \begin{bmatrix} P_{d2}^{-1} & 0 \\ 0 & I \end{bmatrix} \begin{bmatrix} \tilde{A}_F & \tilde{B}_F \\ \tilde{C}_F & 0 \end{bmatrix} \begin{bmatrix} P_{d2}^{-\mathrm{T}} P_{d3} & 0 \\ 0 & I \end{bmatrix}, \tag{2.41}$$

$$P_d \triangleq J_{d0}^{-\mathrm{T}} \begin{bmatrix} \tilde{P}_{d1} & \tilde{P}_{d2} \\ \tilde{P}_{d2} & \tilde{P}_{d2} \end{bmatrix} J_{d0}^{-1}. \tag{2.42}$$

By these definitions, the conditions in (2.33) and (2.34) can be rewritten as

$$J_{d1}^{\mathrm{T}} \begin{bmatrix} P_d & \bar{C}^{\mathrm{T}} \\ \bar{C} & Z_d \end{bmatrix} J_{d1} < 0,$$

$$J_{d2}^{\mathrm{T}} \begin{bmatrix} P_d & P_d\bar{A} & P_d\bar{B} \\ \bar{A}^{\mathrm{T}} P_d & P_d & 0 \\ \bar{B}^{\mathrm{T}} P_d & 0 & I \end{bmatrix} J_{d2} < 0.$$

Since J_{d1} and J_{d2} both are nonsingular, the above two conditions imply that there exist variables P_d in (2.42), Z_d together with filter realization $(A_F,\ B_{F,}\ C_F)$ in (2.41) such that (2.31) and (2.32) hold. Consequently, by Lemma 2.2, (2.30), (2.33), and (2.34) sufficiently ensure the existence of a filter with a guaranteed H_2 performance level bound $\sqrt{\gamma}$.

Substituting the expression of A_F, B_F, and C_F in (2.41) into the transfer function of the filter and performing some matrix manipulations similar to that in the proof of Theorem 2.1, one sees that (2.35) and (2.36) both are admissible realizations for the filter $\mathcal{F}$. The entire proof is completed. □

Remark 2.6 Similar to continuous-time case, it is also valid that some variables in Lemma 2.2 are redundant, and these additional variables are also introduced when $\tilde{P}_{d2}$ is decomposed as $\tilde{P}_{d2} = P_{d2}P_{d3}^{-1}P_{d2}^{\mathrm{T}}$. Hence, with no conservatism introduced, it can be set as $\tilde{P}_{d2} = P_{d2} = P_{d3}$, and P_d in Lemma2.1 reduces to the following form

$$P_d = \begin{bmatrix} P_{d1} & P_{d2} \\ P_{d2} & P_{d2} \end{bmatrix}.$$

By Theorem 2.3, the first method for designing the optimal H_2 filters for discrete-time nominal systems is formulated in Algorithm 3.

Algorithm 3 H_2 Filter Design I (Discrete-Time)

- Solve the minimization problem

$$\min_{\tilde{P}_{d1},\,\tilde{P}_{d2},\,Z_d,\,\tilde{A}_F,\,\tilde{B}_F,\,\tilde{C}_F,\,\gamma} \gamma \quad \text{s.t. (2.30), (2.33) and(2.34)}$$

- Compute filter $\mathcal{F}$ in (2.4) by (2.35) or (2.36).

Note that the LMI in (2.33) has the same form as the one in (2.15). Therefore, Lemma 1.3 can also be applied to (2.33) such that the number of the free variables in Theorem 2.3 can be reduced. Similar to the derivation of Theorem 2.2, the choice for P_{d2} could be $P_{d2} = -\tilde{P}_{d1}$ and the elimination procedure finally results in a fixed value for $\tilde{C}_F$ as $\tilde{C}_F = L$. Accordingly, we obtain another method for H_2 filter design for discrete-time nominal systems.

Theorem 2.4 *Consider the discrete-time system* $\mathcal{S}$ *in* (2.1) *and assume that* $M \in \mathcal{M}$ *is fixed but arbitrary. Given a scalar* $\gamma > 0$*, a filter* $\mathcal{F}$ *in* (2.4) *exists such that the filtering error system* $\mathcal{E}$ *in* (2.5) *is asymptotically stable and satisfies* $\mathbf{E}\left[e(t)^{\mathrm{T}} e(t)\right] < \gamma$ *if and only if the LMIs in (2.30), (2.34) and in the following*

$$\begin{bmatrix} \tilde{P}_{d1} & L^{\mathrm{T}} \\ L & Z_d \end{bmatrix} > 0 \tag{2.43}$$

are feasible in the real matrix variables $\tilde{P}_{d1}$, $\tilde{P}_{d2}$, Z_d, $\tilde{A}_F$, *and* $\tilde{B}_F$. *Moreover, if these conditions are feasible, an admissible state-space realization of the filter* $\mathcal{F}$ *in* (2.4) *is given by*

$$A_F = \tilde{P}_{d1}^{-1} \tilde{A}_F \tilde{P}_{d2}^{-1} \tilde{P}_{d1}, \; B_F = -\tilde{P}_{d1}^{-1} \tilde{B}_F, \; C_F = L. \tag{2.44}$$

Proof The proof is omitted due to its similarity to that of Theorem 2.2. A special point should be pointed out as follows. When Lemma 1.3 is applied to eliminate the variable $\tilde{C}_F$ in (2.33), it reduces to two LMIs:

$$\begin{bmatrix} \tilde{P}_{d1} & \tilde{P}_{d2} \\ \tilde{P}_{d2}^{\mathrm{T}} & \tilde{P}_{d2} \end{bmatrix} > 0, \; \begin{bmatrix} \tilde{P}_{d1} & L^{\mathrm{T}} \\ L & Z_d \end{bmatrix} > 0.$$

Note that the first one has been implied by (2.34). So reserving the second one, i.e., (2.43), suffices to ensure the validity of the results. □

Based on Theorem 2.4, another method for the optimal H_2 filter design for discrete-time nominal systems is given in Algorithm 4.

Algorithm 4 H_2 Filter Design II (Discrete-Time)

- Solve the minimization problem

$$\min_{\tilde{P}_{d1}, \tilde{P}_{d2}, Z_d, \tilde{A}_F, \tilde{B}_F, \gamma} \gamma \;\; \text{s.t.} \; (2.30), \; (2.34), \; \text{and} (2.43)$$

- Compute filter $\mathcal{F}$ in (2.4) by (2.44).

2.1.5 Connection to the Kalman Filtering: Discrete-Time

Following similar discussions as in Sect. 2.1.3, we can also establish the connection between the H_2 filter design methods in Sect. 2.1.4 and the Kalman filtering theory in the discrete-time case. Here, we only provide the main points for illustration.

In order to design the Kalman filter to the system $\mathcal{S}$ in (2.1), similar to the continuous-time case, we need to assume

$$BD^{\mathrm{T}} = 0 \text{ and } DD^{\mathrm{T}} = I.$$

For the system $\mathcal{S}$ in (2.1) with nominal parameters, the discrete-time Kalman filter in the form of the general filter $\mathcal{F}$ in (2.4) is with matrices given by [9]

$$\begin{aligned} A_F &= A - APC^{\mathrm{T}}\left(CPC^{\mathrm{T}} + I\right)^{-1}C, \\ B_F &= APC^{\mathrm{T}}\left(CPC^{\mathrm{T}} + I\right)^{-1}, \\ C_F &= L, \end{aligned}$$

where P is a positive definite matrix that solves the following discrete-time Riccati equation:

$$APA^{\mathrm{T}} - P - APC^{\mathrm{T}}\left(CPC^{\mathrm{T}} + I\right)^{-1}CPA^{\mathrm{T}} + BB^{\mathrm{T}} = 0. \tag{2.45}$$

To relate the Kalman filter gains with the H_2 filter realization obtained by Theorem 2.4, performing a congruence transformation to the inequality in (2.34), we have

$$\begin{aligned} &\begin{bmatrix} I & 0 & 0 & 0 & 0 \\ -I & I & 0 & 0 & 0 \\ 0 & 0 & I & 0 & 0 \\ 0 & 0 & -I & I & 0 \\ 0 & 0 & 0 & 0 & I \end{bmatrix} \begin{bmatrix} \tilde{P}_{d1} & \tilde{P}_{d2} & X_{d1} & \tilde{A}_F & \tilde{P}_{d1}B + \tilde{B}_F D \\ * & \tilde{P}_{d2} & X_{d2} & \tilde{A}_F & \tilde{P}_{d2}B + \tilde{B}_F D \\ * & * & \tilde{P}_{d1} & \tilde{P}_{d2} & 0 \\ * & * & * & \tilde{P}_{d2} & 0 \\ * & * & * & * & I \end{bmatrix} \begin{bmatrix} I & -I & 0 & 0 & 0 \\ 0 & I & 0 & 0 & 0 \\ 0 & 0 & I & -I & 0 \\ 0 & 0 & 0 & I & 0 \\ 0 & 0 & 0 & 0 & I \end{bmatrix} \\ &= \begin{bmatrix} \tilde{P}_{d1} & \tilde{P}_{d2} - \tilde{P}_{d1} & X_{d1} & \tilde{A}_F - X_{d1} & \tilde{P}_{d1}B + \tilde{B}_F D \\ * & \tilde{P}_{d1} - \tilde{P}_{d2} & (\tilde{P}_{d2} - \tilde{P}_{d1})A & (\tilde{P}_{d1} - \tilde{P}_{d2})A & (\tilde{P}_{d2} - \tilde{P}_{d1})B \\ * & * & \tilde{P}_{d1} & \tilde{P}_{d2} - \tilde{P}_{d1} & 0 \\ * & * & * & \tilde{P}_{d1} - \tilde{P}_{d2} & 0 \\ ** & * & * & * & I \end{bmatrix} \\ &> 0. \end{aligned} \tag{2.46}$$

By analogy to the continuous-time case (details are omitted), we can take $\tilde{P}_{d1} - \tilde{P}_{d2} = \epsilon \tilde{P}$, where $\epsilon > 0$ is a sufficient small scalar and $\tilde{P} > 0$ satisfies $\tilde{P} - A^{\mathrm{T}}\tilde{P}A > 0$. Under this choice, the inequality in (2.46) turns to be

$$\begin{bmatrix} \tilde{P}_{d1} & -\epsilon\tilde{P} & X_{d1} & \tilde{A}_F - X_{d1} & \tilde{P}_{d1}B + \tilde{B}_F D \\ * & \epsilon\tilde{P} & -\epsilon\tilde{P}A & \epsilon\tilde{P}A & -\epsilon\tilde{P}B \\ * & * & \tilde{P}_{d1} & -\epsilon\tilde{P} & 0 \\ * & * & * & \epsilon\tilde{P} & 0 \\ * & * & * & * & I \end{bmatrix} > 0. \tag{2.47}$$

To keep the feasibility of this inequality for $\epsilon \to 0^+$, we need to prescribe $\tilde{A}_F - X_{d1} = 0$, under which, the inequality in (2.47), when $\epsilon \to 0^+$, reduces to

$$\begin{bmatrix} \tilde{P}_{d1} & X_{d1} & \tilde{P}_{d1}B + \tilde{B}_F D \\ * & \tilde{P}_{d1} & 0 \\ * & * & I \end{bmatrix} > 0.$$

Using the Schur complement and noting the assumption $BD^{\mathrm{T}} = 0$ and $DD^{\mathrm{T}} = I$, we further get

$$\begin{aligned}
&\tilde{P}_{d1} - \begin{bmatrix} X_{d1} & \tilde{P}_{d1}B + \tilde{B}_F D \end{bmatrix} \begin{bmatrix} \tilde{P}_{d1} & 0 \\ 0 & I \end{bmatrix} \begin{bmatrix} X_{d1} & \tilde{P}_{d1}B + \tilde{B}_F D \end{bmatrix}^{\mathrm{T}} \\
&= \tilde{P}_{d1} - X_{d1}\tilde{P}_{d1}^{-1}X_{d1}^{\mathrm{T}} - \tilde{P}_{d1}BB\tilde{P}_{d1}^{\mathrm{T}} - \tilde{B}_F\tilde{B}_F^{\mathrm{T}} \\
&= \tilde{P}_{d1} - \tilde{P}_{d1}BB\tilde{P}_{d1}^{\mathrm{T}} - \left(\tilde{P}_{d1}A + \tilde{B}_F C\right)\tilde{P}_{d1}^{-1}\left(\tilde{P}_{d1}A + \tilde{B}_F C\right)^{\mathrm{T}} - \tilde{B}_F\tilde{B}_F^{\mathrm{T}} \\
&= \tilde{P}_{d1} - \tilde{P}_{d1}BB\tilde{P}_{d1} - \tilde{P}_{d1}A\tilde{P}_{d1}^{-1}(\tilde{P}_{d1}A)^{\mathrm{T}} - \tilde{P}_{d1}A\tilde{P}_{d1}^{-1}(\tilde{B}_F C)^{\mathrm{T}} \\
&\quad - \tilde{B}_F C\tilde{P}_{d1}^{-1}(\tilde{P}_{d1}A)^{\mathrm{T}} - \tilde{B}_F\left(C\tilde{P}_{d1}^{-1}C + I\right)\tilde{B}_F^{\mathrm{T}} \qquad (2.48) \\
&= \tilde{P}_{d1} - \tilde{P}_{d1}BB\tilde{P}_{d1} - \tilde{P}_{d1}A\tilde{P}_{d1}^{-1}(\tilde{P}_{d1}A)^{\mathrm{T}} \\
&\quad - \left(\tilde{B}_F + \tilde{P}_{d1}A\tilde{P}_{d1}^{-1}C^{\mathrm{T}}\left(C\tilde{P}_{d1}^{-1}C^{\mathrm{T}} + I\right)^{-1}\right)\left(C\tilde{P}_{d1}^{-1}C + I\right) \\
&\quad \times \left(\tilde{B}_F + \tilde{P}_{d1}A\tilde{P}_{d1}^{-1}C^{\mathrm{T}}\left(C\tilde{P}_{d1}^{-1}C^{\mathrm{T}} + I\right)^{-1}\right)^{\mathrm{T}} \\
&\quad + \tilde{P}_{d1}A\tilde{P}_{d1}^{-1}C^{\mathrm{T}}\left(C\tilde{P}_{d1}^{-1}C^{\mathrm{T}} + I\right)^{-1}\left(\tilde{P}_{d1}A\tilde{P}_{d1}^{-1}C^{\mathrm{T}}\right)^{\mathrm{T}} \\
&> 0. \qquad (2.49)
\end{aligned}$$

Due to the fact that

$$\begin{aligned}
&\left(\tilde{B}_F + \tilde{P}_{d1}A\tilde{P}_{d1}^{-1}C^{\mathrm{T}}\left(C\tilde{P}_{d1}^{-1}C^{\mathrm{T}} + I\right)^{-1}\right)\left(C\tilde{P}_{d1}^{-1}C + I\right) \\
&\times \left(\tilde{B}_F + \tilde{P}_{d1}A\tilde{P}_{d1}^{-1}C^{\mathrm{T}}\left(C\tilde{P}_{d1}^{-1}C^{\mathrm{T}} + I\right)^{-1}\right)^{\mathrm{T}} \geq 0
\end{aligned}$$

and the equality holds only for $\tilde{B}_F = -\tilde{P}_{d1}A\tilde{P}_{d1}^{-1}C^{\mathrm{T}}\left(C\tilde{P}_{d1}^{-1}C^{\mathrm{T}}+I\right)^{-1}$, the inequality in (2.47) can be replaced by

$$\tilde{P}_{d1} - \tilde{P}_{d1}BB\tilde{P}_{d1} - \tilde{P}_{d1}A\tilde{P}_{d1}^{-1}(\tilde{P}_{d1}A)^{\mathrm{T}} + \tilde{P}_{d1}A\tilde{P}_{d1}^{-1}C^{\mathrm{T}} \times \left(C\tilde{P}_{d1}^{-1}C^{\mathrm{T}}+I\right)^{-1}\left(\tilde{P}_{d1}A\tilde{P}_{d1}^{-1}C^{\mathrm{T}}\right)^{\mathrm{T}} > 0$$

or equivalently, via performing a congruence transformation, changing the sign, and defining $\tilde{P} \triangleq \tilde{P}_{d1}^{-1}$,

$$A\tilde{P}A^{\mathrm{T}} - \tilde{P} - A\tilde{P}C^{\mathrm{T}}\left(C\tilde{P}C^{\mathrm{T}}+I\right)^{-1}C\tilde{P}A^{\mathrm{T}} + BB < 0. \qquad (2.50)$$

Comparing this inequality with the discrete-time Riccati equation in (2.45) and noting the monotonicity of the solution to the Riccati equation, one sees that the optimal H_2 performance $\sqrt{\gamma}$ computed by Theorem 2.4 can get arbitrarily close to $\mathrm{Tr}\left[LPL^{\mathrm{T}}\right]$ for the H_2 performance of the Kalman filter. What's more, the filter realization computed from (2.44) is given by

$$\begin{aligned}
A_F &= \tilde{P}_{d1}^{-1}\tilde{A}_F\tilde{P}_{d2}^{-1}\tilde{P}_{d1} = \tilde{P}_{d1}^{-1}X_{d1}\\
&= \tilde{P}_{d1}^{-1}(\tilde{P}_{d1}A + \tilde{B}_F C)\\
&= A - \tilde{P}_{d1}^{-1}\tilde{P}_{d1}A\tilde{P}_{d1}^{-1}C^{\mathrm{T}}\left(C\tilde{P}_{d1}^{-1}C^{\mathrm{T}}+I\right)^{-1}C\\
&= A - A\tilde{P}C^{\mathrm{T}}\left(C\tilde{P}C^{\mathrm{T}}+I\right)^{-1}C,\\
B_F &= -\tilde{P}_{d1}^{-1}\tilde{B}_F\\
&= \tilde{P}_{d1}^{-1}\tilde{P}_{d1}A\tilde{P}_{d1}^{-1}C^{\mathrm{T}}\left(C\tilde{P}_{d1}^{-1}C^{\mathrm{T}}+I\right)^{-1}\\
&= A\tilde{P}C^{\mathrm{T}}\left(C\tilde{P}C^{\mathrm{T}}+I\right)^{-1},\\
C_F &= L,
\end{aligned}$$

which are in a form consistent with the discrete-time Kalman filter.

2.1.6 *Quadratic Robust H_2 Filtering*

In Sects. 2.1.2 and 2.1.4, several sufficient and necessary conditions in terms of LMI have been obtained for designing optimal H_2 filters for nominal continuous- or discrete-time systems. The crucial linearization procedures have been established such that the problems of H_2 filter design are cast into the feasibility of the corresponding convex optimization problems. In this subsection, these obtained results will be further extended and applied to design robust H_2 filters.

To handle the robustness of uncertain dynamical systems, a basic notion in control theory is quadratic stability [10–12]. In the framework of quadratic stability, many results regarding robust H_2 or Kalman filtering for systems with norm-bounded uncertainties have been proposed [6, 9, 13–26], where the filtering scheme is often termed as *quadratic guaranteed cost filtering or estimation* [19, 20, 22, 23]. This subsection will employ the notion of quadratic stability to investigate the robust H_2 filtering problem for systems with convex polyhedral uncertainties bounded in $\mathcal{M}$.

When uncertain but fixed parameter λ exists, the filtering error variance of system $\mathcal{E}$ as $t \to +\infty$ satisfies

$$\sup_{\lambda\in\Gamma} \mathbf{E}\left[e(t)^{\mathrm{T}}e(t)\right] = \sup_{\lambda\in\Gamma} \mathrm{Tr}[C(\lambda)P(\lambda)C(\lambda)^{\mathrm{T}}],$$

where parameter-dependent matrix $P(\lambda)$ is calculated from the following parameter-dependent Lyapunov equation

$$\bar{A}(\lambda)P(\lambda) + P(\lambda)\bar{A}(\lambda)^{\mathrm{T}} + \bar{B}(\lambda)\bar{B}(\lambda)^{\mathrm{T}} = 0 \qquad (2.51)$$

for the continuous-time case, or

$$\bar{A}(\lambda)P(\lambda)\bar{A}(\lambda)^{\mathrm{T}} - P(\lambda) + \bar{B}(\lambda)\bar{B}(\lambda)^{\mathrm{T}} = 0 \qquad (2.52)$$

for the discrete-time case.

For a given filter $\mathcal{F}$, suppose there exists a matrix $\bar{P} > 0$ satisfying the parameter-dependent LMI, for the continuous-time case,

$$\bar{A}(\lambda)\bar{P} + \bar{P}\bar{A}(\lambda)^{\mathrm{T}} + \bar{B}(\lambda)\bar{B}(\lambda)^{\mathrm{T}} < 0, \qquad (2.53)$$

or for the discrete-time case,

$$\bar{A}(\lambda)\bar{P}\bar{A}(\lambda)^{\mathrm{T}} - \bar{P} + \bar{B}(\lambda)\bar{B}(\lambda)^{\mathrm{T}} < 0 \qquad (2.54)$$

for all $M(\lambda) \in \mathcal{M}$. From the monotonicity of the solution of matrix $P(\lambda)$ to Lyapunov equations, there holds $P(\lambda) \leq \bar{P}$ for all $\lambda \in \Gamma$, which results in

$$\begin{aligned}\sup_{\lambda\in\Gamma} \mathbf{E}\left[e(t)^{\mathrm{T}}e(t)\right] &= \sup_{\lambda\in\Gamma} \mathrm{Tr}[C(\lambda)P(\lambda)C(\lambda)^{\mathrm{T}}] \\ &\leq \sup_{\lambda\in\Gamma} \mathrm{Tr}[\bar{C}(\lambda)\bar{P}\bar{C}(\lambda)^{\mathrm{T}}].\end{aligned}$$

Therefore for a prescribed robust H_2 performance bound $\sqrt{\gamma}$, a natural idea for solving the robust H_2 filtering problem is to look for a filter satisfying (2.53) or (2.54) together with $\sup_{\lambda\in\Gamma} \mathrm{Tr}[C(\lambda)\bar{P}C(\lambda)^{\mathrm{T}}] \leq \gamma$. Let $(\bar{A}_i, \bar{B}_i, \bar{C}_i)$ be the matrix

$(\bar{A}(\lambda),\ \bar{B}(\lambda),\ \bar{C}(\lambda))$ at the vertices of the polyhedral domain $\mathcal{M}$. Consequently, a quadratic approach to robust H_2 performance analysis for the filtering error system under a given filter can be obtained in the following lemma.

Lemma 2.3 *Given system $\mathcal{S}$ in (2.1), filter $\mathcal{F}$ in (2.4), and a scalar $\gamma > 0$, the filtering error system $\mathcal{E}$ in (2.5) is robustly asymptotically stable and satisfies $\mathbf{E}\left[e(t)^{\mathrm{T}}e(t)\right] < \gamma$ for all $\lambda \in \Gamma$ if the matrix inequalities, for the continuous-time case,*

$$\mathrm{Tr}[Z] < \gamma, \begin{bmatrix} P & \bar{C}_i^{\mathrm{T}} \\ \bar{C}_i & Z \end{bmatrix} > 0, \begin{bmatrix} \bar{A}_i^{\mathrm{T}}P + P\bar{A}_i & P\bar{B}_i \\ \bar{B}_i^{\mathrm{T}}P & -I \end{bmatrix} < 0, \tag{2.55}$$

or for the discrete-time case,

$$\mathrm{Tr}[Z] < \gamma, \begin{bmatrix} P & \bar{C}_i^{\mathrm{T}} \\ \bar{C}_i & Z \end{bmatrix} > 0, \begin{bmatrix} P & P\bar{A}_i & P\bar{B}_i \\ \bar{A}_i^{\mathrm{T}}P & P & 0 \\ \bar{B}_i^{\mathrm{T}}P & 0 & I \end{bmatrix} > 0 \tag{2.56}$$

are feasible in the real matrix variables P and Z for all $i = 1, 2, \ldots, s$.

Proof Suppose that the matrix inequalities in (2.55) and (2.56) hold for some P and Z. Letting $\bar{P} = P^{-1}$, $T_1 = \mathrm{diag}\{\bar{P},\ I_p\}$, $T_2 = \mathrm{diag}\{\bar{P},\ I_l\}$, and $T_3 = \mathrm{diag}\{\bar{P},\ \bar{P},\ I_l\}$ and noting that $\sum_{i=1}^{s}\lambda_i = 1$ and $\lambda_i \geq 0$, we have

$$\begin{aligned}
\begin{bmatrix} \bar{P} & \bar{P}\bar{C}(\lambda)^{\mathrm{T}} \\ \bar{C}(\lambda)\bar{P} & Z \end{bmatrix} &= T_1^{\mathrm{T}}\begin{bmatrix} P & \bar{C}(\lambda)^{\mathrm{T}} \\ \bar{C}(\lambda) & Z \end{bmatrix}T_1 \\
&= T_1^{\mathrm{T}}\left(\sum_{i=1}^{s}\lambda_i\begin{bmatrix} P & \bar{C}_i^{\mathrm{T}} \\ \bar{C}_i & Z \end{bmatrix}\right)T_1 \\
&> 0, \\
\begin{bmatrix} \bar{P}\bar{A}(\lambda)^{\mathrm{T}} + \bar{A}(\lambda)\bar{P} & \bar{B}(\lambda) \\ \bar{B}(\lambda)^{\mathrm{T}} & -I \end{bmatrix} &= T_2^{\mathrm{T}}\begin{bmatrix} \bar{A}(\lambda)^{\mathrm{T}}P + P\bar{A}(\lambda) & P\bar{B}(\lambda) \\ \bar{B}(\lambda)^{\mathrm{T}}P & -I \end{bmatrix}T_2 \\
&= T_2^{\mathrm{T}}\left(\sum_{i=1}^{s}\lambda_i\begin{bmatrix} \bar{A}_i^{\mathrm{T}}P + P\bar{A}_i & P\bar{B}_i \\ \bar{B}_i^{\mathrm{T}}P & -I \end{bmatrix}\right)T_2 \\
&< 0, \\
\begin{bmatrix} \bar{P} & \bar{A}(\lambda)\bar{P} & \bar{B}(\lambda) \\ \bar{P}\bar{A}(\lambda)^{\mathrm{T}} & \bar{P} & 0 \\ \bar{B}(\lambda)^{\mathrm{T}} & 0 & I \end{bmatrix} &= T_3^{\mathrm{T}}\begin{bmatrix} P & P\bar{A}(\lambda) & P\bar{B}(\lambda) \\ \bar{A}(\lambda)^{\mathrm{T}}P & P & 0 \\ \bar{B}(\lambda)^{\mathrm{T}}P & 0 & I \end{bmatrix}T_3 \\
&= T_3^{\mathrm{T}}\left(\sum_{i=1}^{s}\lambda_i\begin{bmatrix} P & P\bar{A}_i & P\bar{B}_i \\ \bar{A}_i^{\mathrm{T}}P & P & 0 \\ \bar{B}_i^{\mathrm{T}}P & 0 & I \end{bmatrix}\right)T_3 \\
&> 0.
\end{aligned}$$

which, by the Schur complement, imply $\bar{C}(\lambda)\bar{P}\bar{C}(\lambda)^{\mathrm{T}} < Z$ and the inequalities in (2.53) and (2.54). Therefore, we have

$$\sup_{\lambda\in\Gamma} \mathbf{E}\left[e(t)^{\mathrm{T}}e(t)\right] \le \sup_{\lambda\in\Gamma} \mathrm{Tr}[\bar{C}(\lambda)\bar{P}\bar{C}(\lambda)^{\mathrm{T}}] < \mathrm{Tr}[Z] < \gamma.$$

□

As the starting point of Lemma 2.2, the inequalities in (2.53) and (2.54) mean that an upper bound of the robust H_2 filtering performance level is evaluated by adopting a single Lyapunov matrix $\bar{P}$ independent of the uncertain parameters, which is essentially the same as the notion of quadratic stability (then the meaning of *quadratic approaches*, as termed in book, is self-explanatory). Compared with Lemmas 2.1 and 2.2, it is easily found that Lemma 2.3 is a direct extension of Lemmas 2.1 and 2.2 to polytopic uncertain systems. Indeed, the last two inequalities in (2.55) and (2.56) include as many LMIs as the number of the vertices of the polytope $\mathcal{M}$ with each LMI being the corresponding one in Lemmas 2.1 and 2.2 at the vertex of $\mathcal{M}$. Since $\mathcal{M}$ is a convex combination of the matrices at the vertices, the conditions in Lemmas 2.1 and 2.2 for arbitrary matrix $M \in \mathcal{M}$ have been automatically guaranteed by those in Lemma 2.3.

The conditions in Lemma 2.3 can also be directly linearized with respect to the unknown filter matrices by the same procedures as in the derivation of H_2 filter design methods for nominal systems. This yields the following quadratic approach to robust H_2 filter design.

Theorem 2.5 *Consider system $\mathcal{S}$ in* (2.1). *Given a scalar $\gamma > 0$, a filter $\mathcal{F}$ in* (2.4) *exists such that the filtering error system $\mathcal{E}$ in* (2.5) *is robustly asymptotically stable and satisfies $\mathbf{E}\left[e(t)^{\mathrm{T}}e(t)\right] < \gamma$ for all $\lambda \in \Gamma$ if the matrix inequalities*

$$\mathrm{Tr}[Z] < \gamma, \tag{2.57}$$

$$\begin{bmatrix} \tilde{P}_1 & \tilde{P}_2 & L_i^{\mathrm{T}} \\ \tilde{P}_2 & \tilde{P}_2 & -\tilde{C}_F^{\mathrm{T}} \\ L_i & -\tilde{C}_F & Z \end{bmatrix} > 0\,(i = 1, 2, \ldots, s), \tag{2.58}$$

and

- *for the continuous-time case,*

$$\begin{bmatrix} X_{1,i} + X_{1,i}^{\mathrm{T}} & \tilde{A}_F + X_{2,i}^{\mathrm{T}} & \tilde{P}_1 B_i + \tilde{B}_F D_i \\ * & \tilde{A}_F + \tilde{A}_F^{\mathrm{T}} & \tilde{P}_2 B_i + \tilde{B}_F D_i \\ * & * & -I \end{bmatrix} < 0\,(i = 1, 2, \ldots, s), \tag{2.59}$$

- *for the discrete-time case,*

$$\begin{bmatrix} \tilde{P}_1 & \tilde{P}_2 & X_{1,i} & \tilde{A}_F & \tilde{P}_1 B_i + \tilde{B}_F D_i \\ * & \tilde{P}_2 & X_{2,i} & \tilde{A}_F & \tilde{P}_2 B_i + \tilde{B}_F D_i \\ * & * & \tilde{P}_1 & \tilde{P}_2 & 0 \\ * & * & * & \tilde{P}_2 & 0 \\ * & * & * & * & I \end{bmatrix} > 0 \, (i = 1, 2, \ldots, s), \tag{2.60}$$

are feasible in the real matrix variables $\tilde{P}_1$, $\tilde{P}_2$, Z, $\tilde{A}_F$, $\tilde{B}_F$, *and* $\tilde{C}_F$, *where*

$$X_{1,i} \triangleq \tilde{P}_1 A_i + \tilde{B}_F C_i, \; X_{2,i} \triangleq \tilde{P}_2 A_i + \tilde{B}_F C_i.$$

Moreover, if these conditions are feasible, an admissible state-space realization of the filter $\mathcal{F}$ *in* (2.4) *is given by*

$$\begin{bmatrix} A_F & B_F \\ C_F & 0 \end{bmatrix} = \begin{bmatrix} \tilde{P}_2^{-1} & 0 \\ 0 & I \end{bmatrix} \begin{bmatrix} \tilde{A}_F & \tilde{B}_F \\ \tilde{C}_F & 0 \end{bmatrix} \tag{2.61}$$

or

$$\begin{bmatrix} A_F & B_F \\ C_F & 0 \end{bmatrix} = \begin{bmatrix} \tilde{A}_F & \tilde{B}_F \\ \tilde{C}_F & 0 \end{bmatrix} \begin{bmatrix} \tilde{P}_2^{-1} & 0 \\ 0 & I \end{bmatrix}. \tag{2.62}$$

Proof For the continuous-time case and the discrete-time case, the proof is quite similar to that of Theorems 2.1 and 2.3, respectively. □

Based on Theorem 2.5, the first quadratic approach to robust H_2 filter design for systems with polytopic uncertainty is summarized in Algorithm 5.

Algorithm 5 Quadratic Approach to Robust H_2 Filter Design

- Solve the minimization problem:
  - Continuous-time case

$$\min_{\tilde{P}_1, \tilde{P}_2, Z, \tilde{A}_F, \tilde{B}_F, \tilde{C}_F, \gamma} \gamma \;\; \text{s.t. } (2.57), \; (2.58), \; (2.59)$$

  - Discrete-time case

$$\min_{\tilde{P}_1, \tilde{P}_2, Z, \tilde{A}_F, \tilde{B}_F, \tilde{C}_F, \gamma} \gamma \;\; \text{s.t. } (2.57), \; (2.58), \; (2.60)$$

- Compute filter $\mathcal{F}$ in (2.4) by (2.61) or (2.62).

According to Theorems 2.2 and 2.4, when matrix L is known, i.e., $L(\lambda) = L = L_1 = \cdots = L_s$, matrix variable $\tilde{C}_F$ in (2.58) can be eliminated and filter matrix C_F can take $C_F = L$ with no loss of generality. In this situation, (2.58) reduces to

$$\begin{bmatrix} \tilde{P}_1 & \tilde{P}_2 \\ \tilde{P}_2 & \tilde{P}_2 \end{bmatrix} > 0, \begin{bmatrix} \tilde{P}_1 & L^{\mathrm{T}} \\ L & Z \end{bmatrix} > 0$$

and an admissible state-space realization of the filter $\mathcal{F}$ in (2.4) is given by

$$A_F = \tilde{P}_1^{-1} \tilde{A}_F \tilde{P}_2^{-1} \tilde{P}_1, \; B_F = -\tilde{P}_1^{-1} \tilde{B}_F, \; C_F = L. \tag{2.63}$$

Consequently, we have Algorithm 6, another quadratic approach to robust H_2 filter design based on the LMI technique.

Algorithm 6 Quadratic Approach to Robust H_2 Filter Design (Known L)

- Solve the minimization problem:
  - Continuous-time case

$$\min_{\tilde{P}_1, \tilde{P}_2, Z, \tilde{A}_F, \tilde{B}_F, \gamma} \gamma \text{ s.t. } \begin{cases} (2.57), \ (2.59) \\ \begin{bmatrix} \tilde{P}_1 & \tilde{P}_2 \\ \tilde{P}_2 & \tilde{P}_2 \end{bmatrix} > 0, \begin{bmatrix} \tilde{P}_1 & L^{\mathrm{T}} \\ L & Z \end{bmatrix} > 0 \end{cases}$$

  - Discrete-time case

$$\min_{\tilde{P}_1, \tilde{P}_2, Z, \tilde{A}_F, \tilde{B}_F, \gamma} \gamma \text{ s.t. } \begin{cases} (2.57), \ (2.60) \\ \begin{bmatrix} \tilde{P}_1 & L^{\mathrm{T}} \\ L & Z \end{bmatrix} > 0 \end{cases}$$

- Compute filter $\mathcal{F}$ in (2.4) by (2.63).

2.2 Quadratic Robust H_∞ Filter Design

In Sect. 2.1.6, quadratic approaches have been proposed for the design of robust H_2 filters for polytopic uncertain systems. In the H_2 setting, the noise process or series is assumed to be Gaussian, which means that the performance of an H_2 filter relies much on the knowledge of the statistic property of the noise input. However, in real situations, this assumption is not always completely satisfied. To overcome this drawback of H_2 filters, other assumptions independent of the statistic information of the noise input are often made and new filtering schemes based on the new assumptions are taken into consideration. The most frequently used assumption of such types is the square integrable function or square summable series with respect to time, that is, the noise $w(t) \in L_2[0, +\infty)$ for the continuous-time case and $w(t) \in l_2[0, +\infty)$ for the discrete-time case. As for the physical meaning of this type of noises, $w(t) \in L_2[0, +\infty)$ or $w(t) \in l_2[0, +\infty)$ implies the energy-bounded property of noises. Obviously, this type of assumption differs from the Gaussian white noise type, and moreover, does not need any information on the statistics of the noise. H_∞ filtering to be considered in this section is one of the filtering schemes that are based on this assumption for noises.

2.2.1 Problem Formulation

Consider a stable uncertain system $\mathcal{S}$ described by the following state-space model:

$$\begin{aligned} \mathcal{S}:\ \delta[x(t)] &= A(\lambda)x(t) + B(\lambda)w(t), \\ y(t) &= C(\lambda)x(t) + D(\lambda)w(t), \\ z(t) &= L(\lambda)x(t) + E(\lambda)w(t), \end{aligned} \tag{2.64}$$

where $x(t) \in \mathbb{R}^n$ is the state vector, $y(t) \in \mathbb{R}^m$ is the measured output, $z(t) \in \mathbb{R}^p$ is the signal to be estimated, and $w(t) \in \mathbb{R}^l$ is the noise. The meaning of $\delta[\cdot]$ is the same as that in (2.1). $A(\lambda)$, $B(\lambda)$, $C(\lambda)$, $D(\lambda)$, $L(\lambda)$, and $E(\lambda)$ are appropriately dimensioned matrices, and are assumed to satisfy

$$R(\lambda) \triangleq (A(\lambda), B(\lambda), C(\lambda), D(\lambda), L(\lambda), E(\lambda)) \in \mathcal{R} \tag{2.65}$$

where $\mathcal{R}$ is a given convex polyhedral domain bounded by s vertices

$$\mathcal{R} \triangleq \left\{ R(\lambda) : R(\lambda) = \sum_{i=1}^{s} \lambda_i R_i;\ \lambda \in \Gamma \right\}$$

with $R_i \triangleq (A_i, B_i, C_i, D_i, L_i, E_i)$ denoting the vertices of the polytope, and Γ denoting the unit simplex defined in (2.3).

In this section, we are interested in designing a nonproper LTI filter as:

$$\begin{aligned} \mathcal{F}:\ \delta[x_F(t)] &= A_F x_F(t) + B_F y(t), \\ z_F(t) &= C_F x_F(t) + D_F y(t), \end{aligned} \tag{2.66}$$

where $x_F(t) \in \mathbb{R}^n$ is the filter vector, and A_F, B_F, C_F, and D_F are appropriately dimensioned real-valued matrices to be determined.

Defining the filtering error signal $e(t) \triangleq z(t) - z_F(t)$ and augmenting the system $\mathcal{S}$ to include the dynamics of the filter $\mathcal{F}$, we have the following state-space representation of the filtering error system:

$$\begin{aligned} \mathcal{E}:\ \delta[\xi(t)] &= \bar{A}(\lambda)\xi(t) + \bar{B}(\lambda)w(t), \\ e(t) &= \bar{C}(\lambda)\xi(t) + \bar{D}(\lambda)w(t) \end{aligned} \tag{2.67}$$

where $\xi(t)$, $\bar{A}(\lambda)$, and $\bar{B}(\lambda)$ are defined in (2.5), and

$$\bar{C}(\lambda) \triangleq \begin{bmatrix} L(\lambda) - D_F C(\lambda) & -C_F \end{bmatrix},\ \bar{D}(\lambda) \triangleq E(\lambda) - D_F D(\lambda).$$

For any fixed $\lambda \in \Gamma$, the transfer function of the filtering error system from the noise input $w(t)$ to the filtering error $e(t)$ can be written as:

$$T(\delta, \lambda) \triangleq \bar{C}(\lambda)\left[\delta I - \bar{A}(\lambda)\right]^{-1}\bar{B}(\lambda) + \bar{D}(\lambda). \tag{2.68}$$

To design filter $\mathcal{F}$ in the H_∞ setting, suppose that the noise $w(t) \in L_2[0, +\infty)$ for the continuous-time case and $w(t) \in l_2[0, +\infty)$ for the discrete-time case. It is still assumed that both system $\mathcal{S}$ and filter $\mathcal{F}$ have zero initial conditions, that is, $x(0) = 0$ and $x_F(0) = 0$. Then, the robust H_∞ filtering problem to be addressed for system $\mathcal{S}$ in this section is formulated as follows.

Robust H_∞ Filtering Problem: Given system $\mathcal{S}$ in (2.64), design a filter $\mathcal{F}$ of the form in (2.66) such that the filtering error system $\mathcal{E}$ in (2.67), for all $\lambda \in \Gamma$, is robustly asymptotically stable, and for all nonzero $w(t) \in L_2[0, +\infty)$ for the continuous-time case or all nonzero $w(t) \in l_2[0, +\infty)$ for the discrete-time case, satisfies

$$\sup_{\lambda \in \Gamma} \frac{\|e\|_2}{\|w\|_2} < \gamma, \quad \left(\text{i.e., } \sup_{\lambda \in \Gamma} \|T(\delta, \lambda)\|_\infty < \gamma\right) \tag{2.69}$$

where $\|e\|_2 = \|e\|_{L_2}$ for the continuous-time case and $\|e\|_2 = \|e\|_{l_2}$ for the discrete-time case, and γ is a given positive constant. Moreover, a filter satisfying the above conditions is referred to as *a filter with a guaranteed robust H_∞ performance bound γ*.

The specification in (2.69) indicates that γ is an upper bound of the worst energy-to-energy gain, i.e., L_2-induced or l_2-induced gain, of the filtering error system. For time-varying systems and nonlinear systems that do not have a transfer function representation, H_∞ filtering is usually termed as *energy-to-energy* filtering; as such, the physical meaning of this filtering scheme is apparent. When uncertainty exists, this bound should be guaranteed for all the uncertainties. If the specification in (2.69) is satisfied, the filtering error energy will be bounded by $\gamma \|w\|_2$ for all energy-bounded noise $w(t)$. Smaller γ means the better noise energy attenuation level of a filter. Hence, if possible, it is desirable to (globally or locally) minimize γ in the robust H_∞ filtering problem.

2.2.2 H_∞ Filtering for Nominal Systems

First, we consider the H_∞ filtering problem for nominal systems. A given upper bound of the H_∞ norm of the filtering error system can be characterized by the so-called bounded real lemma (BRL) in terms of LMI [1, 27–29].

Lemma 2.4 (Continuous-Time [1, 27]) *Consider system $\mathcal{S}$ in* (2.64) *and assume that $R \in \mathcal{R}$ is fixed but arbitrary. Given filter $\mathcal{F}$ in* (2.66) *and a scalar $\gamma > 0$, the filtering error system $\mathcal{E}$ in* (2.67) *is asymptotically stable and satisfies* (2.69) *if and only if the following matrix inequalities*

$$P > 0, \quad \begin{bmatrix} \bar{A}^{\mathrm{T}}P + P\bar{A} & P\bar{B} & \bar{C}^{\mathrm{T}} \\ \bar{B}^{\mathrm{T}}P & -\gamma^2 I & \bar{D}^{\mathrm{T}} \\ \bar{C} & \bar{D} & -I \end{bmatrix} < 0 \tag{2.70}$$

are feasible in the real matrix variable P.

Lemma 2.5 (Discrete-Time [27–29]) *Consider system $\mathcal{S}$ in* (2.64) *and assume that $R \in \mathcal{R}$ is fixed but arbitrary. Given filter $\mathcal{F}$ in* (2.66) *and a scalar $\gamma > 0$, the filtering error system $\mathcal{E}$ in* (2.67) *is asymptotically stable and satisfies* (2.69) *if and only if the following matrix inequality*

$$\begin{bmatrix} -P & P\bar{A} & P\bar{B} & 0 \\ \bar{A}^{\mathrm{T}}P & -P & 0 & \bar{C}^{\mathrm{T}} \\ \bar{B}^{\mathrm{T}}P & 0 & -\gamma^2 I & \bar{D}^{\mathrm{T}} \\ 0 & \bar{C} & \bar{D} & -I \end{bmatrix} < 0 \tag{2.71}$$

is feasible in the matrix variable P.

Since the big inequalities in (2.70) and (2.71) are not linear simultaneously with respect to the filter matrices (A_F, B_F, C_F, D_F) and the Lyapunov matrix P, one cannot directly obtain the filter realization from Lemma 2.4 or 2.5 by solving LMI-based convex optimization problems. To convert the NLMIs in (2.70) and (2.71) into LMIs, we also resort to a linearization procedure similar to the one used in dealing with H_2 filtering problems in Sect. 2.1. Now, we give the following necessary and sufficient condition for the existence of H_∞ filters for both continuous- and discrete-time nominal systems.

Theorem 2.6 *Consider system $\mathcal{S}$ in* (2.64) *and assume that $R \in \mathcal{R}$ is fixed but arbitrary. Given a scalar $\gamma > 0$, a filter $\mathcal{F}$ in* (2.66) *exists such that the filtering error system $\mathcal{E}$ in* (2.67) *is asymptotically stable and satisfies* (2.69) *if and only if the following inequalities*

- *for the continuous-time case,*

$$\begin{bmatrix} \tilde{P}_1 & \tilde{P}_2 \\ \tilde{P}_2 & \tilde{P}_2 \end{bmatrix} > 0, \quad \begin{bmatrix} X_1 + X_1^{\mathrm{T}} & \tilde{A}_F + X_2^{\mathrm{T}} & Y_1 & L^{\mathrm{T}} - C^{\mathrm{T}}\tilde{D}_F^{\mathrm{T}} \\ * & \tilde{A}_F + \tilde{A}_F^{\mathrm{T}} & Y_2 & -\tilde{C}_F^{\mathrm{T}} \\ * & * & -\gamma^2 I & E^{\mathrm{T}} - D^{\mathrm{T}}\tilde{D}_F^{\mathrm{T}} \\ * & * & * & -I \end{bmatrix} < 0 \tag{2.72}$$

- *for the discrete-time case,*

$$\begin{bmatrix} -\tilde{P}_1 & -\tilde{P}_2 & X_1 & \tilde{A}_F & Y_1 & 0 \\ * & -\tilde{P}_2 & X_2 & \tilde{A}_F & Y_2 & 0 \\ * & * & -\tilde{P}_1 & -\tilde{P}_2 & 0 & L^{\mathrm{T}} - C^{\mathrm{T}}\tilde{D}_F^{\mathrm{T}} \\ * & * & * & -\tilde{P}_2 & 0 & -\tilde{C}_F^{\mathrm{T}} \\ * & * & * & * & -\gamma^2 I & E^{\mathrm{T}} - D^{\mathrm{T}}\tilde{D}_F^{\mathrm{T}} \\ * & * & * & * & * & -I \end{bmatrix} < 0 \tag{2.73}$$

are feasible in the real matrix variables $\tilde{P}_1$, $\tilde{P}_2$, $\tilde{A}_F$, $\tilde{B}_F$, $\tilde{C}_F$, and $\tilde{D}_F$, where

$$X_1 \triangleq \tilde{P}_1 A + \tilde{B}_F C,\ X_2 \triangleq \tilde{P}_2 A + \tilde{B}_F C,$$
$$Y_1 \triangleq \tilde{P}_1 B + \tilde{B}_F D,\ Y_2 \triangleq \tilde{P}_2 B + \tilde{B}_F D,$$

Moreover, if these conditions are feasible, an admissible state-space realization of the filter $\mathcal{F}$ in (2.66) *is given by*

$$\begin{bmatrix} A_F & B_F \\ C_F & D_F \end{bmatrix} = \begin{bmatrix} \tilde{P}_2^{-1} & 0 \\ 0 & I \end{bmatrix} \begin{bmatrix} \tilde{A}_F & \tilde{B}_F \\ \tilde{C}_F & \tilde{D}_F \end{bmatrix} \tag{2.74}$$

or

$$\begin{bmatrix} A_F & B_F \\ C_F & D_F \end{bmatrix} = \begin{bmatrix} \tilde{A}_F & \tilde{B}_F \\ \tilde{C}_F & \tilde{D}_F \end{bmatrix} \begin{bmatrix} \tilde{P}_2^{-1} & 0 \\ 0 & I \end{bmatrix}. \tag{2.75}$$

Proof In the first part, we prove that (2.72) ⇔ (2.70) while (2.73) ⇔ (2.71).

(2.70) ⇒ (2.72) *and* (2.71) ⇒ (2.73): Suppose that there exist P such that (2.70) and (2.71) are satisfied. Partition matrix P into four blocks as the following form

$$P = \begin{bmatrix} P_1 & P_2 \\ P_2^{\mathrm{T}} & P_3 \end{bmatrix}$$

with each block being an $n \times n$ matrix. The nonsingularity of P enables us to assume that, by invoking a small perturbation if necessary, P_2 is nonsingular. Define multiplier matrix

$$J_0 \triangleq \begin{bmatrix} I & 0 \\ 0 & P_3^{-1}P_2^{\mathrm{T}} \end{bmatrix} \tag{2.76}$$

and introduce the following matrix transformations

$$\begin{bmatrix} \tilde{A}_F & \tilde{B}_F \\ \tilde{C}_F & \tilde{D}_F \end{bmatrix} \triangleq \begin{bmatrix} P_2 A_F P_3^{-1} P_2^{\mathrm{T}} & P_2 B_F \\ C_F P_3^{-1} P_2^{\mathrm{T}} & D_F \end{bmatrix}, \tag{2.77}$$

$$\begin{bmatrix} \tilde{P}_1 & \tilde{P}_2 \\ \tilde{P}_2 & \tilde{P}_2 \end{bmatrix} \triangleq J_0^{\mathrm{T}} P J_0 = \begin{bmatrix} P_1 & P_2 P_3^{-1} P_2^{\mathrm{T}} \\ P_2 P_3^{-1} P_2^{\mathrm{T}} & P_2 P_3^{-1} P_2^{\mathrm{T}} \end{bmatrix}. \tag{2.78}$$

Then, the following equations can be obtained

$$J_0^{\mathrm{T}} P \bar{A} J_0 = \begin{bmatrix} X_1 & \tilde{A}_F \\ X_2 & \tilde{A}_F \end{bmatrix},$$
$$J_0^{\mathrm{T}} P \bar{B} = \begin{bmatrix} \tilde{P}_1 B + \tilde{B}_F D \\ \tilde{P}_2 B + \tilde{B}_F D \end{bmatrix},$$
$$\bar{C} J_0 = \begin{bmatrix} L - \tilde{D}_F C & -\tilde{C}_F \end{bmatrix}.$$

Furthermore, by defining

$$J_c \triangleq \mathrm{diag}\left\{J_0,\ I_l,\ I_p\right\},\ J_d \triangleq \mathrm{diag}\left\{J_0,\ J_0,\ I_l,\ I_p\right\} \tag{2.79}$$

and noting the equations given above, it readily follows that

$$J_c^{\mathrm{T}} \begin{bmatrix} \bar{A}^{\mathrm{T}} P + P \bar{A} & P\bar{B} & \bar{C}^{\mathrm{T}} \\ \bar{B}^{\mathrm{T}} P & -\gamma^2 I & \bar{D}^{\mathrm{T}} \\ \bar{C} & \bar{D} & -I \end{bmatrix} J_c = \begin{bmatrix} X_1 + X_1^{\mathrm{T}} & \tilde{A}_F + X_2^{\mathrm{T}} & Y_1 & L^{\mathrm{T}} - C^{\mathrm{T}} \tilde{D}_F^{\mathrm{T}} \\ * & \tilde{A}_F + \tilde{A}_F^{\mathrm{T}} & Y_2 & -\tilde{C}_F^{\mathrm{T}} \\ * & * & -\gamma^2 I & E^{\mathrm{T}} - D^{\mathrm{T}} \tilde{D}_F^{\mathrm{T}} \\ * & * & 0 & -I \end{bmatrix},$$

$$J_d^{\mathrm{T}} \begin{bmatrix} -P & P\bar{A} & P\bar{B} & 0 \\ \bar{A}^{\mathrm{T}} P & -P & 0 & \bar{C}^{\mathrm{T}} \\ \bar{B}^{\mathrm{T}} P & 0 & -\gamma^2 I & \bar{D}^{\mathrm{T}} \\ 0 & \bar{C} & \bar{D} & -I \end{bmatrix} J_d = \begin{bmatrix} -\tilde{P}_1 & -\tilde{P}_2 & X_1 & \tilde{A}_F & Y_1 & 0 \\ * & -\tilde{P}_2 & X_2 & \tilde{A}_F & Y_2 & 0 \\ * & * & -\tilde{P}_1 & -\tilde{P}_2 & 0 & L^{\mathrm{T}} - C^{\mathrm{T}} \tilde{D}_F^{\mathrm{T}} \\ * & * & * & -\tilde{P}_2 & 0 & -\tilde{C}_F^{\mathrm{T}} \\ * & * & * & * & -\gamma^2 I & E^{\mathrm{T}} - D^{\mathrm{T}} \tilde{D}_F^{\mathrm{T}} \\ * & * & * & * & * & -I \end{bmatrix},$$

which together with (2.78) imply that (2.70) $\Rightarrow$ (2.72) and (2.71) $\Rightarrow$ (2.73).

(2.70) $\Leftarrow$ (2.72) *and* (2.71) $\Leftarrow$ (2.73): Suppose that there exist $\tilde{P}_1$, $\tilde{P}_2$, $\tilde{A}_F$, $\tilde{B}_F$, $\tilde{C}_F$, and $\tilde{D}_F$ such that (2.72) and (2.73) are satisfied. Obviously, $\tilde{P}_2$ is nonsingular, which means that it is always possible to find two invertible matrices P_2 and P_3 such that $\tilde{P}_2 = P_2 P_3^{-1} P_2^{\mathrm{T}}$. Introduce the following definitions

$$\begin{bmatrix} A_F & B_F \\ C_F & D_F \end{bmatrix} \triangleq \begin{bmatrix} P_2^{-1} & 0 \\ 0 & I \end{bmatrix} \begin{bmatrix} \tilde{A}_F & \tilde{B}_F \\ \tilde{C}_F & \tilde{D}_F \end{bmatrix} \begin{bmatrix} P_2^{-\mathrm{T}} P_3 & 0 \\ 0 & I \end{bmatrix}, \tag{2.80}$$

$$P \triangleq J_0^{-\mathrm{T}} \begin{bmatrix} \tilde{P}_1 & \tilde{P}_2 \\ \tilde{P}_2 & \tilde{P}_2 \end{bmatrix} J_0^{-1} \tag{2.81}$$

with J_0 defined in (2.76). By these definitions, the conditions in (2.72) and (2.73) can be rewritten as

$$J_c^{\mathrm{T}}\begin{bmatrix}\bar{A}^{\mathrm{T}}P+P\bar{A} & P\bar{B} & \bar{C}^{\mathrm{T}}\\ \bar{B}^{\mathrm{T}}P & -\gamma^2 I & \bar{D}^{\mathrm{T}}\\ \bar{C} & \bar{D} & -I\end{bmatrix}J_c<0,$$

$$J_d^{\mathrm{T}}\begin{bmatrix}-P & P\bar{A} & P\bar{B} & 0\\ \bar{A}^{\mathrm{T}}P & -P & 0 & \bar{C}^{\mathrm{T}}\\ \bar{B}^{\mathrm{T}}P & 0 & -\gamma^2 I & \bar{D}^{\mathrm{T}}\\ 0 & \bar{C} & \bar{D} & -I\end{bmatrix}J_d<0,$$

with J_c and J_d defined in (2.79). Since both J_c and J_d are nonsingular, the above two conditions imply that there exist a variable P in (2.81) together with filter realization $(A_F,\ B_F,\ C_F,\ D_F)$ in (2.80) such that (2.70) and (2.71) hold. Hence, we have (2.70) $\Leftarrow$ (2.72) and (2.71) $\Leftarrow$ (2.73). Consequently, by Lemma 2.4 and 2.5, Theorem 2.6 gives a necessary and sufficient condition that can ensure the existence of a filter with a guaranteed H_∞ performance level bound γ.

From the expression of $(A_F,\ B_F,\ C_F,\ D_F)$ in (2.80), admissible state-space realizations of the filter $\mathcal{F}$ can be obtained from (2.74) or (2.75) by following the second part of the proof of Theorem 2.1. The entire proof is completed. □

All the variables, including the square of the H_∞ performance bound γ, are in a linear form in (2.72) and (2.73). Hence, applying Theorem 2.6, we can design the optimal H_∞ filters for nominal systems by the LMI technique, which is summarized in Algorithm 7.

Algorithm 7 H_∞ Filter Design

- Solve the minimization problem:
  - Continuous-time case
$$\min_{\tilde{P}_1,\,\tilde{P}_2,\,\tilde{A}_F,\,\tilde{B}_F,\,\tilde{C}_F,\,\tilde{D}_F,\,\mu} \gamma^2=\mu \quad \text{s.t. (2.72)}$$
  - Discrete-time case
$$\min_{\tilde{P}_1,\,\tilde{P}_2,\,\tilde{A}_F,\,\tilde{B}_F,\,\tilde{C}_F,\,\tilde{D}_F,\,\mu} \gamma^2=\mu \quad \text{s.t. (2.73)}$$
- Compute filter $\mathcal{F}$ in (2.66) by (2.74) or (2.75).

2.2.3 *Quadratic Robust H_∞ Filtering*

It is well-known that, from a time-domain point of view, the sufficiency of the BRLs can be easily verified through the Lyapunov function or functional method [1]. For instance, the conditions in Lemma 2.4 and 2.5 can be established by the following procedures. Suppose (2.70) and (2.71) hold for some $P>0$, then by the Schur complement, the quadratic Lyapunov function

$$V(t)=\bar{x}(t)^{\mathrm{T}}Px(t) \tag{2.82}$$

satisfies, for the continuous-time case,

$$\dot{V}(t) + e(t)^{\mathrm{T}}e(t) - \gamma^2 w(t)^{\mathrm{T}}w(t) = \xi(t)^{\mathrm{T}}\begin{bmatrix} \bar{A}^{\mathrm{T}}P + P\bar{A} + \bar{C}^{\mathrm{T}}\bar{C} & P\bar{B} \\ \bar{B}^{\mathrm{T}}P & -\gamma^2 I \end{bmatrix}\xi(t) < 0,$$

and, for the discrete-time case,

$$\Delta V(t) + e(t)^{\mathrm{T}}e(t) - \gamma^2 w(t)^{\mathrm{T}}w(t) = \xi(t)^{\mathrm{T}}\begin{bmatrix} \bar{A}^{\mathrm{T}}P\bar{A} - P + \bar{C}^{\mathrm{T}}\bar{C} & \bar{A}^{\mathrm{T}}P\bar{B} \\ \bar{B}^{\mathrm{T}}P\bar{A} & \bar{B}^{\mathrm{T}}P\bar{B} - \gamma^2 I \end{bmatrix} \times\xi(t) < 0,$$

with $\xi(t) = \mathrm{col}\{\bar{x}(t),\ w(t)\}$. Noting $V(+\infty) \geq 0$ and integrating or summating the inequalities from 0 to $+\infty$, we have, under the initial condition $\bar{x}(0) = 0$,

$$\begin{aligned} \|e\|_{L_2}^2 - \gamma^2\|w\|_{L_2}^2 &= \int_0^{+\infty}\left(\dot{V}(t) + e(t)^{\mathrm{T}}e(t) - \gamma^2 w(t)^{\mathrm{T}}w(t)\right)\mathrm{d}t - V(+\infty) \\ &\leq \int_0^{+\infty}\left(\dot{V}(t) + e(t)^{\mathrm{T}}e(t) - \gamma^2 w(t)^{\mathrm{T}}w(t)\right)\mathrm{d}t < 0, \\ \|e\|_{l_2}^2 - \gamma^2\|w\|_{l_2}^2 &= \sum_0^{+\infty}\left(\Delta V(t) + e(t)^{\mathrm{T}}e(t) - \gamma^2 w(t)^{\mathrm{T}}w(t)\right) - V(+\infty) \\ &\leq \sum_0^{+\infty}\left(\Delta V(t) + e(t)^{\mathrm{T}}e(t) - \gamma^2 w(t)^{\mathrm{T}}w(t)\right) < 0. \end{aligned}$$

Consequently, the specification in (2.69) is established.

The above derivation shows that the results in Lemma 2.4 and 2.5 are essentially based on the specific *single quadratic* Lyapunov function in (2.82). Moreover, when parameter uncertainties exist in system $\mathcal{S}$, the above derivation is still valid by using this quadratic Lyapunov function, corresponding to the application of the notion of quadratic stability [10–12]. Accordingly, for the case of parametric uncertainty, system matrices in (2.70) and (2.71) should be modified to be in the parameter-dependent form, that is,

$$\begin{bmatrix} \bar{A}(\lambda)^{\mathrm{T}}P + P\bar{A}(\lambda) & P\bar{B}(\lambda) & \bar{C}(\lambda)^{\mathrm{T}} \\ \bar{B}^{\mathrm{T}}P(\lambda) & -\gamma^2 I & \bar{D}(\lambda)^{\mathrm{T}} \\ \bar{C}(\lambda) & 0 & -I \end{bmatrix} < 0,$$

$$\begin{bmatrix} -P & P\bar{A}(\lambda) & P\bar{B}(\lambda) & 0 \\ \bar{A}(\lambda)^{\mathrm{T}}P & -P & 0 & \bar{C}(\lambda)^{\mathrm{T}} \\ \bar{B}(\lambda)^{\mathrm{T}}P & 0 & -\gamma^2 I & \bar{D}(\lambda)^{\mathrm{T}} \\ 0 & \bar{C}(\lambda) & \bar{D}(\lambda) & -I \end{bmatrix} < 0.$$

Due to the convex polytopic assumption of $R(\lambda)$ in (2.65), the above infinite number of parameter-dependent LMIs hold if and only if the corresponding LMIs at the vertices are satisfied. Hence, the following BRL based on a quadratic approach is obtained for the robust H_∞ performance analysis of the filtering error system with polytopic uncertainty.

Lemma 2.6 *Given system $\mathcal{S}$ in* (2.64)*, filter $\mathcal{F}$ in* (2.66)*, and a scalar $\gamma > 0$, the filtering error system $\mathcal{E}$ in* (2.67) *is robustly asymptotically stable and satisfies* (2.69) *for all $\lambda \in \Gamma$ if the following inequalities*

- *for the continuous-time case,*

$$P > 0, \quad \begin{bmatrix} \bar{A}_i^{\mathrm{T}} P + P\bar{A}_i & P\bar{B}_i & \bar{C}_i^{\mathrm{T}} \\ \bar{B}_i^{\mathrm{T}} P & -\gamma^2 I & \bar{D}_i^{\mathrm{T}} \\ \bar{C}_i & \bar{D}_i & -I \end{bmatrix} < 0 \, (i = 1, 2, \ldots, s), \tag{2.83}$$

- *for the discrete-time case,*

$$\begin{bmatrix} -P & P\bar{A}_i & P\bar{B}_i & 0 \\ \bar{A}_i^{\mathrm{T}} P & -P & 0 & \bar{C}_i^{\mathrm{T}} \\ \bar{B}_i^{\mathrm{T}} P & 0 & -\gamma^2 I & \bar{D}_i^{\mathrm{T}} \\ 0 & \bar{C}_i & \bar{D}_i & -I \end{bmatrix} < 0 \, (i = 1, 2, \ldots, s), \tag{2.84}$$

are feasible in the real matrix variable P.

Proof Noting that $\sum_{i=1}^{s} \lambda_i = 1$ and $\lambda_i \geq 0$, we have

$$\begin{bmatrix} \bar{A}(\lambda)^{\mathrm{T}} P + P\bar{A}(\lambda) & P\bar{B}(\lambda) & \bar{C}(\lambda)^{\mathrm{T}} \\ \bar{B}^{\mathrm{T}} P(\lambda) & -\gamma^2 I & 0 \\ \bar{C}(\lambda) & 0 & -I \end{bmatrix} = \sum_{i=1}^{s} \lambda_i \begin{bmatrix} \bar{A}_i^{\mathrm{T}} P + P\bar{A}_i & P\bar{B}_i & \bar{C}_i^{\mathrm{T}} \\ \bar{B}_i^{\mathrm{T}} P & -\gamma^2 I & 0 \\ \bar{C}_i & 0 & -I \end{bmatrix} < 0,$$

$$\begin{bmatrix} -P & P\bar{A}(\lambda) & P\bar{B}(\lambda) & 0 \\ \bar{A}(\lambda)^{\mathrm{T}} P & -P & 0 & \bar{C}(\lambda)^{\mathrm{T}} \\ \bar{B}(\lambda)^{\mathrm{T}} P & 0 & -\gamma^2 I & 0 \\ 0 & \bar{C}(\lambda) & 0 & -I \end{bmatrix} = \sum_{i=1}^{s} \lambda_i \begin{bmatrix} -P & P\bar{A}_i & P\bar{B}_i & 0 \\ \bar{A}_i^{\mathrm{T}} P & -P & 0 & \bar{C}_i^{\mathrm{T}} \\ \bar{B}_i^{\mathrm{T}} P & 0 & -\gamma^2 I & 0 \\ 0 & \bar{C}_i & 0 & -I \end{bmatrix} < 0,$$

which together with the above discussion can complete the proof. □

By applying the linearization procedures in the derivation of Theorem 2.6 to the results in Lemma 2.6, we have the following theorem that provides a quadratic approach to robust H_∞ filter design.

Theorem 2.7 *Consider system $\mathcal{S}$ in* (2.64)*. Given a scalar $\gamma > 0$, a filter $\mathcal{F}$ in* (2.66) *exists such that the filtering error system $\mathcal{E}$ in* (2.67) *is robustly asymptotically stable and satisfies* (2.69) *if the following inequalities*

- *for the continuous-time case,*

$$\begin{bmatrix} \tilde{P}_1 & \tilde{P}_2 \\ \tilde{P}_2 & \tilde{P}_2 \end{bmatrix} > 0, \tag{2.85}$$

$$\begin{bmatrix} X_{1,i} + X_{1,i}^{\mathrm{T}} & \tilde{A}_F + X_{2,i}^{\mathrm{T}} & Y_{1,i} & L_i^{\mathrm{T}} - C_i^{\mathrm{T}} \tilde{D}_F^{\mathrm{T}} \\ * & \tilde{A}_F + \tilde{A}_F^{\mathrm{T}} & Y_{2,i} & -\tilde{C}_F^{\mathrm{T}} \\ * & * & -\gamma^2 I & E_i^{\mathrm{T}} - D_i^{\mathrm{T}} \tilde{D}_F^{\mathrm{T}} \\ * & * & * & -I \end{bmatrix} < 0 \, (i = 1, 2, \ldots, s), \tag{2.86}$$

- *for the discrete-time case,*

$$\begin{bmatrix} -\tilde{P}_1 & -\tilde{P}_2 & X_{1,i} & \tilde{A}_F & Y_{1,i} & 0 \\ * & -\tilde{P}_2 & X_{2,i} & \tilde{A}_F & Y_{2,i} & 0 \\ * & * & -\tilde{P}_1 & -\tilde{P}_2 & 0 & L_i^{\mathrm{T}} - C_i^{\mathrm{T}} \tilde{D}_F^{\mathrm{T}} \\ * & * & * & -\tilde{P}_2 & 0 & -\tilde{C}_F^{\mathrm{T}} \\ * & * & * & * & -\gamma^2 I & E_i^{\mathrm{T}} - D_i^{\mathrm{T}} \tilde{D}_F^{\mathrm{T}} \\ * & * & * & * & * & -I \end{bmatrix} < 0 \, (i = 1, 2, \ldots, s), \tag{2.87}$$

are feasible in the real matrix variables $\tilde{P}_1$, $\tilde{P}_2$, $\tilde{A}_F$, $\tilde{B}_F$, $\tilde{C}_F$, *and* $\tilde{D}_F$, *where*

$$X_{1,i} \triangleq \tilde{P}_1 A_i + \tilde{B}_F C_i, \; X_{2,i} \triangleq \tilde{P}_2 A_i + \tilde{B}_F C_i,$$
$$Y_{1,i} \triangleq \tilde{P}_1 B_i + \tilde{B}_F D_i, \; Y_{2,i} \triangleq \tilde{P}_2 B_i + \tilde{B}_F D_i.$$

Moreover, if these conditions are feasible, an admissible state space realization of the filter $\mathcal{F}$ *in* (2.66) *is given by* (2.74) *or* (2.75).

Based on Theorem 2.7, Algorithm 8 gives a quadratic approach to the design of robust filters with a suboptimal robust H_∞ performance bound γ.

2.3 Quadratic Robust Energy-to-Peak Filter Design

In Sect. 2.2.3, under the energy-bounded assumption for noises, a robust H_∞ filter design method has been proposed for designing filters that guarantee a prescribed noise energy attenuation level γ for all the uncertainties.

In this section, also under the energy-bounded assumption for noises, we consider another filtering scheme, i.e., energy-to-peak filtering (in contrast to the energy-to-energy gain meaning of H_∞ filtering). For this filtering strategy, it is self-explanatory that the worst possible peak value, instead of the energy value, is applied to evaluate the size of the filtering error.

Algorithm 8 Quadratic Approach to Robust H_∞ Filter Design

- Solve the minimization problem:
  - Continuous-time case

$$\min_{\tilde{P}_1, \tilde{P}_2, \tilde{A}_F, \tilde{B}_F, \tilde{C}_F, \tilde{D}_F, \mu} \gamma^2 = \mu \ \text{ s.t. } (2.85)\text{and}(2.86)$$

  - Discrete-time case

$$\min_{\tilde{P}_1, \tilde{P}_2, \tilde{A}_F, \tilde{B}_F, \tilde{C}_F, \tilde{D}_F, \mu} \gamma^2 = \mu \ \text{ s.t. } (2.87)$$

- Compute filter $\mathcal{F}$ in (2.66) by (2.74) or (2.81).

2.3.1 Problem Formulation

Also consider the uncertain stable system $\mathcal{S}$ in (2.1), the full-order proper filter $\mathcal{F}$ in (2.4), and the resulting filtering error system $\mathcal{E}$ in (2.5). To design a filter $\mathcal{F}$ in the energy-to-peak setting, suppose that the noise $w(t) \in L_2[0, +\infty)$ for the continuous-time case and $w(t) \in l_2[0, +\infty)$ for the discrete-time case, and that both system $\mathcal{S}$ and filter $\mathcal{F}$ have zero initial conditions, $x(0) = 0$ and $x_F(0) = 0$. The robust energy-to-peak filtering problem to be addressed for system $\mathcal{S}$ in this section is formulated as follows.

Robust Energy-to-Peak Filtering Problem: Given system $\mathcal{S}$ in (2.1), design a filter $\mathcal{F}$ of the form in (2.4) such that the filtering error system $\mathcal{E}$ in (2.5), for all $\lambda \in \Gamma$, is robustly asymptotically stable, and for all nonzero $w(t) \in L_2[0, +\infty)$ for the continuous-time case or for all nonzero $w(t) \in l_2[0, +\infty)$ for the discrete-time case, satisfies

$$\sup_{\lambda \in \Gamma} \frac{\|e\|_\infty}{\|w\|_2} < \gamma \tag{2.88}$$

where $\|e\|_\infty = \|e\|_{L_\infty}$ and $\|w\|_2 = \|w\|_{L_2}$ for the continuous-time case, $\|e\|_\infty = \|e\|_{l_\infty}$ and $\|w\|_2 = \|w\|_{l_2}$ for the discrete-time case and γ is a given positive constant. Moreover, a filter satisfying the above conditions is referred to as *a filter with a guaranteed robust energy-to-peak performance bound* γ.

The physical meaning of the specification in (2.69) is that the worst energy-to-peak ($L_2 - L_\infty$ or $l_2 - l_\infty$) gain of the filtering error system is bounded by a given level γ for all the uncertainties. If the specification in (2.69) is satisfied, the peak value of the filtering error will not exceed $\gamma\, \|w\|_2$ for all energy-bounded noise $w(t)$. The strictly proper form of the filter $\mathcal{F}$ in (2.4) guarantees that the energy-to-peak gain of the filtering error system is finite, that is, there always exists some sufficiently large γ such that (2.88) can be satisfied. Smaller γ means the better filtering performance of a filter in the energy-to-peak setting. Hence, it is often expected to (globally or locally) minimize γ in the above energy-to-peak filtering problem.

2.3.2 Energy-to-Peak Filtering for Nominal Systems

For nominal systems, the following result provides an energy-to-peak performance criterion in terms of LMI for filtering performance analysis.

Lemma 2.7 ([28, 29]) *Consider system $\mathcal{S}$ in* (2.1) *and assume that $M \in \mathcal{M}$ is fixed but arbitrary. Given filter $\mathcal{F}$ in* (2.4) *and a scalar $\gamma > 0$, the filtering error system $\mathcal{E}$ in* (2.5) *is asymptotically stable and satisfies* (2.88) *if and only if the following matrix inequalities*

- *for the continuous-time case*

$$\begin{bmatrix} -\gamma^2 I & \bar{C} \\ \bar{C}^{\mathrm{T}} & -P \end{bmatrix} < 0, \quad \begin{bmatrix} \bar{A}^{\mathrm{T}} P + P\bar{A} & P\bar{B} \\ \bar{B}^{\mathrm{T}} P & -I \end{bmatrix} < 0 \tag{2.89}$$

- *for the discrete-time case*

$$\begin{bmatrix} -\gamma^2 I & \bar{C} \\ \bar{C}^{\mathrm{T}} & -P \end{bmatrix} < 0, \quad \begin{bmatrix} -P & P\bar{A} & P\bar{B} \\ \bar{A}^{\mathrm{T}} P & -P & 0 \\ \bar{B}^{\mathrm{T}} P & 0 & -I \end{bmatrix} < 0 \tag{2.90}$$

are feasible in the real matrix variable P.

Although the result in Lemma 2.7 does not give the filter realization explicitly, the NLMIs in (2.89) and (2.90) can be converted into LMIs by noting the similarity between the inequalities in (2.89) and (2.90) and those in Lemmas 2.1 and 2.2 for H_2 filtering. Hence, by performing a series of similar matrix transformations, the following necessary and sufficient condition can be obtained for energy-to-peak filter design, leading to an LMI approach to solve the energy-to-peak filtering problem for nominal systems.

Theorem 2.8 *Assume that $M \in \mathcal{M}$ is fixed but arbitrary. Given a scalar $\gamma > 0$, a filter $\mathcal{F}$ in* (2.4) *exists such that the filtering error system $\mathcal{E}$ in* (2.5) *is asymptotically stable and satisfies* (2.88) *if and only if the following inequality*

$$\begin{bmatrix} -\gamma^2 I & L & -\tilde{C}_F \\ L^{\mathrm{T}} & -\tilde{P}_1 & -\tilde{P}_2 \\ -\tilde{C}_F^{\mathrm{T}} & -\tilde{P}_2 & -\tilde{P}_2 \end{bmatrix} < 0 \tag{2.91}$$

and

- *for the continuous-time case*

$$\begin{bmatrix} X_1 + X_1^{\mathrm{T}} & \tilde{A}_F + X_2^{\mathrm{T}} & \tilde{P}_1 B + \tilde{B}_F D \\ * & \tilde{A}_F + \tilde{A}_F^{\mathrm{T}} & \tilde{P}_2 B + \tilde{B}_F D \\ * & * & -I \end{bmatrix} < 0 \tag{2.92}$$

- *for the discrete-time case*

$$\begin{bmatrix} -\tilde{P}_1 & -\tilde{P}_2 & X_1 & \tilde{A}_F & \tilde{P}_1 B + \tilde{B}_F D \\ * & -\tilde{P}_2 & X_2 & \tilde{A}_F & \tilde{P}_2 B + \tilde{B}_F D \\ * & * & -\tilde{P}_1 & -\tilde{P}_2 & 0 \\ * & * & * & -\tilde{P}_2 & 0 \\ * & * & * & * & I \end{bmatrix} < 0 \tag{2.93}$$

are feasible in the real matrix variables $\tilde{P}_1$, $\tilde{P}_2$, $\tilde{A}_F$, $\tilde{B}_F$, *and* $\tilde{C}_F$, *where*

$$X_1 \triangleq \tilde{P}_1 A + \tilde{B}_F C, \ X_2 \triangleq \tilde{P}_2 A + \tilde{B}_F C.$$

Moreover, if these conditions are feasible, an admissible state-space realization of the filter $\mathcal{F}$ *in* (2.4) *is given by* (2.61) *or* (2.62).

Based on Theorem 2.8, we have Algorithm 9 for designing the optimal energy-to-peak filters for nominal systems by the LMI technique.

Algorithm 9 Energy-to-Peak Filter Design I

- Solve the minimization problem:
  - Continuous-time case

$$\min_{\tilde{P}_1, \tilde{P}_2, \tilde{A}_F, \tilde{B}_F, \tilde{C}_F, \mu} \gamma^2 = \mu \quad \text{s.t.} \ (2.91), \ (2.92)$$

  - Discrete-time case

$$\min_{\tilde{P}_1, \tilde{P}_2, \tilde{A}_F, \tilde{B}_F, \tilde{C}_F, \mu} \gamma^2 = \mu \quad \text{s.t.} \ (2.91), \ (2.93)$$

- Compute filter $\mathcal{F}$ in (2.4) by (2.61) or (2.62).

One may note that matrix variable $\tilde{C}_F$ appears only in the inequality in (2.91). Hence, by Lemma 1.3, it is also possible to eliminate $\tilde{C}_F$ from the result in Theorem 2.8, so that the number of free variables can be reduced. The derivation and the result will be similar to the case of Theorems 2.2 and 2.4 dealing with H_2 filtering. With the proof omitted, a modified algorithm (Algorithm 10) is presented for the optimal energy-to-peak filter design for nominal systems, where $\tilde{C}_F$ can be prescribed as $\tilde{C}_F = L$ in advance and no conservatism is introduced.

2.3.3 *Quadratic Robust Energy-to-Peak Filtering*

When there exists uncertain but fixed parameter λ, the energy-to-peak gain of the filtering system satisfies [28–30]

Algorithm 10 Energy-to-Peak Filter Design II

- Solve the minimization problem:
  - Continuous-time case

$$\min_{\tilde{P}_1,\,\tilde{P}_2,\,\tilde{A}_F,\,\tilde{B}_F,\,\mu} \gamma^2 = \mu \quad \text{s.t.} \begin{cases} (2.92) \\ \begin{bmatrix} \tilde{P}_1 & \tilde{P}_2 \\ \tilde{P}_2 & \tilde{P}_2 \end{bmatrix} > 0, \begin{bmatrix} \gamma^2 I & L \\ L^{\mathrm{T}} & \tilde{P}_1 \end{bmatrix} > 0 \end{cases}$$

  - Discrete-time case

$$\min_{\tilde{P}_1,\,\tilde{P}_2,\,\tilde{A}_F,\,\tilde{B}_F,\,\mu} \gamma^2 = \mu \quad \text{s.t.} \begin{cases} (2.93) \\ \begin{bmatrix} \gamma^2 I & L \\ L^{\mathrm{T}} & \tilde{P}_1 \end{bmatrix} > 0 \end{cases}$$

- Compute filter $\mathcal{F}$ in (2.4) by

$$A_F = \tilde{P}_1^{-1}\tilde{A}_F\tilde{P}_2^{-1}\tilde{P}_1,\; B_F = -\tilde{P}_1^{-1}\tilde{B}_F,\; C_F = L.$$

$$\sup_{\lambda\in\Gamma} \frac{\|e\|_\infty}{\|w\|_2} = \sup_{\lambda\in\Gamma} \rho_{\max}^{1/2}[C(\lambda)P(\lambda)C(\lambda)^{\mathrm{T}}],$$

where $\rho_{\max}[\cdot]$ represents the spectral radius of a square matrix, and Lyapunov matrix $P(\lambda)$ is the solution to parameter-dependent Lyapunov equation

$$\bar{A}(\lambda)P(\lambda) + P(\lambda)\bar{A}(\lambda)^{\mathrm{T}} + \bar{B}(\lambda)\bar{B}(\lambda)^{\mathrm{T}} = 0$$

for the continuous-time case, or

$$\bar{A}(\lambda)P(\lambda)\bar{A}(\lambda)^{\mathrm{T}} - P(\lambda) + \bar{B}(\lambda)\bar{B}(\lambda)^{\mathrm{T}} = 0$$

for the discrete-time case. Following the notion of quadratic stability, suppose that (2.89) and (2.90) with matrices $(\bar{A},\ \bar{B},\ \bar{C})$ replaced by its parameter-dependent counterpart $(\bar{A}(\lambda),\ \bar{B}(\lambda),\ \bar{C}(\lambda))$ are still valid for all $M(\lambda) \in \mathcal{M}$. By the Schur complement, the second inequalities in (2.89) and (2.90) imply

$$\begin{aligned} \bar{A}(\lambda)P^{-1} + P^{-1}\bar{A}(\lambda)^{\mathrm{T}} + \bar{B}(\lambda)\bar{B}(\lambda)^{\mathrm{T}} &< 0, \\ \bar{A}(\lambda)P^{-1}\bar{A}(\lambda)^{\mathrm{T}} - P^{-1} + \bar{B}(\lambda)\bar{B}(\lambda)^{\mathrm{T}} &< 0. \end{aligned}$$

From the monotonic property of the solution to Lyapunov equations, there holds $P(\lambda) < P^{-1}$ for all $\lambda \in \Gamma$, which together with the first inequalities in (2.89) and (2.90), respectively, gives rise to

$$\begin{aligned}\sup_{\lambda\in\Gamma}\frac{\|e\|_\infty}{\|w\|_2} &= \sup_{\lambda\in\Gamma}\rho_{\max}^{1/2}[C(\lambda)P(\lambda)C(\lambda)^{\mathrm{T}}]\\ &< \sup_{\lambda\in\Gamma}\rho_{\max}^{1/2}[C(\lambda)P^{-1}C(\lambda)^{\mathrm{T}}]\\ &< \sup_{\lambda\in\Gamma}\rho_{\max}^{1/2}[\gamma^2 I] = \gamma.\end{aligned}$$

Consequently, along the quadratic stability notion, the following lemma provides a quadratic approach to robust energy-to-peak performance analysis for the filtering error system under a given filter.

Lemma 2.8 *Consider system $\mathcal{S}$ in* (2.1) *and assume that $M \in \mathcal{M}$ is fixed but arbitrary. Given filter $\mathcal{F}$ in* (2.4) *and a scalar $\gamma > 0$, the filtering error system $\mathcal{E}$ in* (2.5) *is robustly asymptotically stable and satisfies* (2.88) *if the following matrix inequalities*

- *for the continuous-time case*

$$\begin{bmatrix} -\gamma^2 I & \bar{C}_i \\ \bar{C}_i^{\mathrm{T}} & -P \end{bmatrix} < 0, \quad \begin{bmatrix} \bar{A}_i^{\mathrm{T}} P + P\bar{A}_i & P\bar{B}_i \\ \bar{B}_i^{\mathrm{T}} P & -I \end{bmatrix} < 0 \, (i = 1, 2, \ldots, s)$$

- *for the discrete-time case*

$$\begin{bmatrix} -\gamma^2 I & \bar{C}_i \\ \bar{C}_i^{\mathrm{T}} & -P \end{bmatrix} < 0, \quad \begin{bmatrix} -P & P\bar{A}_i & P\bar{B}_i \\ \bar{A}_i^{\mathrm{T}} P & -P & 0 \\ \bar{B}_i^{\mathrm{T}} P & 0 & -I \end{bmatrix} < 0 \, (i = 1, 2, \ldots, s)$$

are feasible in the real matrix variable P.

Furthermore, linearizing the NLMI conditions in this lemma results in the following quadratic approach to robust energy-to-peak filter design.

Theorem 2.9 *Consider system $\mathcal{S}$ in* (2.1). *Given a scalar $\gamma > 0$, a filter $\mathcal{F}$ in* (2.4) *exists such that the filtering error system $\mathcal{E}$ in* (2.5) *is asymptotically stable and satisfies* (2.88) *if and only if the following inequality*

$$\begin{bmatrix} -\gamma^2 I & L_i & -\tilde{C}_F \\ L_i^{\mathrm{T}} & -\tilde{P}_1 & -\tilde{P}_2 \\ -\tilde{C}_F^{\mathrm{T}} & -\tilde{P}_2 & -\tilde{P}_2 \end{bmatrix} < 0 \, (i = 1, 2, \ldots, s) \tag{2.94}$$

and

- *for the continuous-time case*

$$\begin{bmatrix} X_{1,i} + X_{1,i}^{\mathrm{T}} & \tilde{A}_F + X_{2,i}^{\mathrm{T}} & \tilde{P}_1 B_i + \tilde{B}_F D_i \\ * & \tilde{A}_F + \tilde{A}_F^{\mathrm{T}} & \tilde{P}_2 B_i + \tilde{B}_F D_i \\ * & * & -I \end{bmatrix} < 0 \, (i = 1, 2, \ldots, s) \tag{2.95}$$

• *for the discrete-time case*

$$\begin{bmatrix} -\tilde{P}_1 & -\tilde{P}_2 & X_{1,i} & \tilde{A}_F & \tilde{P}_1 B_i + \tilde{B}_F D_i \\ * & -\tilde{P}_2 & X_{2,i} & \tilde{A}_F & \tilde{P}_2 B_i + \tilde{B}_F D_i \\ * & * & -\tilde{P}_1 & -\tilde{P}_2 & 0 \\ * & * & * & -\tilde{P}_2 & 0 \\ * & * & * & * & I \end{bmatrix} < 0 \, (i = 1, 2, \ldots, s) \tag{2.96}$$

are feasible in the real matrix variables $\tilde{P}_1$, $\tilde{P}_2$, $\tilde{A}_F$, $\tilde{B}_F$, *and* $\tilde{C}_F$, *where*

$$X_{1,i} \triangleq \tilde{P}_1 A_i + \tilde{B}_F C_i, \; X_{2,i} \triangleq \tilde{P}_2 A_i + \tilde{B}_F C_i.$$

Moreover, if these conditions are feasible, an admissible state-space realization of the filter $\mathcal{F}$ *in* (2.4) *is given by* (2.61) *or* (2.62).

Based on Theorem 2.9, two quadratic approaches (Algorithms 11 and 12) to design suboptimal robust energy-to-peak filters can be obtained for unknown and known L, respectively.

Algorithm 11 Quadratic Approach to Robust Energy-to-Peak Filter Design

---

• Solve the minimization problem:
  - Continuous-time case

$$\min_{\tilde{P}_1, \tilde{P}_2, \tilde{A}_F, \tilde{B}_F, \tilde{C}_F, \mu} \gamma^2 = \mu \quad \text{s.t. (2.94), (2.95)}$$

  - Discrete-time case

$$\min_{\tilde{P}_1, \tilde{P}_2, \tilde{A}_F, \tilde{B}_F, \tilde{C}_F, \mu} \gamma^2 = \mu \quad \text{s.t. (2.94), (2.96)}$$

• Compute filter $\mathcal{F}$ in (2.4) by (2.61) or (2.62).

Finally, it deserves pointing out that, for single-input-single-output (SISO) systems, the energy-to-peak performance of a transfer function coincides with its H_2 norm. According to [31], the energy-to-peak performance of a continuous-time transfer function can be defined in the frequency-domain as

$$\|T(\mathrm{j}\omega)\|_{L_2 - L_\infty} = \sqrt{\frac{1}{2\pi} \lambda_{\max} \left[\int_{-\infty}^{\infty} T(\mathrm{j}\omega) T(\mathrm{j}\omega)^* \mathrm{d}\omega \right]}$$

where $\lambda_{\max}[\cdot]$ denotes the maximum eigenvalue of a matrix, while in the frequency-domain, the H_2 norm of a transfer function is given by [1].

Algorithm 12 Quadratic Approach to Robust Energy-to-Peak Filter Design (Known L)

- Solve the minimization problem:
  - Continuous-time case

$$\min_{\tilde{P}_1, \tilde{P}_2, \tilde{A}_F, \tilde{B}_F, \mu} \gamma^2 = \mu \text{ s.t. } \begin{cases} (2.95) \\ \begin{bmatrix} \tilde{P}_1 & \tilde{P}_2 \\ \tilde{P}_2 & \tilde{P}_2 \end{bmatrix} > 0, \begin{bmatrix} \gamma^2 I & L \\ L^{\mathrm{T}} & \tilde{P}_1 \end{bmatrix} > 0 \end{cases}$$

  - Discrete-time case

$$\min_{\tilde{P}_1, \tilde{P}_2, \tilde{A}_F, \tilde{B}_F, \mu} \gamma^2 = \mu \text{ s.t. } \begin{cases} (2.96) \\ \begin{bmatrix} \gamma^2 I & L \\ L^{\mathrm{T}} & \tilde{P}_1 \end{bmatrix} > 0 \end{cases}$$

- Compute filter $\mathcal{F}$ in (2.4) by

$$A_F = \tilde{P}_1^{-1} \tilde{A}_F \tilde{P}_2^{-1} \tilde{P}_1, \; B_F = -\tilde{P}_1^{-1} \tilde{B}_F, \; C_F = L.$$

$$\|T(\mathrm{j}\omega)\|_2 = \sqrt{\frac{1}{2\pi} \mathrm{Tr}\left[\int_{-\infty}^{\infty} T(\mathrm{j}\omega) T(\mathrm{j}\omega)^* \mathrm{d}\omega\right]}$$

Obviously, for SISO systems, there holds $\|T(\mathrm{j}\omega)\|_{L_2-L_\infty} = \|T(\mathrm{j}\omega)\|_2$. On the other hand, this fact can also be obtained from Lemma 2.1 and the continuous-time part of Lemma 2.7, since they are obviously equivalent to each other for SISO systems. Hence, for SISO systems, the quadratic approach to robust energy-to-peak filter design in this section is actually that for robust H_2 filter design in Sect. 2.1.

2.4 Examples

In this section, several numerical examples are provided to demonstrate the effectiveness of the presented filter design methods.

Example 2.1 (*H_2 filter design, continuous-time* [4, 6, 32–35]) Consider an uncertain continuous-time system given by

$$\begin{aligned} \dot{x}(t) &= \begin{bmatrix} 0 & -1 + 0.3\alpha \\ 1 & -0.5 \end{bmatrix} x(t) + \begin{bmatrix} -2 & 0 \\ 1 & 0 \end{bmatrix} w(t), \\ y(t) &= \begin{bmatrix} -100 + 10\beta & 100 \end{bmatrix} x(t) + \begin{bmatrix} 0 & 1 \end{bmatrix} w(t), \\ z(t) &= \begin{bmatrix} 1 & 0 \end{bmatrix} x(t), \end{aligned} \tag{2.97}$$

where α and β are unknown but bounded constant parameters. It is assumed that $\alpha = \beta = 0$ corresponds to the nominal system of (2.97). The goal of this example is to design H_2 filters in the form in (2.5) by Algorithms 1 and 2 for the nominal system of (2.97) and by Algorithms 5 and 6 (continuous-time case) for the uncertain system of (2.97).

(1) *Nominal system with* $\alpha = \beta = 0$

Solving the minimization problems in Algorithms 1 and 2 yields the same minimum upper bound of the error variance as

$$\gamma^* = 0.0266.$$

The decision matrix variables in Algorithm 1 are also obtained as

$$\begin{aligned}
\tilde{P}_{c1} &= \begin{bmatrix} 41.7362 & -16.6945 \\ -16.6945 & 66.7779 \end{bmatrix}, \\
\tilde{P}_{c2} &= \begin{bmatrix} 41.6376 & -16.6605 \\ -16.6605 & 66.6418 \end{bmatrix}, \\
\tilde{A}_F &= \begin{bmatrix} -9,994 & 9,946 \\ 10,044 & -9,996 \end{bmatrix}, \\
\tilde{B}_F &= \begin{bmatrix} 99.9999 \\ -99.9999 \end{bmatrix}, \\
\tilde{C}_F &= \begin{bmatrix} -0.9976 & 0 \end{bmatrix},
\end{aligned}$$

which, from (2.17), generate a filter $\mathcal{F}$ with the state-space realization given by

$$\left[\begin{array}{cc} A_F & B_F \\ C_F & 0 \end{array}\right] = \left[\begin{array}{cc|c} -199.6846 & 198.7382 & 2.0015 \\ 100.7881 & -100.3155 & -1.0002 \\ \hline -0.9976 & 0 & 0 \end{array}\right], \tag{2.98}$$

and, from (2.18), generate another filter $\mathcal{F}$ given by

$$\left[\begin{array}{cc} A_F & B_F \\ C_F & 0 \end{array}\right] = \left[\begin{array}{cc|c} -199.6846 & 99.1667 & 99.9999 \\ 201.3334 & -99.6667 & -99.9999 \\ \hline -0.0266 & -0.0067 & 0 \end{array}\right]. \tag{2.99}$$

The matrix variables in Algorithm 2 are obtained as

$$\begin{aligned}
\tilde{P}_{c1} &= \begin{bmatrix} 41.7362 & -16.6945 \\ -16.6945 & 66.7779 \end{bmatrix}, \\
\tilde{P}_{c2} &= \begin{bmatrix} 41.6361 & -16.6595 \\ -16.6595 & 66.6389 \end{bmatrix}, \\
\tilde{A}_F &= \begin{bmatrix} -9,993 & 9,946 \\ 10,043 & -9,996 \end{bmatrix},
\end{aligned}$$

$$\tilde{B}_F = \begin{bmatrix} 100.002 \\ -100.002 \end{bmatrix}.$$

Accordingly, the filter $\mathcal{F}$ generated from (2.21) is given by

$$\begin{bmatrix} A_F & B_F \\ C_F & 0 \end{bmatrix} = \left[\begin{array}{cc|c} -199.6673 & 198.6673 & -1.9967 \\ 100.8336 & -100.3336 & 0.9983 \\ \hline 1 & 0 & 0 \end{array}\right]. \quad (2.100)$$

Note that $BD^{\mathrm{T}} = 0$ and $DD^{\mathrm{T}} = I$. Thus, for the nominal system of this example, the well-known Kalman filter (see Sect. 2.1.3) can be utilized to estimate $z(t)$. To design the Kalman filter for (2.97), we solve the following ARE:

$$AP + PA^{\mathrm{T}} - PC^{\mathrm{T}}CP + BB^{\mathrm{T}} = 0,\ P \geq 0$$

and obtain the unique solution P as

$$P = \begin{bmatrix} 0.0266 & 0.0067 \\ 0.0067 & 0.0166 \end{bmatrix}.$$

Then, the Kalman filter gain matrices in (2.23) are obtained from

$$\begin{aligned} A_F &= A - PC^{\mathrm{T}}C = \begin{bmatrix} -199.6669 & 198.6669 \\ 100.8335 & -100.3335 \end{bmatrix}, \\ B_F &= PC^{\mathrm{T}} = \begin{bmatrix} -1.9967 \\ 0.9983 \end{bmatrix}, \\ C_F &= L = \begin{bmatrix} 1 & 0 \end{bmatrix}, \end{aligned} \quad (2.101)$$

with the optimal filtering error variance bound given by

$$\mathbf{E}\left[e(t)^{\mathrm{T}}e(t)\right] \leq \mathrm{Tr}\left[LPL^{\mathrm{T}}\right] = 0.0266.$$

It is shown that the filtering error variance bound of the filters in (2.98), (2.99), and (2.100) designed by the optimal H_2 approaches presented in this chapter is equal to that of the Kalman filter in (2.101). Furthermore, comparing the filter parameters presented above, it is realized that, if omitting the numerically computational error, the filter state-space realization in (2.100) is "equivalent" to the Kalman filter in (2.101). Indeed, according to the discussions in Sect. 2.1.3, these results are not surprising due to the fact that the objective of the optimal H_2 filter for the nominal system is the same as that of the Kalman filter.

(2) *Polytopic system with* $|\alpha| \leq 1$ *and* $|\beta| \leq 1$

For the uncertain case, system (2.97) can be represented by the form of polytopic system $\mathcal{S}$ in (2.1) with four vertices. Applying the quadratic approaches, Algorithms

5 and 6 (continuous-time case), we obtain the same minimum filtering error variance upper bound

$$\gamma^* = 5.7281.$$

Accordingly, the filter state-space realizations obtained from (2.61), (2.62), and (2.63) are, respectively, given by

$$\begin{bmatrix} A_F & B_F \\ C_F & 0 \end{bmatrix} = \left[\begin{array}{cc|c} -0.8748 & 0.0946 & 0.0129 \\ 1.8493 & -1.3812 & -0.0095 \\ \hline -0.7698 & 0.0746 & 0 \end{array}\right], \tag{2.102}$$

$$\begin{bmatrix} A_F & B_F \\ C_F & 0 \end{bmatrix} = \left[\begin{array}{cc|c} -2.2372 & -0.3924 & 0.0036 \\ 2.5257 & -0.0189 & -0.0035 \\ \hline -5.7281 & -2.8533 & 0 \end{array}\right], \tag{2.103}$$

$$\begin{bmatrix} A_F & B_F \\ C_F & 0 \end{bmatrix} = \left[\begin{array}{cc|c} -1.0428 & 0.1279 & -0.0107 \\ 1.8127 & -1.2135 & 0.0082 \\ \hline 1 & 0 & 0 \end{array}\right]. \tag{2.104}$$

To illustrate the effectiveness of these filters in (2.102) and (2.104) (we omit (2.103) because its transfer function is the same as the one of the filter in (2.102)), the actual values of the H_2 norm square of the filtering error systems under different parameters α and β are depicted in Fig. 2.1, where $T(\delta, \alpha, \beta)$ denotes the transfer function of the filtering error system. It is shown that within the parameter uncertainty domain, the actual values of the H_2 norm square are obviously smaller than the obtained minimum error variance upper bound $\gamma^* = 5.7281$.

In addition, to illustrate the robustness difference of the designed filters against parameter uncertainties, Table 2.2 involves the actual values of the H_2 norm square of the filtering error systems generated from the Kalman filter in (2.101) and the robust H_2 filters in (2.102) and (2.104). Although the Kalman filter can guarantee the optimal filtering error variance level $\gamma^* = 0.0266$ for the nominal system, its performance deteriorates greatly when encountering parameter uncertainties. For instance, the actual filtering error variance level of the Kalman filter for $(\alpha, \beta) = (1, 1)$ is 31.4293, 1181 times larger than the value for the nominal $(\alpha, \beta) = (0, 0)$. The results in Table 2.2 display that the robustness of the Kalman filter against parameter uncertainties is the weakest among the listed filters. Thus, it is necessary to take uncertainties into consideration in the process of designing filters, so that the designed filters are robust enough in practical applications.

Example 2.2 (*H_2 filter design, discrete-time,* [7, 8, 33, 34, 36, 37]). Consider the following uncertain discrete-time system:

$$\begin{aligned} x(t+1) &= \begin{bmatrix} 0.9 & 0.1 + 0.06\alpha \\ 0.01 + 0.05\beta & 0.9 \end{bmatrix} x(t) + \begin{bmatrix} 1 & 0 & 0 \\ 0 & 1 & 0 \end{bmatrix} w(t), \\ y(t) &= \begin{bmatrix} 1 & 0 \end{bmatrix} x(t) + \begin{bmatrix} 0 & 0 & \sqrt{2} \end{bmatrix} w(t), \end{aligned}$$

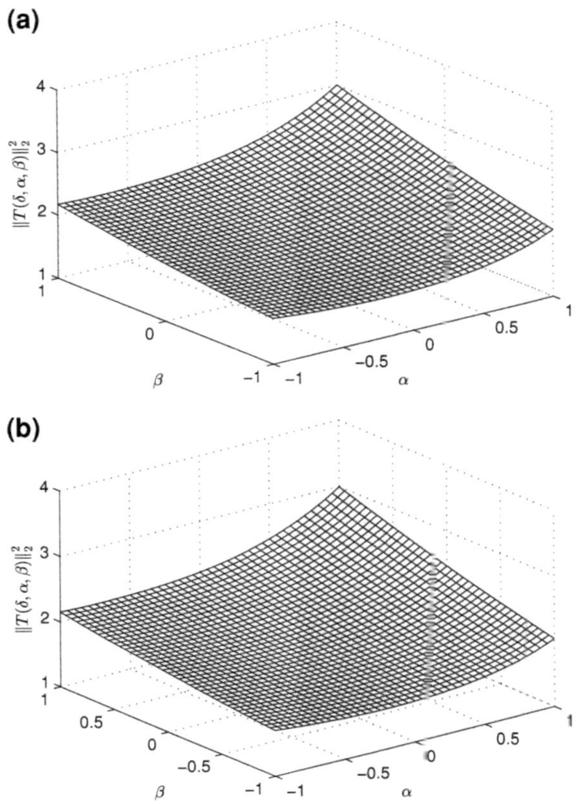

Fig. 2.1 Actual performance $\|T(\delta, \alpha, \beta)\|_2^2$ with different α and β in Example 2.1 [**a** the filter in Eq. (2.102); **b** the filter in Eq. (2.104)]

Table 2.2 Robustness comparison in Example 2.1 ((α, β) are in the bracket)

		$\|T(\delta, \alpha, \beta)\|_2^2$				
Filter	Bound γ^*	(0, 0)	$(-1, -1)$	$(-1, 1)$	$(1, -1)$	(1, 1)
Filter in (2.101)	—	0.0266	10.7343	0.2963	2.8822	31.4293
Filter in (2.102)	5.7281	1.8922	1.7217	2.1682	2.0528	3.0028
Filter in (2.104)	5.7281	1.8771	1.7062	2.1492	2.0452	2.9954

$$z(t) = \begin{bmatrix} 1 & 1 \end{bmatrix} x(t), \tag{2.105}$$

where α and β are uncertain parameters satisfying $|\alpha| \leq 1$ and $|\beta| \leq 1$. For the nominal system, it is assumed that $\alpha = \beta = 0$. We will design H_2 filters in the form (2.5) for the nominal system of (2.105) by Algorithms 3 and 4 and for the uncertain system of (2.105) by Algorithms 5 and 6 (discrete-time case), respectively.

(1) *Nominal system with* $\alpha = \beta = 0$

By Algorithms 1 and 2, the same minimum upper bound of the error variance can be obtained as

$$\gamma^* = 8.0759.$$

And the state-space realizations of the filters generated from (2.35), (2.36), and (2.44) are, respectively, given by

$$\begin{bmatrix} A_F & B_F \\ C_F & 0 \end{bmatrix} = \left[\begin{array}{cc|c} 0.4514 & 0.0616 & -0.5375 \\ -0.1996 & 0.8914 & -0.2509 \\ \hline -0.9705 & -0.4273 & 0 \end{array}\right], \tag{2.106}$$

$$\begin{bmatrix} A_F & B_F \\ C_F & 0 \end{bmatrix} = \left[\begin{array}{cc|c} 0.4811 & 0.0894 & -0.2405 \\ -0.0010 & 0.8616 & 0.0005 \\ \hline -2.6250 & -5.4509 & 0 \end{array}\right], \tag{2.107}$$

$$\begin{bmatrix} A_F & B_F \\ C_F & 0 \end{bmatrix} = \left[\begin{array}{cc|c} 0.4427 & 0.1000 & 0.4573 \\ -0.1615 & 0.9000 & 0.1715 \\ \hline 1 & 1 & 0 \end{array}\right]. \tag{2.108}$$

(2) *Uncertain system with* $|\alpha| \leq 1$ *and* $|\beta| \leq 1$

The uncertainty polytope of the system in (2.105) consists of four vertices. By Algorithms 5 and 6 (discrete-time case) with $s = 4$, we get the optimal filtering error variance upper bound as

$$\gamma^* = 100.0276,$$

and the state-space realizations of the filters computed from (2.61), (2.62), and (2.63) are, respectively, given by

$$\begin{bmatrix} A_F & B_F \\ C_F & 0 \end{bmatrix} = \left[\begin{array}{cc|c} 0.0365 & 0.0640 & -0.9239 \\ -0.5890 & 0.9004 & -0.6506 \\ \hline -1.1228 & -0.2914 & 0 \end{array}\right], \tag{2.109}$$

$$\begin{bmatrix} A_F & B_F \\ C_F & 0 \end{bmatrix} = \left[\begin{array}{cc|c} 0.0826 & -0.0768 & -0.0413 \\ -0.0002 & 0.8543 & 0.0001 \\ \hline -29.8402 & -70.1874 & 0 \end{array}\right], \tag{2.110}$$

$$\begin{bmatrix} A_F & B_F \\ C_F & 0 \end{bmatrix} = \left[\begin{array}{cc|c} 0.0383 & 0.0982 & 0.8490 \\ -0.3681 & 0.8986 & 0.3779 \\ \hline 1 & 1 & 0 \end{array}\right]. \tag{2.111}$$

For each filter, the actual values of the H_2 norm square of the filtering error system $T(\delta, \alpha, \beta)$ with respect to uncertain parameters α and β are shown in Fig. 2.2. Note that the actual values of the H_2 norm square are far beneath the obtained guaranteed filtering error variance bound $\gamma^* = 100.0276$. Thus, the designed robust H_2 filters in (2.109) and (2.111) are effective.

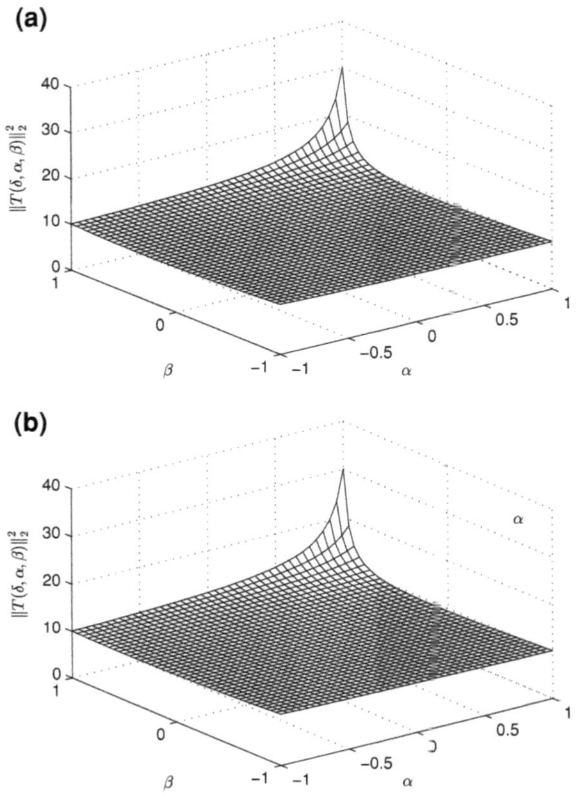

Fig. 2.2 Actual performance $\|T(\delta, \alpha, \beta)\|_2^2$ with different α and β in Example 2.2 [**a** the filter in Eq. (2.109); **b** the filter in Eq. (2.111)]

Table 2.3 Robustness comparison in Example 2.2

Filters	Eq. (2.106)	Eq. (2.108)	Eq. (2.109)	Eq. (2.111)
$\sup_{\|\alpha\|\leq 1, \|\beta\|\leq 1} \|T(\delta, \alpha, \beta)\|_2^2$	49.4385	49.5109	30.0769	30.1377

Furthermore, Table 2.3 collects results on the robustness comparison between the filters in (2.106), (2.108), (2.109), and (2.111). Although the filters in (2.106) and (2.108) are the optimal with $\gamma^* = 8.0759$ for the nominal system, they are less robust against parameter uncertainties than the robust H_2 filters in (2.109) and (2.111).

Example 2.3 (*H_∞ filter design, continuous-time*). Consider an uncertain continuous-time system given by

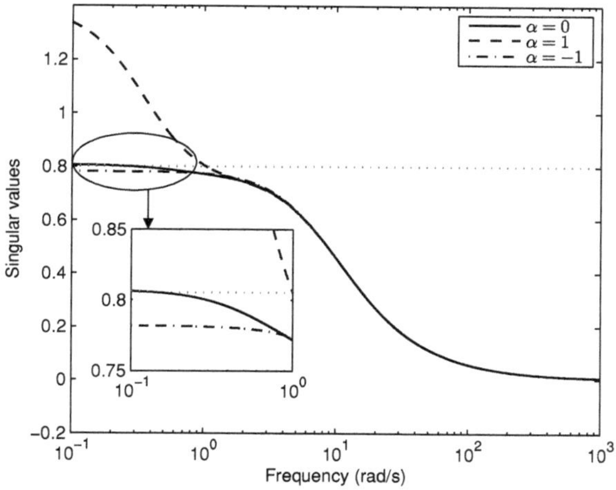

Fig. 2.3 Singular value curves with the filter in Eq. (2.113) under $\alpha = 0$, 1 and -1 in Example 2.3

$$\begin{aligned}\dot{x}(t) &= \begin{bmatrix} -2 & 0 \\ -1 & -0.7+0.4\alpha \end{bmatrix} x(t) + \begin{bmatrix} -0.5 & 0 \\ 2 & 0 \end{bmatrix} w(t),\\ y(t) &= \begin{bmatrix} 0 & 1 \end{bmatrix} x(t) + \begin{bmatrix} 0 & 1 \end{bmatrix} w(t),\\ z(t) &= \begin{bmatrix} 2 & 1 \end{bmatrix} x(t),\end{aligned} \tag{2.112}$$

where parameter α satisfies $|\alpha| \le 1$ for the uncertain system and $\alpha = 0$ for the nominal system. Design H_∞ filters of the form in (2.5), respectively, for the nominal system of (2.112) by Algorithm 7 (continuous-time case) and for the uncertain system of (2.112) by Algorithm 8 (continuous-time case).

For the nominal system, Algorithm 7 provides the optimal H_∞ filtering performance upper bound

$$\gamma^* = 0.8063$$

and yields filters with the following state-space realizations:

$$\begin{bmatrix} A_F & B_F \\ C_F & D_F \end{bmatrix} = \left[\begin{array}{cc|c} -3.1789 & 0.6728 & 0.8479 \\ -2.4339 & -8.2371 & -7.3671 \\ \hline -1.9077 & -0.9881 & 0.0039 \end{array}\right], \tag{2.113}$$

$$\begin{bmatrix} A_F & B_F \\ C_F & D_F \end{bmatrix} = \left[\begin{array}{cc|c} -5.1363 & 14.4562 & -0.0623 \\ 0.3066 & -6.2797 & -0.6517 \\ \hline 0.8361 & -8.7680 & 0.0039 \end{array}\right], \tag{2.114}$$

which are obtained from (2.74) and (2.75), respectively.

Connecting the filter in (2.113) to the original system in (2.112), we plot the singular values of the filtering error system for different parameter α, shown in Fig. 2.3. On one hand, one can see that the designed optimal H_∞ filter in (2.113) is effective since the maximum singular value of the filtering error system with respect

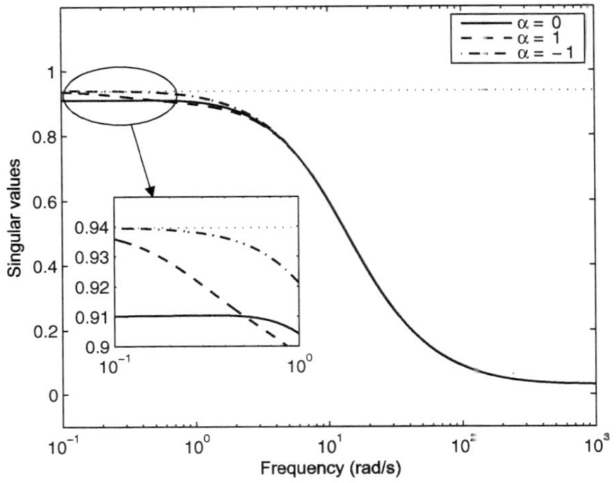

Fig. 2.4 Singular value curves with the filter in Eq. (2.115) under $\alpha = 0$, 1 and -1 in Example 2.3

to frequency ω is tightly bounded by the obtained optimal H_∞ filtering performance upper bound $\gamma^* = 0.8063$ for the nominal system. On the other hand, Fig. 2.3 clearly shows that the filter cannot guarantee the optimal H_∞ performance level when parameter uncertainty exists. Hence, in the sequel, we focus on design robust H_∞ filters for the system in (2.112).

For the uncertain system, applying Algorithm 8 to design robust H_∞ filters, we obtain the minimum H_∞ filtering performance upper bound as $\gamma^* = 0.9397$ and the following two filters with following state-space realizations:

$$\begin{bmatrix} A_F & B_F \\ C_F & D_F \end{bmatrix} = \left[\begin{array}{cc|c} -4.1377 & 0.6820 & 0.8987 \\ 0.5488 & -10.1544 & -10.1829 \\ \hline -2.0160 & -0.9713 & 0.0296 \end{array}\right], \tag{2.115}$$

$$\begin{bmatrix} A_F & B_F \\ C_F & D_F \end{bmatrix} = \left[\begin{array}{cc|c} -5.0838 & 11.4492 & 0.0930 \\ 0.4517 & -9.2084 & -0.8533 \\ \hline 0.4929 & -9.4139 & 0.0296 \end{array}\right]. \tag{2.116}$$

We still connect the filter in (2.115) to the original system in (2.112) and plot the singular values of the filtering error system for different parameter α, shown in Fig. 2.4. Figure 2.4 demonstrates that all the maximum singular values of the filtering error system at the two vertices ($\alpha = 1, -1$) and at the nominal parameter ($\alpha = 0$) are smaller than the guaranteed H_∞ filtering performance upper bound $\gamma^* = 0.9397$. Thus, the effectiveness of the filter in (2.115) (as well as the one in (2.116)) is validated.

Moreover, comparing the results in Figs. 2.3 and 2.4, one clearly sees that the maximum singular value under the filter in (2.113) over the uncertainty domain is larger than 1.2 (seeing the curve for $\alpha = -1$ in Fig. 2.3), which is beyond the guaranteed H_∞ performance upper bound $\gamma^* = 0.9397$ under the filter in (2.115). This demonstrates that the filter in (2.115) designed by the quadratic approach Algorithm 8 for the uncertain system is more robust than the one in (2.113) designed by the optimal approach Algorithm 7 for the nominal system.

Example 2.4 (*H_∞ filter design, discrete-time*) Consider an uncertain discrete-time system given by

$$\begin{aligned} x(t+1) &= \begin{bmatrix} \alpha & -0.5 \\ 1 & 1 \end{bmatrix} x(t) + \begin{bmatrix} -6 & 0 \\ 2 & 0 \end{bmatrix} w(t), \\ y(t) &= \begin{bmatrix} -100 & 10 \end{bmatrix} x(t) + \begin{bmatrix} 0 & 1 \end{bmatrix} w(t), \\ z(t) &= \begin{bmatrix} 1 & 0 \end{bmatrix} x(t). \end{aligned} \tag{2.117}$$

Uncertain parameter α is assumed as $\alpha = 0$ for the nominal system and $|\alpha| \le 0.25$ for the uncertain system. We apply Algorithm 7 (discrete-time case) and Algorithm 8 (discrete-time case) to design H_∞ filters for the nominal system and the uncertain system of (2.117), respectively.

For the nominal system, the optimal H_∞ performance upper bound obtained by Algorithm 7 is $\gamma^* = 1.0000$, which can be guaranteed by a filter with the following two state-space realizations:

$$\begin{bmatrix} A_F & B_F \\ C_F & D_F \end{bmatrix} = \left[\begin{array}{cc|c} -0.8726 & -0.3981 & 9.2822 \times 10^{-3} \\ 1.9190 & 0.8756 & 9.9001 \times 10^{-3} \\ \hline -0.1680 & 7.8408 \times 10^{-2} & 8.2567 \times 10^{-3} \end{array}\right], \tag{2.118}$$

$$\begin{bmatrix} A_F & B_F \\ C_F & D_F \end{bmatrix} = \left[\begin{array}{cc|c} 4.1549 \times 10^{-3} & 1.0249 \times 10^{-2} & 1.4551 \times 10^{-3} \\ 1.2184 \times 10^{-2} & -1.1731 \times 10^{-3} & 6.5640 \times 10^{-4} \\ \hline -0.8725 & -10.0037 & 8.2567 \times 10^{-3} \end{array}\right] \tag{2.119}$$

which are generated from (2.74) and (2.75), respectively.

Connect the obtained filter in (2.118) to the original system in (2.117) and depict the singular value curves of the resulting filtering error system with $\alpha = 0, 0.25, -0.25$. The results are displayed in Fig. 2.5, which clearly demonstrates the effectiveness of the filter in (2.118) in guaranteeing the optimal H_∞ performance upper bound $\gamma^* = 1.0000$ when $\alpha = 0$. Figure 2.5 also exposes that the filter cannot well endure the parameter uncertainty. For instance, the maximum singular value under $\alpha = 0.25$ is about 2, almost twice of $\gamma^* = 1.0000$ under $\alpha = 0$.

Next, robust H_∞ filters are designed by Algorithm 8 for the uncertain system of (2.117). The obtained minimum H_∞ filtering performance upper bound and the corresponding filter realizations are given as

$$\gamma^* = 1.2492$$

and

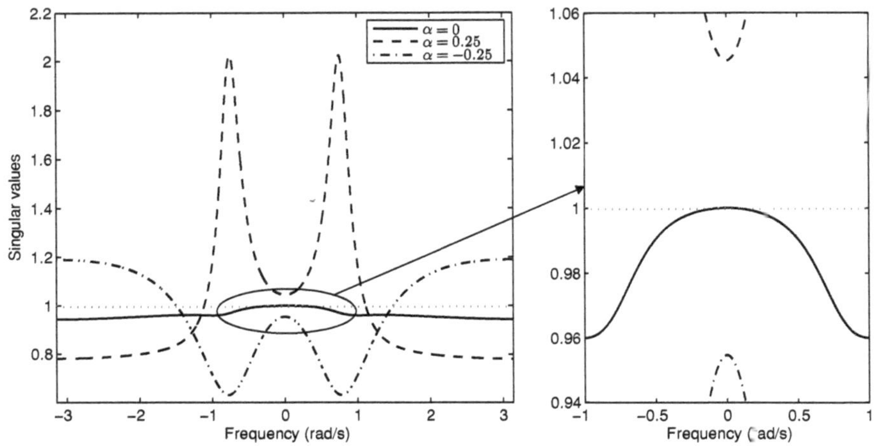

Fig. 2.5 Singular value curves with the filter in (2.118) under $\alpha = 0$, 0.25 and -0.25 in Example 2.4

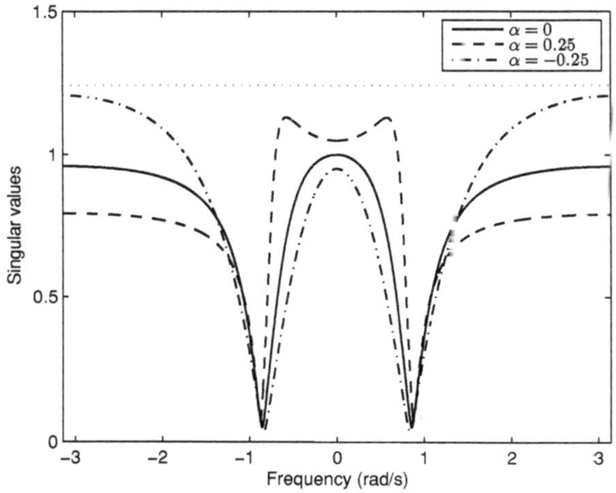

Fig. 2.6 Singular value curves with the filter in Eq. (2.120) under $\alpha = 0$, 0.25 and -0.25 in Example 2.4

$$\begin{bmatrix} A_F & B_F \\ C_F & D_F \end{bmatrix} = \left[\begin{array}{cc|c} -1.0426 & -0.6039 & 1.2911 \times 10^{-2} \\ 1.6584 & 0.9605 & -7.0043 \times 10^{-3} \\ -0.1372 & -8.6277 \times 10^{-2} & -8.6277 \times 10^{-3} \end{array}\right], \tag{2.120}$$

$$\begin{bmatrix} A_F & B_F \\ C_F & D_F \end{bmatrix} = \left[\begin{array}{cc|c} 3.4435 \times 10^{-2} & -5.4779 \times 10^{-2} & 1.6021 \times 10^{-4} \\ 7.3264 \times 10^{-2} & -0.1165 & 1.0072 \times 10^{-4} \\ -5.9589 \times 10^{-2} & -11.4952 & -8.6277 \times 10^{-3} \end{array}\right] \tag{2.121}$$

Figure 2.6 depicts the singular value curves of the filtering error system resulting from the filter in (2.120) with $\alpha = 0,\ 0.25,\ -0.25$. The effectiveness of the filter is apparent. Moreover, for the uncertain system of this example, the minimum H_∞ filtering performance upper bound $\gamma^* = 1.2492$ guaranteed by the filter in (2.120) is obviously smaller than the maximum singular values for the case of $\alpha = 0.25$ in Fig. 2.5. Thus, the robustness of the filter in (2.120) is better than that of the filter in (2.118).

2.5 Summary and Notes

2.5.1 Summary

This chapter has addressed the quadratic approaches to robust filter design for polytopic uncertain systems. By a linearization procedure, filter design methods are firstly derived for nominal systems and then the notion of quadratic stability is employed to extend the design results to the polytopic uncertainty case. Systematic results on the commonly used H_2, H_∞, and energy-to-peak filtering schemes have been established in Sects. 2.1, 2.2, and 2.3, respectively. The relationship between the optimal H_2 filtering and the Kalman filtering has been revealed. All the design methods have been formulated in the form of solving a set of LMIs. Numerical examples have been presented to illustrate the effectiveness of the design methods.

For nominal systems, the derived filter design methods can design the optimal filters in the corresponding performance; for uncertain systems, suboptimal robust filters can be effectively designed by the quadratic approaches. However, it may have been noted from, e.g., Table 2.2 or 2.3, that the guaranteed performance upper bound of each filter designed by the quadratic approaches is still far from the actual optimal performance bound of this filter. This implies that the quadratic approaches are too conservative and still leave much room that deserves further investigation. In Chap. 3, the parameter-dependent idea will be employed to reduce this conservatism and improve the filter design methods in the quadratic framework.

2.5.2 Notes

For robust H_2 filtering subject to polytopic uncertainty, quadratic approaches in the LMI framework were investigated in [4, 5, 8, 38, 39], among which [4, 5, 38] are focused on continuous-time systems and [8, 38, 39] on discrete-time systems, respectively. For robust H_∞ filtering, related results are reported in [5, 8, 38–43], where [5, 38, 40, 41, 43] are concerned with polytopic uncertain continuous-time systems, and [8, 39, 41, 42] are about polytopic uncertain discrete-time systems, respectively. Quadratic approaches to robust energy-to-peak filtering can be found in [41, 44]. Among these mentioned results, except [39] where observer-type filters are designed for state estimation, all other references design linear filters with the general state-space form.

For converting the conditions for filter design into the LMI form, the change-of-variable procedures in (2.11)–(2.14) are based on the results reported in [35]. Connections between the optimal H_2 filtering and the Kalman filtering can be found in [8, 38], where the relationship between the optimal H_∞ filter and the central H_∞ filter is also established.

References

1. Boyd, S., El Ghaoui, L., Feron, E., Balakrishnan, V.: Linear Matrix Inequalities in Systems and Control Theory. SIAM, Philadelphia (1994)
2. Colaneri, P., Geromel, J.C., Locatelli, A.: Control Theory and Design-An RH_2 and RH_∞ Viewpoint. Academic, New York (1997)
3. Zhou, K., Doyle, J.C., Glover, K.: Robust and Optimal Control. Prentice-Hall, New Jersey (1996)
4. Geromel, J.C.: Optimal linear filtering under parameter uncertainty. IEEE Trans. Signal Process. **47**(1), 168–175 (1999)
5. Geromel, J.C., de Oliveira, M.C.: H_2 and H_∞ robust filtering for convex bounded uncertain systems. IEEE Trans. Autom. Control **46**(1), 100–107 (2001)
6. Shaked, U., de Souza, C.E.: Robust minimum variance filtering. IEEE Trans. Signal Process. **43**(11), 2474–2483 (1995)
7. Geromel, J.C., Bernussou, J., Garcia, G., de Oliveira, M.C.: H_2 and H_∞ robust filtering for discrete-time linear systems. SIAM J. Control Optim. **38**, 1353–1368 (2000)
8. Geromel, J.C., Bernussou, J., Garcia, G., de Oliveira, M.C.: H_2 and H_∞ robust filtering for discrete-time linear systems. In: Proceedings of the 37th IEEE Conference on Decision and Control, pp. 632–637. Tampa, Florida USA (1998)
9. Theodor, Y., Shaked, U.: Robust discrete-time minimum-variance filtering. IEEE Trans. Signal Process. **44**(2), 181–189 (1996)
10. Barmish, B.R.: Necessary and sufficient conditions for quadratic stabilizability of an uncertain linear systems. J. Optim. Theory Appl. **46**(4), 399–408 (1985)
11. Khargonekar, P.P., Petersen, I.R., Zhou, K.: Robust stabilization of uncertain linear systems: Quadratic stabilizability and H_∞ control theory. IEEE Trans. Autom. Control **35**(3), 356–361 (1990)
12. Petersen, I.R.: A stabilization algorithm for a class of uncertain linear systems. Syst. Control Lett. **8**, 351–357 (1987)
13. Jain, B.N.: Guaranteed error estimation in uncertain systems. IEEE Trans. Autom. Control **AC-20**, 230–232 (1975)
14. Haddad, W.M., Bernstein, D.S.: Robust, reduced-order, nonstrictly proper state estimation via the optimal projection equations with Petersen-Hollot bounds. Syst. Control Lett. **9**(5), 423–431 (1987)
15. Haddad, W.M., Bernstein, D.S.: Robust, reduced-order, nonstrictly proper state estimation via the optimal projection equations with guaranteed cost bounds. IEEE Trans. Autom. Control **33**(6), 591–595 (1988)
16. Haddad, W.M., Bernstein, D.S.: The optimal projection equations for reduced-order, discrete-time state estimation for linear systems with multiplicativewhite noise. Syst. Control Lett. **8**(4), 381–388 (1987)
17. Bernstein, D.S., Haddad, W.M.: Steady-state Kalman filtering with an H_∞ error bound. Syst. Control Lett. **12**(1), 9–16 (1989)
18. Bernstein, D.S., Haddad, W.M., Mustafa, D.: Mixed-norm H_2/H_∞ regulation and estimation: the discrete-time case. Syst. Control Lett. **16**(4), 235–247 (1991)

19. Petersen, I.R., McFarlane, D.C.: Robust state estimation for uncertain systems. In: Proceedings the 30th IEEE Conference on Decision and Control, pp. 2630–2631. Brighton, England (1991)
20. Petersen, I.R., McFarlane, D.C.: Optimal guaranteed cost control and filtering for uncertain linear systems. IEEE Trans. Autom. Control **39**(9), 1971–1977 (1994)
21. Petersen, I.R., McFarlane, D.C.: Optimal guaranteed cost filtering for uncertain discrete-time systems. Int. J. Robust Nonlinear Control **6**, 267–280 (1996)
22. Xie, L., Soh, Y.C., de Souza, C.E.: Robust Kalman filtering for uncertain discrete-time systems. IEEE Trans. Autom. Control **39**(6), 1310–1314 (1994)
23. Xie, L., Soh, Y.C.: Robust Kalman filtering for uncertain systems. Syst. Control Lett. **22**, 123–129 (1994)
24. Bolzern, P., Colaneri, P., Nicolao, G.D.: Optimal robust filtering with time-varying parameter uncertainty. Int. J. Control **63**(3), 557–576 (1996)
25. Bolzern, P., Colaneri, P., Nicolao, G.D.: Optimal design of robust predictors for linear discrete-time systems. Syst. Control Lett. **26**(1), 25–31 (1995)
26. Bolzern, P., Colaneri, P., Nicolao, G.D.: Guaranteed-cost prediction of discrete-time systems: the finite- and infinite-horizon case. In: Proceedings of the 2nd IFAC Symposium on Robust Control Design, pp. 471–476. Budapest, Hungary (1997)
27. Gahinet, P., Apkarian, P.: A linear matrix inequality approach to H_∞ control. Int. J. Robust Nonlinear Control **4**(4), 421–448 (1994)
28. Grigoriadis, K.M., Watson, J.T.: Reduced-order H_∞ and L_2-L_∞ filtering via linear matrix inequalities. IEEE Trans. Aerosp. Electr. Syst. **33**(4), 1326–1338 (1997)
29. Skelton, R.E., Iwasaki, T., Grigoriadis, K.: A Unified Algebraic Approach to Control Design. Taylor & Francis, London (1997)
30. Wilson, D.A.: Convolution and Hankel operator norms for linear systems. IEEE Trans. Autom. Control **34**(1), 94–97 (1989)
31. Rotea, M.A.: The generalized H_2 control problem. Automatica **29**(2), 373–385 (1993)
32. Barbosa, K.A., de Souza, C.E., Trofino, A.: Robust H_2 filtering for uncertain linear systems: LMI based methods with parametric Lyapunov functions. Syst. Control Lett. **54**(3), 251–262 (2005)
33. Duan, Z., Zhang, J., Zhang, C.Z., Mosca, E.: Robust H_2 and H_∞ filtering for uncertain linear systems. Automatica **42**(11), 1919–1926 (2006)
34. Gao, H., Meng, X., Chen, T.: A new design of robust H_2 filters for uncertain systems. Syst. Control Lett. **57**(7), 585–593 (2008)
35. Tuan, H.D., Apkarian, P., Nguyen, T.Q.: Robust and reduced-order filtering: new LMI-based characterizations and methods. IEEE Trans. Signal Process. **49**(12), 2975–2984 (2001)
36. Geromel, J.C., De Oliveira, M.C., Bernussou, J.: Robust filtering of discrete-time linear systems with parameter dependent Lyapunov functions. SIAM J. Control Optim. **41**(3), 700–711 (2002)
37. Xie, L., Lu, L., Zhang, H.: Improved robust H_2 and H_∞ filtering for uncertain discrete-time systerms. Automatica **40**, 873–880 (2004)
38. Geromel, J.C., de Oliveira, M.C.: H_2 and H_∞ robust filtering for convex bounded uncertain systems. In: Proceedings of the 37th IEEE Conference on Decision and Control, pp. 146–151. Tampa, Florida USA (1998)
39. Palhares, R.M., Peres, P.L.D.: Optimal filtering schemes for linear discrete-time systems: a linear matrix inequality approach. Int. J. Syst. Sci. **29**(6), 587–593 (1998)
40. Palhares, R.M., Peres, P.L.D.: Robust H_∞-filtering design with pole placement constraint via linear matrix inequalities. J. Optim. Theory Appl. **102**(2), 239–261 (1999)
41. Palhares, R.M., Peres, P.L.D.: Mixed L_2-L_∞/H_∞ filtering for uncertain linear systems: a linear matrix inequality approach. Int. J. Syst. Sci. **31**(9), 1091–1098 (2000)
42. Palhares, R.M., Peres, P.L.D.: Robust H_∞ filter design with pole constraints for discrete-time systems. J. Frankl. Inst. **337**(6), 713–723 (2000)
43. Jin, S.H., Park, J.B.: Robust H_∞ filter for polytopic uncertain systems via convex optimization. IEE Proc. Control Theory Appl. **148**(1), 55–59 (2001)
44. Palhares, R.M., Peres, P.L.D.: Robust filtering with guaranteed energy-to-peak performance-an LMI approach. Automatica **36**(6), 851–858 (2000)

Chapter 3
Parameter-Dependent Robust Filter Design

In Chap. 2, quadratic approaches to the design of robust filters have been presented. Numerical results have shown that the quadratic approaches are able to design filters with a guaranteed H_2, H_∞ or energy-to-peak filtering performance level. However, as pointed out at the end of Chap. 2, for a filter designed by the quadratic approach, the gap of the guaranteed filtering performance cost and the actual filtering performance bound is sometimes tremendous. This results from the inherent conservatism of the notion of quadratic stability, that is, a *single* Lyapunov matrix has to be fixed over the entire uncertainty domain.

To reduce the conservatism of the quadratic approaches, in this chapter, we will consider parameter-dependent approaches to robust H_2 and H_∞ filters design. New H_2 and H_∞ performance criteria are first developed by introducing slack matrices into the standard criteria presented in Chap. 2. Filter realization conditions in the generic parameter-dependent form are then derived by imposing special structure on the slack matrices in the new filtering analysis results. In the sequel, we will focus on two special parameter-dependence forms such that the parameter-dependent conditions for filter realizations are well converted into LMIs. Finally, several numerical examples will be provided to demonstrate the effectiveness and advantages of the parameter-dependent methods.

In the first type of the parameter-dependent form, slack matrices are fixed over the entire uncertainty domain while "Lyapunov" matrices are linearly parameter-dependent, which results in the linearly parameter-dependent approaches to robust filter design. To make full use of the parameter-dependent idea, in the second type of the parameter-dependence form, all the "Lyapunov" matrices and part of the slack matrices are in the general homogeneous polynomial parameter-dependence form with a prescribed degree, which yields a series of polynomial parameter-dependent approaches to robust filter design.

H. Gao and X. Li, *Robust Filtering for Uncertain Systems*,
Communications and Control Engineering, DOI: 10.1007/978-3-319-05903-7_3,

3.1 Slack Matrix

In this section, we re-consider the criteria for the H_2 and H_∞ performances. The criteria stated in Chap. 2 will be further developed by introducing additional *slack matrix* or *slack matrices*, with the aid of which, new criteria for the concerned performance are proposed.

3.1.1 H_2 Performance Criterion

The following lemma gives equivalent LMI characterizations of the conditions in Lemmas 2.1 and 2.2 for the H_2 filtering performance of nominal systems.

Lemma 3.1 *The conditions in Lemmas* 2.1 *and* 2.2 *are feasible if and only if the following matrix inequalities*

$$\mathrm{Tr}[Z] < \gamma, \ \begin{bmatrix} P & \bar{C}^{\mathrm{T}} \\ \bar{C} & Z \end{bmatrix} > 0 \tag{3.1}$$

and

- *for the continuous-time case*

$$\begin{bmatrix} 0 & P & 0 \\ P & 0 & 0 \\ 0 & 0 & -I \end{bmatrix} + \mathrm{sym}\left\{\begin{bmatrix} -W & W\bar{A} & W\bar{B} \\ -F & F\bar{A} & F\bar{B} \\ -G & G\bar{A} & G\bar{B} \end{bmatrix}\right\} < 0 \tag{3.2}$$

- *for the discrete-time case*

$$\begin{bmatrix} P & 0 & 0 \\ 0 & -P & 0 \\ 0 & 0 & -I \end{bmatrix} + \mathrm{sym}\left\{\begin{bmatrix} -W & W\bar{A} & W\bar{B} \\ -F & F\bar{A} & F\bar{B} \\ -G & G\bar{A} & G\bar{B} \end{bmatrix}\right\} < 0 \tag{3.3}$$

are feasible in the symmetric matrix variables P and Z and the general matrix variables W, F and G.

Proof Let $P = P_c$, $Z = Z_c$ for the continuous-time case and $P = P_d$, $Z = Z_d$ for the discrete-time case, respectively. Note that the inequalities in (2.8), (2.9) or (2.30), (2.31) are the ones in (3.1). It only needs to prove the conditions in (3.2) and (3.3) are, respectively, equivalent to the ones in (2.10) and (2.32).

For the continuous-time case, set

$$\Phi_c = \begin{bmatrix} 0 & P & 0 \\ P & 0 & 0 \\ 0 & 0 & -I \end{bmatrix}, \ Y = \begin{bmatrix} W \\ F \\ G \end{bmatrix}, \ X = I_{4n+l}, \ \bar{Z} = \begin{bmatrix} -I & \bar{A} & \bar{B} \end{bmatrix}.$$

Under this setting, (3.2) can be rewritten as

$$\Phi_c + X^{\mathrm{T}} Y Z + Z^{\mathrm{T}} Y^{\mathrm{T}} X < 0,$$

and moreover, we can choose

$$\mathcal{N}_X = 0,\ \mathcal{N}_{\bar{Z}} = \begin{bmatrix} \bar{A} & \bar{B} \\ I & 0 \\ 0 & I \end{bmatrix}.$$

Using Lemma 1.2 with $\Phi = \Phi_c$ and noting that $\mathcal{N}_X^{\mathrm{T}} \Phi_c \mathcal{N}_X$ vanishes, we have (3.2) is equivalent to

$$\mathcal{N}_{\bar{Z}}^{\mathrm{T}} \Phi_c \mathcal{N}_{\bar{Z}} < 0.$$

This is exactly the one in (2.10) since

$$\begin{bmatrix} \bar{A}^{\mathrm{T}} P + P\bar{A} & P\bar{B} \\ \bar{B}^{\mathrm{T}} & -I \end{bmatrix} = \begin{bmatrix} \bar{A} & \bar{B} \\ I & 0 \\ 0 & I \end{bmatrix}^{\mathrm{T}} \begin{bmatrix} 0 & P & 0 \\ P & 0 & 0 \\ 0 & 0 & -I \end{bmatrix} \begin{bmatrix} \bar{A} & \bar{B} \\ I & 0 \\ 0 & I \end{bmatrix} = \mathcal{N}_{\bar{Z}}^{\mathrm{T}} \Phi_c \mathcal{N}_{\bar{Z}}.$$

For the discrete-time case, reset

$$\Phi_d = \begin{bmatrix} P & 0 & 0 \\ 0 & -P & 0 \\ 0 & 0 & -I \end{bmatrix},\ Y = \begin{bmatrix} W \\ F \\ G \end{bmatrix},\ X = I_{4n+l},\ \bar{Z} = \begin{bmatrix} -I & \bar{A} & \bar{B} \end{bmatrix}.$$

The equivalence between (3.3) and (2.32) can be similarly established by applying Lemma 1.2 and noting the following relationship:

$$\begin{aligned} & \begin{bmatrix} P & P\bar{A} & P\bar{B} \\ \bar{A}^{\mathrm{T}} P & P & 0 \\ \bar{B}^{\mathrm{T}} P & 0 & I \end{bmatrix} > 0 \\ \Leftrightarrow \quad & \begin{bmatrix} -P & -P\bar{A} & -P\bar{B} \\ -\bar{A}^{\mathrm{T}} P & -P & 0 \\ -\bar{B}^{\mathrm{T}} P & 0 & -I \end{bmatrix} < 0 \\ \xLeftrightarrow{\text{Lemma 1.1}} \quad & \begin{bmatrix} \bar{A}^{\mathrm{T}} P \bar{A} - P & \bar{A}^{\mathrm{T}} P \bar{B} \\ \bar{B}^{\mathrm{T}} P \bar{A} & \bar{B}^{\mathrm{T}} P \bar{B} - I \end{bmatrix} = \mathcal{N}_Z^{\mathrm{T}} \Phi_d \mathcal{N}_Z < 0. \end{aligned}$$

The proof is completed. □

Remark 3.1 For the energy-to-peak performance, the corresponding results can be easily obtained from Lemmas 2.7 and 2.8. Compared with the case of Lemma 3.1, the only modification is to replace the inequalities in (3.1) by

$$\begin{bmatrix} -\gamma^2 I & \bar{C} \\ \bar{C}^{\mathrm{T}} & -P \end{bmatrix} < 0.$$

The results corresponding to energy-to-peak filtering can be obtained similarly and thus are omitted. Especially, as commented at the end of Sect. 2.3, H_2 filtering and energy-to-peak filtering is the same problem for SISO systems.

Compared with Lemmas 2.1 and 2.2 in Chap. 2, three extra slack matrices W, F and H are involved in Lemma 3.1. Note that for nominal systems, Lemmas 2.1 and 2.2 are already necessary and sufficient conditions for the H_2 performance. Hence, these slack matrices will not further reduce the conservatism of the conditions for nominal systems, that is, they are redundant for nominal systems. For the discrete-time case, this fact can be elucidated in a straightforward way of specifying the slack matrices in (3.3). Let

$$W = P, \ F = 0, \ G = 0,$$

and then (3.3) reduces to

$$\begin{bmatrix} P & 0 & 0 \\ 0 & -P & 0 \\ 0 & 0 & -I \end{bmatrix} + \mathrm{sym}\left\{\begin{bmatrix} -W & W\bar{A} & W\bar{B} \\ -F & F\bar{A} & F\bar{B} \\ -G & G\bar{A} & G\bar{B} \end{bmatrix}\right\} = \begin{bmatrix} -P & P\bar{A} & P\bar{B} \\ \bar{A}^{\mathrm{T}}P & -P & 0 \\ \bar{B}^{\mathrm{T}}P & 0 & -I \end{bmatrix} < 0.$$

The right hand side (RHS) of "=" is the one in (2.32).

However, the most attractive merit of (3.2) and (3.3) is that the product terms between Lyapunov matrix P and system matrices have been eliminated, which is crucial in treating parameter-dependent synthesis problems of uncertain systems. A ready-made example is the parameter-dependent filter design problem to be addressed in the sequel.

Based on Lemma 3.1, the following condition for robust H_2 performance analysis for the filtering error system $\mathcal{S}$ in (2.1) can be obtained in terms of parameter-dependent LMI. This result can also be found in [1].

Lemma 3.2 *Given system $\mathcal{S}$ in* (2.1), *filter $\mathcal{F}$ in* (2.4) *and a scalar $\gamma > 0$, the filtering error system $\mathcal{E}$ in* (2.5) *is robustly asymptotically stable and satisfies* (2.7) *for all $\lambda \in \Gamma$ if and only if the following matrix inequalities*

$$\mathrm{Tr}[Z(\lambda)] < \gamma, \tag{3.4}$$

$$\begin{bmatrix} P(\lambda) & \bar{C}(\lambda)^{\mathrm{T}} \\ \bar{C}(\lambda) & Z(\lambda) \end{bmatrix} > 0, \tag{3.5}$$

$$\Phi(\lambda) + \Psi(\lambda) + \Psi(\lambda)^{\mathrm{T}} < 0, \tag{3.6}$$

where

$$\Psi(\lambda) = \begin{bmatrix} -W(\lambda) & W(\lambda)\bar{A}(\lambda) & W(\lambda)\bar{B}(\lambda) \\ -F(\lambda) & F(\lambda)\bar{A}(\lambda) & F(\lambda)\bar{B}(\lambda) \\ -G(\lambda) & G(\lambda)\bar{A}(\lambda) & G(\lambda)\bar{B}(\lambda) \end{bmatrix}$$

and

- *for the continuous-time case*

$$\Phi(\lambda) = \Phi_c(\lambda) \triangleq \begin{bmatrix} 0 & P(\lambda) & 0 \\ P(\lambda) & 0 & 0 \\ 0 & 0 & -I \end{bmatrix}$$

- *for the discrete-time case*

$$\Phi(\lambda) = \Phi_d(\lambda) \triangleq \begin{bmatrix} P(\lambda) & 0 & 0 \\ 0 & -P(\lambda) & 0 \\ 0 & 0 & -I \end{bmatrix},$$

are feasible in the real matrix functions $P(\lambda) = P(\lambda)^{\mathrm{T}}$, $Z(\lambda) = Z(\lambda)^{\mathrm{T}}$, $W(\lambda)$, $F(\lambda)$ *and* $G(\lambda)$.

3.1.2 H_∞ Performance Criterion

By introducing slack matrices for Lemmas 2.4 and 2.5, the following alternative BRLs can be obtained for the H_∞ filtering performance of nominal systems.

Lemma 3.3 *The conditions in Lemmas 2.4 and 2.5 are feasible if and only if the following matrix inequalities*

- *for the continuous-time case*

$$P > 0, \quad \begin{bmatrix} 0 & P & 0 & 0 \\ P & 0 & 0 & \bar{C}^{\mathrm{T}} \\ 0 & 0 & -\gamma^2 I & \bar{D}^{\mathrm{T}} \\ 0 & \bar{C} & \bar{D} & -I \end{bmatrix} + \mathrm{sym}\left\{\begin{bmatrix} -W & W\bar{A} & W\bar{B} & 0 \\ -F & F\bar{A} & F\bar{B} & 0 \\ -G & G\bar{A} & G\bar{B} & 0 \\ 0 & 0 & 0 & 0 \end{bmatrix}\right\} < 0, \tag{3.7}$$

- *for the discrete-time case*

$$P > 0, \quad \begin{bmatrix} P & 0 & 0 & 0 \\ 0 & -P & 0 & \bar{C}^{\mathrm{T}} \\ 0 & 0 & -\gamma^2 I & \bar{D}^{\mathrm{T}} \\ 0 & \bar{C} & \bar{D} & -I \end{bmatrix} + \mathrm{sym}\left\{\begin{bmatrix} -W & W\bar{A} & W\bar{B} & 0 \\ -F & F\bar{A} & F\bar{B} & 0 \\ -G & G\bar{A} & G\bar{B} & 0 \\ 0 & 0 & 0 & 0 \end{bmatrix}\right\} < 0, \tag{3.8}$$

are feasible in the symmetric matrix variable P *and the general matrix variables* W, F *and* G.

Proof For the continuous-time case, let

$$\Xi_c = \begin{bmatrix} 0 & P & 0 & 0 \\ P & 0 & 0 & \bar{C}^{\mathrm{T}} \\ 0 & 0 & -\gamma^2 I & \bar{D}^{\mathrm{T}} \\ 0 & \bar{C} & \bar{D} & -I \end{bmatrix}, \; Y = \begin{bmatrix} W \\ F \\ G \end{bmatrix}, \; X = \begin{bmatrix} I & 0 & 0 \\ 0 & I & 0 \\ 0 & 0 & I \\ 0 & 0 & 0 \end{bmatrix}, \; Z = \begin{bmatrix} -I & \bar{A} & \bar{B} & 0 \end{bmatrix},$$

while for the discrete-time case, let

$$\Xi_d = \begin{bmatrix} P & 0 & 0 & 0 \\ 0 & -P & 0 & \bar{C}^{\mathrm{T}} \\ 0 & 0 & -\gamma^2 I & \bar{D}^{\mathrm{T}} \\ 0 & \bar{C} & \bar{D} & -I \end{bmatrix}, \; Y = \begin{bmatrix} W \\ F \\ G \end{bmatrix}, \; X = \begin{bmatrix} I & 0 & 0 \\ 0 & I & 0 \\ 0 & 0 & I \\ 0 & 0 & 0 \end{bmatrix}, \; Z = \begin{bmatrix} -I & \bar{A} & \bar{B} & 0 \end{bmatrix}.$$

Under these settings, we can select

$$\mathcal{N}_X = \begin{bmatrix} 0 \\ 0 \\ 0 \\ I \end{bmatrix}, \; \mathcal{N}_Z = \begin{bmatrix} \bar{A} & \bar{B} & 0 \\ I & 0 & 0 \\ 0 & I & 0 \\ 0 & 0 & I \end{bmatrix}.$$

The inequalities in (3.7) with $\Phi = \Xi_c$ and (3.8) with $\Phi = \Xi_d$ can be rewritten as

$$\Phi + X^{\mathrm{T}} Y Z + Z^{\mathrm{T}} Y^{\mathrm{T}} X < 0,$$

which, from Lemma 1.2, is equivalent to

$$\mathcal{N}_Z^{\mathrm{T}} \Phi \mathcal{N}_Z < 0$$

since $\mathcal{N}_X^{\mathrm{T}} \Phi \mathcal{N}_X = -I < 0$ is automatically satisfied. By calculation, we have, for the continuous-time case,

$$\begin{bmatrix} \bar{A}^{\mathrm{T}} P + P\bar{A} & P\bar{B} & \bar{C}^{\mathrm{T}} \\ \bar{B}^{\mathrm{T}} P & -\gamma^2 I & \bar{D}^{\mathrm{T}} \\ \bar{C} & \bar{D} & -I \end{bmatrix} = \mathcal{N}_Z^{\mathrm{T}} \Phi_c \mathcal{N}_Z < 0,$$

and for the discrete-time case,

$$\begin{bmatrix} -P & P\bar{A} & P\bar{B} & 0 \\ \bar{A}^{\mathrm{T}} P & -P & 0 & \bar{C}^{\mathrm{T}} \\ \bar{B}^{\mathrm{T}} P & 0 & -\gamma^2 I & \bar{D}^{\mathrm{T}} \\ 0 & \bar{C} & 0 & -I \end{bmatrix} < 0$$

$$\overset{\text{Lemma 1.1}}{\Longleftrightarrow} \begin{bmatrix} \bar{A}^{\mathrm{T}} P \bar{A} - P & \bar{A}^{\mathrm{T}} P \bar{B} & \bar{C}^{\mathrm{T}} \\ \bar{B}^{\mathrm{T}} P \bar{A} & \bar{B}^{\mathrm{T}} P \bar{B} - \gamma^2 I & \bar{D}^{\mathrm{T}} \\ \bar{C} & \bar{D} & -I \end{bmatrix} = \mathcal{N}_Z^{\mathrm{T}} \Phi_d \mathcal{N}_Z < 0.$$

The proof is completed. □

The following BRL extends Lemma 3.3 to the parameter uncertainty case, which gives necessary and sufficient conditions for robust H_∞ performance analysis for the filtering error system $\mathcal{S}$ in (2.64) in terms of parameter-dependent LMI.

Lemma 3.4 *Given system $\mathcal{S}$ in* (2.64), *filter $\mathcal{F}$ in* (2.66) *and a scalar $\gamma > 0$, the filtering error system $\mathcal{E}$ in* (2.67) *is robustly asymptotically stable and satisfies* (2.69) *for all $\lambda \in \Gamma$ if and only if the following matrix inequalities*

$$P(\lambda) > 0, \tag{3.9}$$

$$\Xi(\lambda) + \Upsilon(\lambda) + \Upsilon(\lambda)^{\mathrm{T}} < 0, \tag{3.10}$$

where

$$\Upsilon(\lambda) = \begin{bmatrix} -W(\lambda) & W(\lambda)\bar{A}(\lambda) & W(\lambda)\bar{B}(\lambda) & 0 \\ -F(\lambda) & F(\lambda)\bar{A}(\lambda) & F(\lambda)\bar{B}(\lambda) & 0 \\ -G(\lambda) & G(\lambda)\bar{A}(\lambda) & G(\lambda)\bar{B}(\lambda) & 0 \\ 0 & 0 & 0 & 0 \end{bmatrix}$$

and

- *for the continuous-time case*

$$\Xi(\lambda) = \Xi_c(\lambda) \triangleq \begin{bmatrix} 0 & P(\lambda) & 0 & 0 \\ P(\lambda) & 0 & 0 & \bar{C}(\lambda)^{\mathrm{T}} \\ 0 & 0 & -\gamma^2 I & \bar{D}(\lambda)^{\mathrm{T}} \\ 0 & \bar{C}(\lambda) & \bar{D}(\lambda) & -I \end{bmatrix},$$

- *for the discrete-time case*

$$\Xi(\lambda) = \Xi_d(\lambda) \triangleq \begin{bmatrix} P(\lambda) & 0 & 0 & 0 \\ 0 & -P(\lambda) & 0 & \bar{C}(\lambda)^{\mathrm{T}} \\ 0 & 0 & -\gamma^2 I & \bar{D}(\lambda)^{\mathrm{T}} \\ 0 & \bar{C}(\lambda) & \bar{D}(\lambda) & -I \end{bmatrix},$$

are feasible in the real matrix functions $P(\lambda) = P(\lambda)^{\mathrm{T}}$, $W(\lambda)$, $F(\lambda)$ and $G(\lambda)$.

3.2 Filter Realization

Lemmas 3.2 and 3.4 do not tell us how to get the state space realizations of filters. In what follows, we further deal with these results such that filter realizations can be explicitly obtained from the derived conditions.

3.2.1 H_2 Filter Realization

For the H_2 filtering condition, Lemma 3.2, we select the following structure for the slack matrices:

$$\begin{aligned} W(\lambda) &= \begin{bmatrix} W_1(\lambda) & W_3 \\ W_2(\lambda) & W_4 \end{bmatrix}, \\ F(\lambda) &= \begin{bmatrix} F_1(\lambda) & \alpha W_3 \\ F_2(\lambda) & \alpha W_4 \end{bmatrix}, \\ G(\lambda) &= \begin{bmatrix} G_1(\lambda) & 0_{l\times n} \end{bmatrix}, \end{aligned} \tag{3.11}$$

where scalar $\alpha = 1$ for the continuous-time case and $\alpha = 0$ for the discrete-time case. The above structural specifications mean that the (1,1) and (2,1) blocks of $W(\lambda)$ and $F(\lambda)$ and the (1,1) block of $G(\lambda)$ are dependent on the parameter λ, the (1,2) and (2,2) blocks of $W(\lambda)$ and $F(\lambda)$ are assumed to be independent of the parameter λ, and the (1,2) block of $G(\lambda)$ are set to be zero. Similar treatment can also be found in [1, 2]. Let $P(\lambda)$ be partitioned as

$$P(\lambda) = \begin{bmatrix} P_1(\lambda) & P_2(\lambda) \\ P_2(\lambda)^{\mathrm{T}} & P_3(\lambda) \end{bmatrix}.$$

Without loss of generality, $W_3(\lambda)$ and $W_4(\lambda)$ can be assumed to be invertible. Define matrices

$$\begin{aligned} \phi &\triangleq \begin{bmatrix} I & 0 \\ 0 & W_4^{-\mathrm{T}} W_3^{\mathrm{T}} \end{bmatrix}, \\ \tilde{P}(\lambda) &\triangleq \begin{bmatrix} \tilde{P}_1(\lambda) & \tilde{P}_2(\lambda) \\ * & \tilde{P}_3(\lambda) \end{bmatrix} = \phi^{\mathrm{T}} P(\lambda) \phi, \\ \tilde{W}_1(\lambda) &\triangleq W_1(\lambda), \\ \tilde{F}_1(\lambda) &\triangleq F_1(\lambda), \\ \tilde{W}_2(\lambda) &\triangleq W_3 W_4^{-1} W_2(\lambda), \\ \tilde{F}_2(\lambda) &\triangleq W_3 W_4^{-1} F_2(\lambda), \\ \tilde{W}_3 &\triangleq W_3 W_4^{-\mathrm{T}} W_3^{\mathrm{T}}, \end{aligned} \tag{3.12}$$

and

$$\begin{bmatrix} \tilde{A}_F & \tilde{B}_F \\ \tilde{C}_F & 0 \end{bmatrix} \triangleq \begin{bmatrix} W_3 & 0 \\ 0 & I \end{bmatrix} \begin{bmatrix} A_F & B_F \\ C_F & 0 \end{bmatrix} \begin{bmatrix} W_4^{-\mathrm{T}} W_3^{\mathrm{T}} & 0 \\ 0 & I \end{bmatrix}. \tag{3.13}$$

Suppose that (3.5) and (3.6) are satisfied for some $P(\lambda)$, $Z(\lambda)$, $W(\lambda)$, $F(\lambda)$ and $G(\lambda)$. Then applying congruence transformations to (3.5) and (3.6) by

$$\phi_1 \triangleq \mathrm{diag}\{\phi,\ I_p\} \text{ and } \phi_2 \triangleq \mathrm{diag}\{\phi,\ \phi,\ I_l\}, \tag{3.14}$$

respectively, and substituting the state space matrices of the filtering error system in (2.5) into (3.5) and (3.6), we obtain

$$\Pi(\lambda) \triangleq \begin{bmatrix} \tilde{P}_1(\lambda) & \tilde{P}_2(\lambda) & L(\lambda)^{\mathrm{T}} \\ * & \tilde{P}_3(\lambda) & -\tilde{C}_F^{\mathrm{T}} \\ * & * & Z(\lambda) \end{bmatrix} > 0, \tag{3.15}$$

$$\tilde{\Phi}(\lambda) + \tilde{\Psi}(\lambda) + \tilde{\Psi}(\lambda)^{\mathrm{T}} < 0, \tag{3.16}$$

where

- for the continuous-time case

$$\tilde{\Phi}(\lambda) \triangleq \begin{bmatrix} 0 & \tilde{P}(\lambda) & 0 \\ \tilde{P}(\lambda) & 0 & 0 \\ 0 & 0 & -I \end{bmatrix},$$

$$\tilde{\Psi}(\lambda) \triangleq \left[\begin{array}{cc|cc|c} -\tilde{W}_1(\lambda) & -\tilde{W}_3 & \tilde{W}_1(\lambda)A(\lambda) + \tilde{B}_F C(\lambda) & \tilde{A}_F & \tilde{W}_1(\lambda)B(\lambda) + \tilde{B}_F D(\lambda) \\ -\tilde{W}_2(\lambda) & -\tilde{W}_3 & \tilde{W}_2(\lambda)A(\lambda) + \tilde{B}_F C(\lambda) & \tilde{A}_F & \tilde{W}_2(\lambda)B(\lambda) + \tilde{B}_F D(\lambda) \\ \hline -\tilde{F}_1(\lambda) & -\tilde{W}_3 & \tilde{F}_1(\lambda)A(\lambda) + \tilde{B}_F C(\lambda) & \tilde{A}_F & \tilde{F}_1(\lambda)B(\lambda) + \tilde{B}_F D(\lambda) \\ -\tilde{F}_2(\lambda) & -\tilde{W}_3 & \tilde{F}_2(\lambda)A(\lambda) + \tilde{B}_F C(\lambda) & \tilde{A}_F & \tilde{F}_2(\lambda)B(\lambda) + \tilde{B}_F D(\lambda) \\ \hline -\tilde{G}_1(\lambda) & 0 & \tilde{G}_1(\lambda)A(\lambda) & 0 & \tilde{G}_1(\lambda)B(\lambda) \end{array}\right],$$

- for the discrete-time case

$$\tilde{\Phi}(\lambda) \triangleq \begin{bmatrix} \tilde{P}(\lambda) & 0 & 0 \\ 0 & -\tilde{P}(\lambda) & 0 \\ 0 & 0 & -I \end{bmatrix},$$

$$\tilde{\Psi}(\lambda) \triangleq \left[\begin{array}{cc|cc|c} -\tilde{W}_1(\lambda) & -\tilde{W}_3 & \tilde{W}_1(\lambda)A(\lambda) + \tilde{B}_F C(\lambda) & \tilde{A}_F & \tilde{W}_1(\lambda)B(\lambda) + \tilde{B}_F D(\lambda) \\ -\tilde{W}_2(\lambda) & -\tilde{W}_3 & \tilde{W}_2(\lambda)A(\lambda) + \tilde{B}_F C(\lambda) & \tilde{A}_F & \tilde{W}_2(\lambda)B(\lambda) + \tilde{B}_F D(\lambda) \\ \hline -\tilde{F}_1(\lambda) & 0 & \tilde{F}_1(\lambda)A(\lambda) & 0 & \tilde{F}_1(\lambda)B(\lambda) \\ -\tilde{F}_2(\lambda) & 0 & \tilde{F}_2(\lambda)A(\lambda) & 0 & \tilde{F}_2(\lambda)B(\lambda) \\ \hline -\tilde{G}_1(\lambda) & 0 & \tilde{G}_1(\lambda)A(\lambda) & 0 & \tilde{G}_1(\lambda)B(\lambda) \end{array}\right].$$

The above transformations result in the following lemma for exploring filter realizations by checking the feasibility of parameter-dependent LMIs.

Lemma 3.5 *Consider system $\mathcal{S}$ in* (2.1). *Given a scalar $\gamma > 0$, a filter $\mathcal{F}$ in* (2.4) *exists such that the filtering error system $\mathcal{E}$ in* (2.5) *is robustly asymptotically stable and satisfies* (2.7) *for all $\lambda \in \Gamma$ if*

$$\mathrm{Tr}[Z(\lambda)] - \gamma < 0 \tag{3.17}$$

and the matrix inequalities in (3.15) *and* (3.16) *are feasible in the real matrix functions $\tilde{P}(\lambda) = \begin{bmatrix} \tilde{P}_1(\lambda) & \tilde{P}_2(\lambda) \\ * & \tilde{P}_3(\lambda) \end{bmatrix} = \tilde{P}(\lambda)^{\mathrm{T}}$, $Z(\lambda) = Z(\lambda)^{\mathrm{T}}$, $\tilde{W}_1(\lambda)$, $\tilde{W}_2(\lambda)$, $\tilde{F}_1(\lambda)$, $\tilde{F}_2(\lambda)$ and $\tilde{G}_1(\lambda)$, and the matrices $\tilde{W}_3$, $\tilde{A}_F$, $\tilde{B}_F$ and $\tilde{C}_F$.*

Moreover, if these conditions are feasible, an admissible state space realization of the filter $\mathcal{F}$ in (2.4) *is given by*

$$\begin{bmatrix} A_F & B_F \\ C_F & 0 \end{bmatrix} = \begin{bmatrix} \tilde{W}_3^{-1} & 0 \\ 0 & I \end{bmatrix} \begin{bmatrix} \tilde{A}_F & \tilde{B}_F \\ \tilde{C}_F & 0 \end{bmatrix} \tag{3.18}$$

or

$$\begin{bmatrix} A_F & B_F \\ C_F & 0 \end{bmatrix} = \begin{bmatrix} \tilde{A}_F & \tilde{B}_F \\ \tilde{C}_F & 0 \end{bmatrix} \begin{bmatrix} \tilde{W}_3^{-1} & 0 \\ 0 & I \end{bmatrix}. \tag{3.19}$$

Proof Suppose that (3.15)–(3.17) are feasible in the matrix functions $\tilde{P}(\lambda)$, $Z(\lambda)$, $\tilde{W}_1(\lambda)$, $\tilde{W}_2(\lambda)$, $\tilde{F}_1(\lambda)$, $\tilde{F}_2(\lambda)$ and $\tilde{G}_1(\lambda)$, and the matrices $\tilde{W}_3$, $\tilde{A}_F$, $\tilde{B}_F$ and $\tilde{C}_F$. Note that (3.16) implies $-\tilde{W}_3 - \tilde{W}_3^{\mathrm{T}} < 0$ for the continuous-time case and $-\tilde{W}_3 - \tilde{W}_3^{\mathrm{T}} < -P_3(\lambda) < 0$ for the discrete-time case, respectively, which both guarantee that $\tilde{W}_3$ is nonsingular. Thus, one can always find square and invertible matrices W_3 and W_4 satisfying $\tilde{W}_3 = W_3 W_4^{-\mathrm{T}} W_3^{\mathrm{T}}$. Now define the nonsingular matrix ϕ as in (3.12), ϕ_1 and ϕ_2 as in (3.14),

$$\begin{aligned} P(\lambda) &\triangleq \phi^{-\mathrm{T}} \tilde{P}(\lambda) \phi^{-1}, \\ W(\lambda) &\triangleq \begin{bmatrix} \tilde{W}_1(\lambda) & W_3 \\ W_4 W_3^{-1} \tilde{W}_2(\lambda) & W_4 \end{bmatrix}, \\ F(\lambda) &\triangleq \begin{bmatrix} \tilde{F}_1(\lambda) & \alpha W_3 \\ W_4 W_3^{-1} \tilde{F}_2(\lambda) & \alpha W_4 \end{bmatrix}, \\ G(\lambda) &\triangleq \begin{bmatrix} \tilde{G}_1(\lambda) & 0_{l \times n} \end{bmatrix}, \end{aligned} \tag{3.20}$$

and

$$\begin{bmatrix} A_F & B_F \\ C_F & 0 \end{bmatrix} \triangleq \begin{bmatrix} W_3^{-1} & 0 \\ 0 & I \end{bmatrix} \begin{bmatrix} \tilde{A}_F & \tilde{B}_F \\ \tilde{C}_F & 0 \end{bmatrix} \begin{bmatrix} W_3^{-\mathrm{T}} W_4^{\mathrm{T}} & 0 \\ 0 & I \end{bmatrix}. \tag{3.21}$$

By some algebraic manipulations, (3.15) and (3.16) can be rewritten as

$$\phi_1^{\mathrm{T}} \begin{bmatrix} P(\lambda) & \bar{C}(\lambda)^{\mathrm{T}} \\ * & Z(\lambda) \end{bmatrix} \phi_1 > 0,$$
$$\phi_2^{\mathrm{T}} \left(\Phi(\lambda) + \Psi(\lambda) + \Psi(\lambda)^{\mathrm{T}} \right) \phi_2 < 0.$$

Then performing congruence transformations to the above inequalities by ϕ_1^{-1} and ϕ_2^{-1}, respectively, one can get (3.5) and (3.6). In addition, (3.17) is just the one in (3.4). Hence, (3.4)–(3.6) are assured to be feasible for some $P(\lambda)$, $Z(\lambda)$, $W(\lambda)$, $F(\lambda)$, $G(\lambda)$, A_F, B_F and C_F defined in (3.20), (3.21). Consequently, we conclude from Lemma 3.2 that the filter with a state space realization in (3.21) guarantees the filtering error system $\mathcal{E}$ in (2.5) to be robustly asymptotically stable and satisfy (2.7) for all $\lambda \in \Gamma$.

To prove that (3.18) and (3.19) are admissible filter realizations, considering the relationship $\tilde{W}_3 = W_3 W_4^{-\mathrm{T}} W_3^{\mathrm{T}}$ and (3.21) as well as the filter transfer function, we have

$$\begin{aligned} T_F(\delta) &= C_F \left(\delta I - A_F\right)^{-1} B_F \\ &= \left(\tilde{C}_F W_3^{-\mathrm{T}} W_4^{\mathrm{T}}\right) \left(\delta I - W_3^{-1} \tilde{A}_F W_3^{-\mathrm{T}} W_4^{\mathrm{T}}\right)^{-1} \left(W_3^{-1} \tilde{B}_F\right) \\ &= \tilde{C}_F \left(\delta I - \tilde{W}_3^{-1} \tilde{A}_F\right)^{-1} \left(\tilde{W}_3^{-1} \tilde{B}_F\right) \end{aligned}$$

and

$$\begin{aligned} T_F(\delta) &= C_F \left(\delta I - A_F\right)^{-1} B_F \\ &= \left(\tilde{C}_F W_3^{-\mathrm{T}} W_4^{\mathrm{T}}\right) \left(\delta I - W_3^{-1} \tilde{A}_F W_3^{-\mathrm{T}} W_4^{\mathrm{T}}\right)^{-1} \left(W_3^{-1} \tilde{B}_F\right) \\ &= \left(\tilde{C}_F \tilde{W}_3^{-1}\right) \left(\delta I - \tilde{A}_F \tilde{W}_3^{-1}\right)^{-1} \tilde{B}_F. \end{aligned}$$

This completes the proof. □

Remark 3.2 Recall (2.59) and (2.60) in Chap. 2 and (3.16) in this chapter. For convenience, we also call $\tilde{P}$ and $\tilde{P}(\lambda)$ *Lyapunov matrices* although they are transformed congruently from P and $P(\lambda)$. It should be emphasized that the quadratic results *cannot* be extended to the parameter-dependent case by directly replacing $\tilde{P}$ by its parameter-dependent form $\tilde{P}(\lambda)$. This is because it is generally impossible to compute the parameter-independent filter matrices (A_F, B_F, C_F) from the parameter-dependent matrix $\tilde{P}_2(\lambda)$ as in (2.61) or (2.62). In Lemma 3.5, the filter realization is independent of the Lyapunov matrix $\tilde{P}(\lambda)$, so that $\tilde{P}(\lambda)$ can be parameter-dependent, which is the role of slack matrices in relaxing synthesis conditions.

3.2.2 H_∞ Filter Realization

In this section, we look for H_∞ filter realizations from Lemma 3.4. By performing similar manipulations as in the derivation of Lemma 3.5, we have the following existence condition for H_∞ filter realizations in terms of solving parameter-dependent LMIs.

Lemma 3.6 *Consider system $\mathcal{S}$ in* (2.64). *Given a scalar $\gamma > 0$, a filter $\mathcal{F}$ in* (2.66) *exists such that the filtering error system $\mathcal{E}$ in* (2.67) *is robustly asymptotically stable and satisfies* (2.69) *for all $\lambda \in \Gamma$ if*

$$\tilde{P}(\lambda) > 0, \tag{3.22}$$

$$\tilde{\Xi}(\lambda) + \tilde{\Upsilon}(\lambda) + \tilde{\Upsilon}(\lambda)^{\mathrm{T}} < 0, \tag{3.23}$$

where

- *for the continuous-time case*

$$\tilde{\Xi}(\lambda) \triangleq \begin{bmatrix} 0 & \tilde{P}(\lambda) & 0 & 0 \\ \tilde{P}(\lambda) & 0 & 0 & \tilde{C}(\lambda)^{\mathrm{T}} \\ 0 & 0 & -\gamma^2 I & \tilde{D}(\lambda)^{\mathrm{T}} \\ 0 & \tilde{C}(\lambda) & \tilde{D}(\lambda) & -I \end{bmatrix},$$

$$\tilde{\Upsilon}(\lambda) \triangleq \left[\begin{array}{cc|cc|c|c} -\tilde{W}_1(\lambda) & -\tilde{W}_3 & \tilde{W}_1(\lambda)A(\lambda) + \tilde{B}_F C(\lambda) & \tilde{A}_F & \tilde{W}_1(\lambda)B(\lambda) + \tilde{B}_F D(\lambda) & 0 \\ -\tilde{W}_2(\lambda) & -\tilde{W}_3 & \tilde{W}_2(\lambda)A(\lambda) + \tilde{B}_F C(\lambda) & \tilde{A}_F & \tilde{W}_2(\lambda)B(\lambda) + \tilde{B}_F D(\lambda) & 0 \\ \hline -\tilde{F}_1(\lambda) & -\tilde{W}_3 & \tilde{F}_1(\lambda)A(\lambda) + \tilde{B}_F C(\lambda) & \tilde{A}_F & \tilde{F}_1(\lambda)B(\lambda) + \tilde{B}_F D(\lambda) & 0 \\ -\tilde{F}_2(\lambda) & -\tilde{W}_3 & \tilde{F}_2(\lambda)A(\lambda) + \tilde{B}_F C(\lambda) & \tilde{A}_F & \tilde{F}_2(\lambda)B(\lambda) + \tilde{B}_F D(\lambda) & 0 \\ \hline -\tilde{G}_1(\lambda) & 0 & \tilde{G}_1(\lambda)A(\lambda) & 0 & \tilde{G}_1(\lambda)B(\lambda) & 0 \\ \hline 0 & 0 & 0 & 0 & 0 & 0 \end{array}\right],$$

$$\tilde{C}(\lambda) \triangleq \begin{bmatrix} L(\lambda) - \tilde{D}_F C(\lambda) & -\tilde{C}_F \end{bmatrix}, \ \tilde{D}(\lambda) \triangleq E(\lambda) - \tilde{D}_F D(\lambda),$$

- *for the discrete-time case*

$$\tilde{\Xi}(\lambda) \triangleq \begin{bmatrix} \tilde{P}(\lambda) & 0 & 0 & 0 \\ 0 & -\tilde{P}(\lambda) & 0 & \tilde{C}(\lambda)^{\mathrm{T}} \\ 0 & 0 & -\gamma^2 I & \tilde{D}(\lambda)^{\mathrm{T}} \\ 0 & \tilde{C}(\lambda) & \tilde{D}(\lambda) & -I \end{bmatrix},$$

$$\tilde{\Upsilon}(\lambda) \triangleq \left[\begin{array}{cc|cc|c|c} -\tilde{W}_1(\lambda) & -\tilde{W}_3 & \tilde{W}_1(\lambda)A(\lambda) + \tilde{B}_F C(\lambda) & \tilde{A}_F & \tilde{W}_1(\lambda)B(\lambda) + \tilde{B}_F D(\lambda) & 0 \\ -\tilde{W}_2(\lambda) & -\tilde{W}_3 & \tilde{W}_2(\lambda)A(\lambda) + \tilde{B}_F C(\lambda) & \tilde{A}_F & \tilde{W}_2(\lambda)B(\lambda) + \tilde{B}_F D(\lambda) & 0 \\ \hline -\tilde{F}_1(\lambda) & 0 & \tilde{F}_1(\lambda)A(\lambda) & 0 & \tilde{F}_1(\lambda)B(\lambda) & 0 \\ -\tilde{F}_2(\lambda) & 0 & \tilde{F}_2(\lambda)A(\lambda) & 0 & \tilde{F}_2(\lambda)B(\lambda) & 0 \\ \hline -\tilde{G}_1(\lambda) & 0 & \tilde{G}_1(\lambda)A(\lambda) & 0 & \tilde{G}_1(\lambda)B(\lambda) & 0 \\ \hline 0 & 0 & 0 & 0 & 0 & 0 \end{array}\right],$$

$$\tilde{C}(\lambda) \triangleq \begin{bmatrix} L(\lambda) - \tilde{D}_F C(\lambda) & -\tilde{C}_F \end{bmatrix}, \ \tilde{D}(\lambda) \triangleq E(\lambda) - \tilde{D}_F D(\lambda)$$

are feasible in the real matrix functions $\tilde{P}(\lambda) = \begin{bmatrix} \tilde{P}_1(\lambda) & \tilde{P}_2(\lambda) \\ * & \tilde{P}_3(\lambda) \end{bmatrix} = \tilde{P}(\lambda)^{\mathrm{T}}$, $\tilde{W}_1(\lambda)$, $\tilde{W}_2(\lambda)$, $\tilde{F}_1(\lambda)$, $\tilde{F}_2(\lambda)$ *and* $\tilde{G}_1(\lambda)$, *and the matrices* $\tilde{W}_3$, $\tilde{A}_F$, $\tilde{B}_F$, $\tilde{C}_F$ *and* $\tilde{D}_F$.

Moreover, if these conditions are feasible, an admissible state space realization of the filter $\mathcal{F}$ *in* (2.66) *is given by*

$$\begin{bmatrix} A_F & B_F \\ C_F & D_F \end{bmatrix} = \begin{bmatrix} \tilde{W}_3^{-1} & 0 \\ 0 & I \end{bmatrix} \begin{bmatrix} \tilde{A}_F & \tilde{B}_F \\ \tilde{C}_F & \tilde{D}_F \end{bmatrix} \tag{3.24}$$

or

$$\begin{bmatrix} A_F & B_F \\ C_F & D_F \end{bmatrix} = \begin{bmatrix} \tilde{A}_F & \tilde{B}_F \\ \tilde{C}_F & \tilde{D}_F \end{bmatrix} \begin{bmatrix} \tilde{W}_3^{-1} & 0 \\ 0 & I \end{bmatrix}. \tag{3.25}$$

Proof Suppose that (3.22) and (3.23) are feasible in the matrix functions $\tilde{P}(\lambda)$, $Z(\lambda)$, $\tilde{W}_1(\lambda)$, $\tilde{W}_2(\lambda)$, $\tilde{F}_1(\lambda)$, $\tilde{F}_2(\lambda)$ and $\tilde{G}_1(\lambda)$, and the matrices $\tilde{W}_3$, $\tilde{A}_F$, $\tilde{B}_F$, $\tilde{C}_F$ and $\tilde{D}_F$. Both of the continuous- and discrete-time cases of (3.23) imply $-\tilde{W}_3 - \tilde{W}_3^{\mathrm{T}} < 0$. Thus, $\tilde{W}_3$ is nonsingular and one can always find square and invertible matrices W_3 and W_4 satisfying $\tilde{W}_3 = W_3 W_4^{-\mathrm{T}} W_3^{\mathrm{T}}$. Define the nonsingular matrix ϕ as in (3.12), $P(\lambda)$, $W(\lambda)$, $F(\lambda)$ and $G(\lambda)$ as in (3.20), A_F, B_F, C_F as in (3.21) together with

$$D_F \triangleq \tilde{D}_F, \ \psi \triangleq \mathrm{diag}\{\phi, \ \phi, \ I_l, \ I_p\}. \tag{3.26}$$

By some algebraic manipulations, (3.22) and (3.23) can be rewritten as

$$\phi^{\mathrm{T}} P(\lambda)\phi > 0,$$
$$\psi^{\mathrm{T}} \left(\Xi(\lambda) + \Upsilon(\lambda) + \Upsilon(\lambda)^{\mathrm{T}} \right) \psi < 0,$$

where $\Xi(\lambda)$ and $\Upsilon(\lambda)$ are defined in (3.10). Then performing congruence transformations to the above inequalities by ϕ^{-1} and ψ^{-1}, respectively, one can get (3.9) and (3.10). This means that (3.9) and (3.10) are feasible for some $P(\lambda)$, $W(\lambda)$, $F(\lambda)$, $G(\lambda)$, A_F, B_F, C_F and D_F defined in (3.20), (3.21) and (3.26). Consequently, we conclude from Lemma 3.4 that the filter with a state space realization in (3.24) or (3.25) guarantees the filtering error system $\mathcal{E}$ in (2.67) to be robustly asymptotically stable and satisfy (2.69) for all $\lambda \in \Gamma$.

The reminder of the proof is almost the same as that in the proof of Lemma 3.5. The proof is completed. □

The comments for Lemma 3.5 in Remark 3.2 are also suitable for Lemma 3.6. In the next two sections, our attention will be focused on appropriately choosing the parameter-dependence form for $\tilde{P}(\lambda)$ such that (1) all the blocks of $\tilde{P}(\lambda)$ must be parameter-dependent and (2) the derived conditions must be numerically tractable by the LMI technique.

3.3 Linearly Parameter-Dependent Approaches

State space realizations of filters have been explicitly found in Lemmas 3.5 and 3.6 However, due to the infinite dimension property of the uncertain parameters λ, it is difficult to solve a feasibility (or an optimization) problem with infinite number of LMIs in (3.15)–(3.17) or (3.22), (3.23). Hence, it is still inconvenient to apply Lemmas 3.5 and 3.6 to design robust filters. It is necessary to develop more numerically tractable results. To this end, we consider two special cases of Lemmas 3.5 and 3.6, where all the conditions will be guaranteed by a finite number of LMIs only. This section is concerned about the first special case, arriving at the *linearly parameter-dependence* approaches to robust filter design.

3.3.1 H_2 Filter Design I

In this section, the following special case is taken into consideration. The slack matrices in Lemma 3.5 are assumed to be parameter-independent, that is, they are fixed over the entire uncertainty domain as

$$\begin{aligned}&\tilde{W}_1(\lambda)=\tilde{W}_1,\ \tilde{W}_2(\lambda)=\tilde{W}_2,\ \tilde{G}_1(\lambda)=\tilde{G}_1,\\&\tilde{F}_1(\lambda)=\tilde{F}_1,\ \tilde{F}_2(\lambda)=\tilde{F}_2.\end{aligned}\tag{3.27}$$

Lyapunov matrix $\tilde{P}(\lambda)$ and symmetric matrix $Z(\lambda)$ in Lemma 3.5 are selected to be of the following linearly parameter-dependence form:

$$\tilde{P}(\lambda)=\sum_{i=1}^{s}\lambda_i\tilde{P}_i,\ \tilde{Z}(\lambda)=\sum_{i=1}^{s}\lambda_i\tilde{Z}_i,\tag{3.28}$$

where $\tilde{P}_i$ and $\tilde{Z}_i$ are fixed matrices to be determined. Since $\tilde{P}(\lambda)$ and $Z(\lambda)$ in Eq. (3.28) are linearly parameter-dependent, they can also be regarded as uncertain matrices in polytopes with s vertices, if $\tilde{P}_i$ and $\tilde{Z}_i$ are known. Under the settings in (3.27) and (3.28), the LHSs of both (3.15) and (3.16) can be rewritten as the linear combination of s matrices as follows:

$$\begin{aligned}\Pi(\lambda)&=\sum_{i=1}^{s}\lambda_i\Pi_i,\\\tilde{\Phi}(\lambda)+\tilde{\Psi}(\lambda)+\tilde{\Psi}(\lambda)^{\mathrm{T}}&=\sum_{i=1}^{s}\lambda_i\left(\tilde{\Phi}_i+\tilde{\Psi}_i+\tilde{\Psi}_i^{\mathrm{T}}\right),\end{aligned}$$

where Π_i, $\tilde{\Phi}_i$ and $\tilde{\Psi}_i$ are $\Pi(\lambda)$, $\tilde{\Phi}(\lambda)$ and $\tilde{\Psi}(\lambda)$, respectively, with the parameter-dependent terms replaced by the corresponding matrices at the ith vertex of the polytope. According to the above discussion, under the settings in (3.27) and (3.28), testing the infinite number of LMIs in (3.15) and (3.16) can be equivalently completed by testing a finite number of LMIs, which is numerically solvable.

Thus, we have the following linearly parameter-dependent approach to robust H_2 filter design.

Theorem 3.1 *Consider system $\mathcal{S}$ in (2.1). Given a scalar $\gamma > 0$, a filter $\mathcal{F}$ in (2.4) exists such that the filtering error system $\mathcal{E}$ in (2.5) is robustly asymptotically stable and satisfies $\mathbf{E}\left[e(t)^{\mathrm{T}}e(t)\right] < \gamma$ for all $\lambda \in \Gamma$ if the matrix inequalities*

$$\mathrm{Tr}[Z_i] - \gamma < 0 \, (i = 1, 2, \ldots, s), \tag{3.29}$$

$$\begin{bmatrix} \tilde{P}_{1,i} & \tilde{P}_{2,i} & L_i^{\mathrm{T}} \\ \tilde{P}_{2,i}^{\mathrm{T}} & \tilde{P}_{3,i} & -\tilde{C}_F^{\mathrm{T}} \\ L_i & -\tilde{C}_F & Z_i \end{bmatrix} > 0 \, (i = 1, 2, \ldots, s), \tag{3.30}$$

$$\tilde{\Phi}_i + \tilde{\Psi}_i + \tilde{\Psi}_i^{\mathrm{T}} < 0 \, (i = 1, 2, \ldots, s), \tag{3.31}$$

where

- *for the continuous-time case*

$$\tilde{\Phi}_i \triangleq \begin{bmatrix} 0 & \tilde{P}_i & 0 \\ \tilde{P}_i & 0 & 0 \\ 0 & 0 & -I \end{bmatrix},$$

$$\tilde{\Psi}_i \triangleq \left[\begin{array}{cc|cc|c} -\tilde{W}_1 & -\tilde{W}_3 & \tilde{W}_1 A_i + \tilde{B}_F C_i & \tilde{A}_F & \tilde{W}_1 B_i + \tilde{B}_F D_i \\ -\tilde{W}_2 & -\tilde{W}_3 & \tilde{W}_2 A_i + \tilde{B}_F C_i & \tilde{A}_F & \tilde{W}_2 B_i + \tilde{B}_F D_i \\ \hline -\tilde{F}_1 & -\tilde{W}_3 & \tilde{F}_1 A_i + \tilde{B}_F C_i & \tilde{A}_F & \tilde{F}_1 B_i + \tilde{B}_F D_i \\ -\tilde{F}_2 & -\tilde{W}_3 & \tilde{F}_2 A_i + \tilde{B}_F C_i & \tilde{A}_F & \tilde{F}_2 B_i + \tilde{B}_F D_i \\ \hline -\tilde{G}_1 & 0 & \tilde{G}_1 A_i & 0 & \tilde{G}_1 B_i \end{array}\right],$$

- *for the discrete-time case*

$$\tilde{\Phi}_i \triangleq \begin{bmatrix} \tilde{P}_i & 0 & 0 \\ 0 & -\tilde{P}_i & 0 \\ 0 & 0 & -I \end{bmatrix},$$

$$\tilde{\Psi}_i \triangleq \left[\begin{array}{cc|cc|c} -\tilde{W}_1 & -\tilde{W}_3 & \tilde{W}_1 A_i + \tilde{B}_F C_i & \tilde{A}_F & \tilde{W}_1 B_i + \tilde{B}_F D_i \\ -\tilde{W}_2 & -\tilde{W}_3 & \tilde{W}_2 A_i + \tilde{B}_F C_i & \tilde{A}_F & \tilde{W}_2 B_i + \tilde{B}_F D_i \\ \hline -\tilde{F}_1 & 0 & \tilde{F}_1 A_i & 0 & \tilde{F}_1 B_i \\ -\tilde{F}_2 & 0 & \tilde{F}_2 A_i & 0 & \tilde{F}_2 B_i \\ \hline -\tilde{G}_1 & 0 & \tilde{G}_1 A_i & 0 & \tilde{G}_1 B_i \end{array}\right],$$

are feasible in the real matrix variables $\tilde{P}_i = \begin{bmatrix} \tilde{P}_{1,i} & \tilde{P}_{2,i} \\ * & \tilde{P}_{3,i} \end{bmatrix} = \tilde{P}_i^{\mathrm{T}}$, $Z_i = Z_i^{\mathrm{T}}$, $\tilde{W}_1$, $\tilde{W}_2$, $\tilde{F}_1$, $\tilde{F}_2$, $\tilde{G}_1$, $\tilde{W}_3$, $\tilde{A}_F$, $\tilde{B}_F$ *and* $\tilde{C}_F$.

Moreover, if these conditions are feasible, an admissible state space realization of the filter $\mathcal{F}$ in (2.4) *is given by* (3.18) *or* (3.19).

The linearly parameter-dependent approach based on Theorem 3.1 is summarized in Algorithm 13 for robust H_2 filter design.

Algorithm 13 Linearly Parameter-Dependent Approach to Robust H_2 Filter Design

- Solve the minimization problem:

$$\min_{\substack{\tilde{P}_i = \begin{bmatrix} \tilde{P}_{1i} & \tilde{P}_{2i} \\ * & \tilde{P}_{3i} \end{bmatrix} = \tilde{P}_i^{\mathrm{T}}, \\ Z_i = Z_i^{\mathrm{T}},\ i=1,2,\ldots,s,\ \tilde{W}_1,\ \tilde{W}_2,\ \tilde{F}_1, \\ \tilde{F}_2,\ \tilde{G}_1,\ \tilde{W}_3,\ \tilde{A}_F,\ \tilde{B}_F,\ \tilde{C}_F,\ \gamma}} \gamma \quad \text{s.t. (3.29), (3.30), (3.31)}$$

- Compute filter in (2.4) by (3.18) or (3.19).

3.3.2 H_∞ Filter Design I

In this section, we consider the linearly parameter-dependent approach to robust H_∞ filter design. Similar to the H_2 case in Sect. 3.3.1, the matrix functions in the conditions of Lemma 3.6 are assigned as follows:

$$\begin{aligned} \tilde{W}_1(\lambda) &= \tilde{W}_1,\ \tilde{W}_2(\lambda) = \tilde{W}_2, \\ \tilde{F}_1(\lambda) &= \tilde{F}_1,\ \tilde{F}_2(\lambda) = \tilde{F}_2, \\ \tilde{G}_1(\lambda) &= \tilde{G}_1,\ \tilde{P}(\lambda) = \sum_{i=1}^{s} \lambda_i \tilde{P}_i. \end{aligned}$$

These settings also mean that the slack matrices in Lemma 3.6 are assumed to be parameter-independent while Lyapunov matrix $\tilde{P}(\lambda)$ is selected to be of the linearly parameter-dependence form. This yields the following result on a linearly parameter-dependent approach to robust H_∞ filter design.

Theorem 3.2 *Consider system $\mathcal{S}$ in* (2.64). *Given a scalar $\gamma > 0$, a filter $\mathcal{F}$ in* (2.66) *exists such that the filtering error system $\mathcal{E}$ in* (2.67) *is robustly asymptotically stable and satisfies* (2.69) *for all $\lambda \in \Gamma$ if*

$$\tilde{P}_i > 0\,(i = 1, 2, \ldots, s), \tag{3.32}$$

$$\tilde{\Xi}_i + \tilde{\Upsilon}_i + \tilde{\Upsilon}_i^{\mathrm{T}} < 0\,(i = 1, 2, \ldots, s), \tag{3.33}$$

where

- *for the continuous-time case*

$$\tilde{\Xi}_i \triangleq \begin{bmatrix} 0 & \tilde{P}_i & 0 & 0 \\ \tilde{P}_i & 0 & 0 & \tilde{C}_i^{\mathrm{T}} \\ 0 & 0 & -\gamma^2 I & \tilde{D}_i^{\mathrm{T}} \\ 0 & \tilde{C}_i & \tilde{D}_i & -I \end{bmatrix},$$

$$\tilde{\Upsilon}_i \triangleq \left[\begin{array}{cc|cc|c|c} -\tilde{W}_1 & -\tilde{W}_3 & \tilde{W}_1 A_i + \tilde{B}_F C_i & \tilde{A}_F & \tilde{W}_1 B_i + \tilde{B}_F D_i & 0 \\ -\tilde{W}_2 & -\tilde{W}_3 & \tilde{W}_2 A_i + \tilde{B}_F C_i & \tilde{A}_F & \tilde{W}_2 B_i + \tilde{B}_F D_i & 0 \\ \hline -\tilde{F}_1 & -\tilde{W}_3 & \tilde{F}_1 A_i + \tilde{B}_F C_i & \tilde{A}_F & \tilde{F}_1 B_i + \tilde{B}_F D_i & 0 \\ -\tilde{F}_2 & -\tilde{W}_3 & \tilde{F}_2 A_i + \tilde{B}_F C_i & \tilde{A}_F & \tilde{F}_2 B_i + \tilde{B}_F D_i & 0 \\ \hline -\tilde{G}_1 & 0 & \tilde{G}_1 A_i & 0 & \tilde{G}_1 B_i & 0 \\ \hline 0 & 0 & 0 & 0 & 0 & 0 \end{array}\right],$$

$$\tilde{C}_i \triangleq \left[\, L_i - \tilde{D}_F C_i \ \ -\tilde{C}_F \,\right],$$
$$\tilde{D}_i \triangleq E_i - \tilde{D}_F D_i,$$

- *for the discrete-time case*

$$\tilde{\Xi}_i \triangleq \begin{bmatrix} \tilde{P}_i & 0 & 0 & 0 \\ 0 & -\tilde{P}_i & 0 & \tilde{C}_i^{\mathrm{T}} \\ 0 & 0 & -\gamma^2 I & \tilde{D}_i^{\mathrm{T}} \\ 0 & \tilde{C}_i & \tilde{D}_i & -I \end{bmatrix},$$

$$\tilde{\Upsilon}_i \triangleq \left[\begin{array}{cc|cc|c|c} -\tilde{W}_1 & -\tilde{W}_3 & \tilde{W}_1 A_i + \tilde{B}_F C_i & \tilde{A}_F & \tilde{W}_1 B_i + \tilde{B}_F D_i & 0 \\ -\tilde{W}_2 & -\tilde{W}_3 & \tilde{W}_2 A_i + \tilde{B}_F C_i & \tilde{A}_F & \tilde{W}_2 B_i + \tilde{B}_F D_i & 0 \\ \hline -\tilde{F}_1 & 0 & \tilde{F}_1 A_i & 0 & \tilde{F}_1 B_i & 0 \\ -\tilde{F}_2 & 0 & \tilde{F}_2 A_i & 0 & \tilde{F}_2 B_i & 0 \\ \hline -\tilde{G}_1 & 0 & \tilde{G}_1 A_i & 0 & \tilde{G}_1 B_i & 0 \\ \hline 0 & 0 & 0 & 0 & 0 & 0 \end{array}\right],$$

$$\tilde{C}_i \triangleq \left[\, L_i - \tilde{D}_F C_i \ \ -\tilde{C}_F \,\right],$$
$$\tilde{D}_i \triangleq E_i - \tilde{D}_F D_i,$$

are feasible in the real matrices $\tilde{P}_i = \begin{bmatrix} \tilde{P}_{1,i} & \tilde{P}_{2,i} \\ * & \tilde{P}_{3,i} \end{bmatrix} = \tilde{P}_i^{\mathrm{T}}$, $\tilde{W}_1$, $\tilde{W}_2$, $\tilde{F}_1$, $\tilde{F}_2$, $\tilde{G}_1$, $\tilde{W}_3$, $\tilde{A}_F$, $\tilde{B}_F$, $\tilde{C}_F$ *and* $\tilde{D}_F$.

Moreover, if these conditions are feasible, an admissible state space realization of the filter $\mathcal{F}$ *in* (2.4) *is given by* (3.24) *or* (3.25).

This linearly parameter-dependent approach based on Theorem 3.2 to robust H_∞ filter design is summarized in Algorithm 14.

Algorithm 14 Linearly Parameter-Dependent Approach to Robust H_∞ Filter Design

- Solve the minimization problem:

$$\min_{\substack{\tilde{P}_i = \begin{bmatrix} \tilde{P}_{1i} & \tilde{P}_{2i} \\ * & \tilde{P}_{3i} \end{bmatrix} = \tilde{P}_i^{\mathrm{T}},\ i=1,2,\ldots,s, \\ \tilde{W}_1,\ \tilde{W}_2,\ \tilde{F}_1,\ \tilde{F}_2,\ \tilde{G}_1,\ \tilde{W}_3, \\ \tilde{A}_F,\ \tilde{B}_F,\ \tilde{C}_F,\ \tilde{D}_F,\ \mu}} \mu = \gamma^2 \quad \text{s.t. (3.32), (3.33)}$$

- Compute filter in (2.66) by (3.24) or (3.25).

3.4 Polynomially Parameter-Dependent Approaches

In Sect. 3.3, linearly parameter-dependent approaches to robust H_2 and H_∞ filter design have been proposed. Although the matrix specifications employed there lead simple and numerically tractable results in terms of LMIs, they are inevitably too restrictive. Exactly speaking, further improvements can be oriented to the following two aspects.

1. Lemmas 3.5 and 3.6 show that since filter realizations given by (3.18) or (3.19) are only dependent on $\tilde{W}_3$, the slack matrices $\tilde{W}_1(\lambda)$, $\tilde{W}_2(\lambda)$, $\tilde{F}_1(\lambda)$, $\tilde{F}_2(\lambda)$, $\tilde{G}_1(\lambda)$ together with $\tilde{H}_1(\lambda)$ could be parameter-dependent. To derive Theorems 3.1 and 3.2, they are all assumed to be fixed, which is the first limitation of Theorems 3.1 and 3.2. Thus, the first idea for possible improvement is to relax these slack matrices to be parameter-dependent.
2. As is termed, Theorems 3.1 and 3.2 are linearly parameter-dependent approaches to filter design. Lyapunov matrix function $\tilde{P}(\lambda)$ of the linear dependence form leads to tractable and simple conditions. Between the adopted $\tilde{P}(\lambda)$ of the linear dependence form and an ideal $\tilde{P}(\lambda)$ such that the resulting conditions have no conservatism, there is a gap that can be reduced. It is unknown what form such an ideal $\tilde{P}(\lambda)$ should exactly have, and any specific choice of the parameter-dependence form for $\tilde{P}(\lambda)$ is only an approximation of the ideal one. Hence, another idea for reducing this gap and thus the conservatism is to select other more complicated but still numerically tractable parameter-dependence form for $\tilde{P}(\lambda)$ and other parameter-dependent matrices in Lemmas 3.5 and 3.6.

In this section, applying the above two ideas, we assume the parameter-dependent matrices in Lemmas 3.5 and 3.6 all to be *polynomially* dependent on the parameter λ, which encompasses the linearly parameter-dependence as a special case. Thus, the results in Sect. 3.3 is expected to be further improved.

3.4.1 Homogeneous Polynomial

In this section, let the matrices $\tilde{P}(\lambda)$, $Z(\lambda)$, $\tilde{W}_1(\lambda)$, $\tilde{W}_2(\lambda)$, $\tilde{F}_1(\lambda)$, $\tilde{F}_2(\lambda)$, $\tilde{G}_1(\lambda)$ and $\tilde{H}_1(\lambda)$ take the homogeneous polynomial of arbitrary degree g and depend polynomially on the uncertain parameters λ_i, $i = 1, 2, \ldots, s$. That is,

$$
\begin{aligned}
\tilde{P}(\lambda) &= \sum_{j=1}^{J(s,g)} \lambda_1^{k_1}\lambda_2^{k_2}\cdots\lambda_s^{k_s}\,\tilde{P}_{K_j(s,g)},\\
Z(\lambda) &= \sum_{j=1}^{J(s,g)} \lambda_1^{k_1}\lambda_2^{k_2}\cdots\lambda_s^{k_s}\,Z_{K_j(s,g)},\\
\tilde{W}_i(\lambda) &= \sum_{j=1}^{J(s,g)} \lambda_1^{k_1}\lambda_2^{k_2}\cdots\lambda_s^{k_s}\,\tilde{W}_{i,K_j(s,g)},\ i = 1, 2\\
\tilde{F}_i(\lambda) &= \sum_{j=1}^{J(s,g)} \lambda_1^{k_1}\lambda_2^{k_2}\cdots\lambda_s^{k_s}\,\tilde{F}_{i,K_j(s,g)},\ i = 1, 2\\
\tilde{G}_1(\lambda) &= \sum_{j=1}^{J(s,g)} \lambda_1^{k_1}\lambda_2^{k_2}\cdots\lambda_s^{k_s}\,\tilde{G}_{1,K_j(s,g)},\\
K_j(s,g) &= (k_1, k_2, \ldots, k_s) \in K(s,g).
\end{aligned}
\tag{3.34}
$$

The notations in the above are explained as follows:

- $K(s, g)$ the set of s-tuples $(k_1, k_2, \ldots, k_s)$ obtained as all possible combination of $k_1, k_2, \ldots, k_s$ with $k_i \in \mathbb{N}_0^+$ and $k_1 + k_2 + \cdots + k_s = g$;
- $K_j(s, g)$ the jth s-tuples of set $K(s, g)$;
- $J(s, g)$ the number of elements in set $K(s, g)$, that is, $J(s, g) = \frac{(s+g-1)!}{g!\,(s-1)!}$;
- $\tilde{P}_{K_j(s,g)}$ constant matrices with subscripts identified by the elements of $K_j(s, g)$, that is, $\tilde{P}_{K_j(s,g)} = \tilde{P}_{(k_1,k_2,\ldots,k_s)}$, and other notations defined in a similar way.

The notations in (3.34) mean that the parameter-dependent matrices $\tilde{P}(\lambda)$, $Z(\lambda)$, $\tilde{W}_1(\lambda)$, $\tilde{W}_2(\lambda)$, $\tilde{F}_1(\lambda)$, $\tilde{F}_2(\lambda)$ and $\tilde{G}_1(\lambda)$ are constructed by the sum of constant matrices with coefficients formed by homogeneous polynomials of λ. Note that,

- when $g = 0$, we have $J(s, g) = 1$ and

$$
\begin{aligned}
\tilde{P}(\lambda) &= \tilde{P}_{K_1(s,0)},\\
Z(\lambda) &= Z_{K_1(s,0)},\\
\tilde{W}_i(\lambda) &= \tilde{W}_{i,K_1(s,0)},\ i = 1, 2\\
\tilde{F}_i(\lambda) &= \tilde{F}_{i,K_1(s,0)},\ i = 1, 2
\end{aligned}
$$

$$\tilde{G}_1(\lambda) = \tilde{G}_{1,K_1(s,0)},$$

where all the matrices are fixed over the uncertainty domain, leading to the standard filtering result in the quadratic framework;

- when $g = 1$, we have $J(s, g) = s$ and

$$\tilde{P}(\lambda) = \sum_{j=1}^{s} \lambda_{s+1-j} \tilde{P}_{K_j(s,1)},$$

$$Z(\lambda) = \sum_{j=1}^{s} \lambda_{s+1-j} Z_{K_j(s,1)},$$

$$\tilde{W}_i(\lambda) = \sum_{j=1}^{s} \lambda_{s+1-j} \tilde{W}_{i,K_j(s,1)}, \ i = 1, 2,$$

$$\tilde{F}_i(\lambda) = \sum_{j=1}^{s} \lambda_{s+1-j} \tilde{F}_{i,K_j(s,1)}, \ i = 1, 2,$$

$$\tilde{G}_1(\lambda) = \sum_{j=1}^{s} \lambda_{s+1-j} \tilde{G}_{1,K_j(s,1)},$$

which is another version of linear parameter-dependence, different from the one employed in the last section. Some more specific comments will be given for the case of $g = 1$ in Sect. 3.4.2.

In addition, it is also worth noting that since $\lambda_i \geq 0$, $i = 1, 2, \ldots, s$, a simple way to ensure $\tilde{P}(\lambda) > 0$ and $Z(\lambda) > 0$ is to impose $\tilde{P}_{K_j(s,g)} > 0$ and $Z_{K_j(s,g)} > 0$ for all $j \in \{1, 2, \ldots, J(s, g)\}$.

For further deriving our results, the following notations are needed:

$I(s, g)$	the set with elements $I_j(s, g)$ given by subsets of $i \in \{1, 2, \ldots, s\}$, associated to the s-tuples $K_j(s, g)$ whose k_i's are nonzero;
$K_j^i(s, g)$	the s-tuples equal to $K_j(s, g)$ but with $k_i > 0$ replaced by $k_i - 1$;
$\beta_j^i(s, g+1)$	the scalar constant coefficients defined as $\beta_j^i(s, g+1) = \frac{g!}{k_1! k_2! \cdots k_s!}$ with $(k_1, k_2, \ldots, k_s) = K_j^i(s, g+1)$

Note that the s-tuples $K_j^i(s, g)$ are defined only in the cases when the corresponding k_i is positive. Note also that, when applied to the elements of $K(s, g+1)$, the s-tuples $K_j^i(s, g+1)$ define subscripts $(k_1, k_2, \ldots, k_s)$ of matrices

$$\begin{gathered} \tilde{P}_{(k_1,k_2,\ldots,k_s)}, \ Z_{(k_1,k_2,\ldots,k_s)}, \ \tilde{G}_{1,(k_1,k_2,\ldots,k_s)}, \\ \tilde{W}_{i,(k_1,k_2,\ldots,k_s)}, \ \tilde{F}_{i,(k_1,k_2,\ldots,k_s)}, \ i = 1, 2 \end{gathered}$$

associated to homogeneously polynomially parameter-dependent matrices of degree g.

As an example, consider a polytope $\mathcal{M}$ with $s = 2$ vertices and let the degree $g = 3$. Then, $J(2,3) = \frac{4!}{3!\,1!} = 4$ and

$$\begin{aligned} K(2,3) &= \{(0,3),(1,2),(2,1),(3,0)\}, \\ \tilde{P}(\lambda) &= \lambda_2^3 \tilde{P}_{(0,3)} + \lambda_1 \lambda_2^2 \tilde{P}_{(1,2)} + \lambda_1^2 \lambda_2 \tilde{P}_{(2,1)} + \lambda_1^3 \tilde{P}_{(3,0)}, \\ I(2,3) &= \{(2),(1,2),(1,2),(1)\}. \end{aligned} \tag{3.35}$$

Moreover,

$$\begin{aligned} K_1^2(2,3) &= (0,2),\ \beta_1^2(2,3) = \frac{2!}{0!\,2!} = 1, \\ K_2^1(2,3) &= (0,2),\ \beta_2^1(2,3) = \frac{2!}{0!\,2!} = 1, \\ K_2^2(2,3) &= (1,1),\ \beta_2^2(2,3) = \frac{2!}{1!\,1!} = 2, \\ K_3^1(2,3) &= (1,1),\ \beta_3^1(2,3) = \frac{2!}{1!\,1!} = 2, \\ K_3^2(2,3) &= (2,0),\ \beta_3^2(2,3) = \frac{2!}{2!\,0!} = 1, \\ K_4^1(2,3) &= (2,0),\ \beta_4^1(2,3) = \frac{2!}{2!\,0!} = 1 \end{aligned}$$

are the only possible triples $K_j^i(2,3)$, $j = 1, \ldots, 4$ associated to $K(2,3)$ and the corresponding value of $\beta_j^i(2,3)$.

3.4.2 H_2 Filter Design II

In this section, we derive new robust H_2 filter design methods by virtue of the polynomially parameter-dependent idea. The following notations are first defined for future use:

$$\begin{aligned} \tilde{\Psi}_{1,i,j} &\triangleq \tilde{W}_{1,K_j^i(s,g+1)} A_i + \beta_j^i(s,g+1) \tilde{B}_F C_i, \\ \tilde{\Psi}_{2,i,j} &\triangleq \tilde{W}_{1,K_j^i(s,g+1)} B_i + \beta_j^i(s,g+1) \tilde{B}_F D_i, \\ \tilde{\Psi}_{3,i,j} &\triangleq \tilde{W}_{2,K_j^i(s,g+1)} A_i + \beta_j^i(s,g+1) \tilde{B}_F C_i, \\ \tilde{\Psi}_{4,i,j} &\triangleq \tilde{W}_{2,K_j^i(s,g+1)} B_i + \beta_j^i(s,g+1) \tilde{B}_F D_i, \\ \tilde{\Psi}_{9,i,j} &\triangleq \tilde{G}_{1,K_j^i(s,g+1)} A_i, \\ \tilde{\Psi}_{10,i,j} &\triangleq \tilde{G}_{1,K_j^i(s,g+1)} B_i. \end{aligned} \tag{3.36}$$

With the above preparation, we are in a position to give one of the main results of this section, that is, a polynomially parameter-dependent approach to robust H_2 filter design.

Theorem 3.3 *If there exist real matrices* $\tilde{P}_{K_j(s,g)} = \begin{bmatrix} \tilde{P}_{1,K_j(s,g)} & \tilde{P}_{2,K_j(s,g)} \\ * & \tilde{P}_{3,K_j(s,g)} \end{bmatrix} = \tilde{P}^{\mathrm{T}}_{K_j(s,g)} > 0$, $Z_{K_j(s,g)} = Z^{\mathrm{T}}_{K_j(s,g)} > 0$, $\tilde{W}_{1,K_j(s,g)}$, $\tilde{W}_{2,K_j(s,g)}$, $\tilde{F}_{1,K_j(s,g)}$, $\tilde{F}_{2,K_j(s,g)}$, $\tilde{G}_{1,K_j(s,g)}$, $j = 1, 2, \ldots, J(s, g)$, $\tilde{W}_3$, $\tilde{A}_F$, $\tilde{B}_F$ *and* $\tilde{C}_F$ *such that the following LMIs hold for all* $j = 1, 2, \ldots, J(s, g+1)$:

$$\Delta_{1,j} \triangleq \sum_{i \in I_j(s,g+1)} \left(\mathrm{Tr}[Z_{K^i_j(s,g+1)}] - \beta^i_j(s, g+1)\gamma \right) < 0, \tag{3.37}$$

$$\Delta_{2,j} \triangleq \sum_{i \in I_j(s,g+1)} \begin{bmatrix} \tilde{P}_{1,K^i_j(s,g+1)} & \tilde{P}_{2,K^i_j(s,g+1)} & \beta^i_j(s, g+1)L^{\mathrm{T}}_i \\ * & \tilde{P}_{3,K^i_j(s,g+1)} & -\beta^i_j(s, g+1)\tilde{C}^{\mathrm{T}}_F \\ * & * & Z_{K^i_j(s,g+1)} \end{bmatrix} > 0, \tag{3.38}$$

$$\Delta_{3,j} \triangleq \sum_{i \in I_j(s,g+1)} \left(\tilde{\Phi}_{i,j} + \tilde{\Psi}_{i,j} + \tilde{\Psi}^{\mathrm{T}}_{i,j} \right) < 0, \tag{3.39}$$

where

- *for the continuous-time case*

$$\tilde{\Phi}_{i,j} \triangleq \begin{bmatrix} 0 & \tilde{P}_{K^i_j(s,g+1)} & 0 \\ \tilde{P}_{K^i_j(s,g+1)} & 0 & 0 \\ 0 & 0 & -\beta^i_j(s, g+1)I \end{bmatrix},$$

$$\tilde{\Psi}_{i,j} \triangleq \left[\begin{array}{cc|cc|c} -\tilde{W}_{1,K^i_j(s,g+1)} & -\beta^i_j(s, g+1)\tilde{W}_3 & \tilde{\Psi}_{1,i,j} & \beta^i_j(s, g+1)\tilde{A}_F & \tilde{\Psi}_{2,i,j} \\ -\tilde{W}_{2,K^i_j(s,g+1)} & -\beta^i_j(s, g+1)\tilde{W}_3 & \tilde{\Psi}_{3,i,j} & \beta^i_j(s, g+1)\tilde{A}_F & \tilde{\Psi}_{4,i,j} \\ \hline -\tilde{F}_{1,K^i_j(s,g+1)} & -\beta^i_j(s, g+1)\tilde{W}_3 & \tilde{\Psi}_{5,i,j} & \beta^i_j(s, g+1)\tilde{A}_F & \tilde{\Psi}_{6,i,j} \\ -\tilde{F}_{2,K^i_j(s,g+1)} & -\beta^i_j(s, g+1)\tilde{W}_3 & \tilde{\Psi}_{7,i,j} & \beta^i_j(s, g+1)\tilde{A}_F & \tilde{\Psi}_{8,i,j} \\ \hline -\tilde{G}_{1,K^i_j(s,g+1)} & 0 & \tilde{\Psi}_{9,i,j} & 0 & \tilde{\Psi}_{10,i,j} \end{array} \right],$$

$$\tilde{\Psi}_{5,i,j} \triangleq \tilde{F}_{1,K^i_j(s,g+1)}A_i + \beta^i_j(s, g+1)\tilde{B}_F C_i,$$

$$\tilde{\Psi}_{6,i,j} \triangleq \tilde{F}_{1,K^i_j(s,g+1)}B_i + \beta^i_j(s, g+1)\tilde{B}_F D_i,$$

$$\tilde{\Psi}_{7,i,j} \triangleq \tilde{F}_{2,K^i_j(s,g+1)}A_i + \beta^i_j(s, g+1)\tilde{B}_F C_i,$$

$$\tilde{\Psi}_{8,i,j} \triangleq \tilde{F}_{2,K^i_j(s,g+1)}B_i + \beta^i_j(s, g+1)\tilde{B}_F D_i,$$

- *for the discrete-time case*

$$\tilde{\Phi}_{i,j} \triangleq \begin{bmatrix} \tilde{P}_{K_j^i(s,g+1)} & 0 & 0 \\ 0 & -\tilde{P}_{K_j^i(s,g+1)} & 0 \\ 0 & 0 & -\beta_j^i(s,g+1)I \end{bmatrix},$$

$$\tilde{\Psi}_{i,j} \triangleq \left[\begin{array}{cc|cc|c} -\tilde{W}_{1,K_j^i(s,g+1)} & -\beta_j^i(s,g+1)\tilde{W}_3 & \tilde{\Psi}_{1,i,j} & \beta_j^i(s,g+1)\tilde{A}_F & \tilde{\Psi}_{2,i,j} \\ -\tilde{W}_{2,K_j^i(s,g+1)} & -\beta_j^i(s,g+1)\tilde{W}_3 & \tilde{\Psi}_{3,i,j} & \beta_j^i(s,g+1)\tilde{A}_F & \tilde{\Psi}_{4,i,j} \\ \hline -\tilde{F}_{1,K_j^i(s,g+1)} & 0 & \tilde{\Psi}_{5,i,j} & 0 & \tilde{\Psi}_{6,i,j} \\ -\tilde{F}_{2,K_j^i(s,g+1)} & 0 & \tilde{\Psi}_{7,i,j} & 0 & \tilde{\Psi}_{8,i,j} \\ \hline -\tilde{G}_{1,K_j^i(s,g+1)} & 0 & \tilde{\Psi}_{9,i,j} & 0 & \tilde{\Psi}_{10,i,j} \end{array}\right],$$

$$\tilde{\Psi}_{5,i,j} \triangleq \tilde{F}_{1,K_j^i(s,g+1)}A_i,\ \tilde{\Psi}_{6,i,j} \triangleq \tilde{F}_{1,K_j^i(s,g+1)}B_i,$$

$$\tilde{\Psi}_{7,i,j} \triangleq \tilde{F}_{2,K_j^i(s,g+1)}A_i,\ \tilde{\Psi}_{8,i,j} \triangleq \tilde{F}_{2,K_j^i(s,g+1)}B_i,$$

and other notations are defined in (3.36). *Then, the homogeneous polynomial matrices given by* (3.34) *ensure* (3.15)–(3.17) *for all* $\lambda \in \Gamma$.

Moreover, if the LMIs of (3.37)–(3.39) *are satisfied for a given degree* $\tilde{g}$*, then the LMIs corresponding to any degree* $g > \tilde{g}$ *are also satisfied.*

Proof Since $\tilde{P}_{K_j(s,g)} > 0$ and $Z_{K_j(s,g)} > 0$, $j = 1, 2, \ldots, J(s,g)$, we know that $\tilde{P}(\lambda)$ and $Z(\lambda)$ defined in (3.34) are both positive definite for all $\lambda \in \Gamma$. Now, note that the LHSs of the inequalities in (3.15)–(3.17) all are homogeneous polynomials of degree $g+1$ that can be written as

$$\mathrm{Tr}[Z(\lambda)] - \gamma \triangleq \sum_{j=1}^{J(s,g+1)} \left(\lambda_1^{k_1}\lambda_2^{k_2}\cdots\lambda_s^{k_s}\Delta_{1,j}\right),$$

$$\Pi(\lambda) \triangleq \sum_{j=1}^{J(s,g+1)} \left(\lambda_1^{k_1}\lambda_2^{k_2}\cdots\lambda_s^{k_s}\Delta_{2,j}\right),$$

$$\tilde{\Phi}(\lambda) + \tilde{\Psi}(\lambda) + \tilde{\Psi}(\lambda)^{\mathrm{T}} \triangleq \sum_{j=1}^{J(s,g+1)} \left(\lambda_1^{k_1}\lambda_2^{k_2}\cdots\lambda_s^{k_s}\Delta_{3,j}\right),$$

with $(k_1, k_2, \ldots, k_s) = K_j(s, g+1)$. Conditions (3.37)–(3.39) imposed for all $l \in \{1, 2, \ldots, J(s, g+1)\}$ ensure that the inequalities in (3.15)–(3.17) hold for all $\lambda \in \Gamma$, and thus the first part of Theorem 3.3 is proved.

Suppose that the LMIs in (3.37)–(3.39) are satisfied for a certain degree $\tilde{g}$, that is, there exist $J(s,\tilde{g})$ symmetric positive definite matrices $\tilde{P}_{K_j(s,g)}$, $Z_{K_j(s,g)}$ and matrices $\tilde{W}_{1,K_j(s,g)}$, $\tilde{W}_{2,K_j(s,g)}$, $\tilde{F}_{1,K_j(s,g)}$, $\tilde{F}_{2,K_j(s,g)}$, $\tilde{G}_{1,K_j(s,g)}$, defined in (3.34), that are homogeneously polynomially parameter-dependent matrices ensuring $\Delta_{1,j} < 0$, $\Delta_{2,j} > 0$ and $\Delta_{3,j} < 0$. Then, the terms of the polynomial matrices

$$
\begin{aligned}
\bar{P}(\lambda) &= (\lambda_1 + \lambda_2 + \cdots + \lambda_s)\,\tilde{P}(\lambda),\\
\bar{Z}(\lambda) &= (\lambda_1 + \lambda_2 + \cdots + \lambda_s)\,Z(\lambda),\\
\bar{W}_1(\lambda) &= (\lambda_1 + \lambda_2 + \cdots + \lambda_s)\,\tilde{W}_1(\lambda),\\
\bar{W}_2(\lambda) &= (\lambda_1 + \lambda_2 + \cdots + \lambda_s)\,\tilde{W}_2(\lambda),\\
\bar{F}_1(\lambda) &= (\lambda_1 + \lambda_2 + \cdots + \lambda_s)\,\tilde{F}_1(\lambda),\\
\bar{F}_2(\lambda) &= (\lambda_1 + \lambda_2 + \cdots + \lambda_s)\,\tilde{F}_2(\lambda),\\
\bar{G}_1(\lambda) &= (\lambda_1 + \lambda_2 + \cdots + \lambda_s)\,\tilde{G}_1(\lambda)
\end{aligned}
$$

with $\tilde{P}(\lambda)$, $Z(\lambda)$, $\tilde{W}_1(\lambda)$, $\tilde{W}_2(\lambda)$, $\tilde{F}_1(\lambda)$, $\tilde{F}_2(\lambda)$ and $\tilde{G}_1(\lambda)$ defined in (3.34) also satisfy the LMIs in (3.37)–(3.39) corresponding to the degree $\tilde{g} + 1$, which can be obtained in this case by linear combination of the LMIs of Theorem 3.3 corresponding to $\tilde{g}$. □

Theorem 3.3 shows that using the polynomially parameter-dependent matrices in (3.34), the infinite number of inequalities in Lemma 3.5 can be guaranteed by finite number of LMIs in Theorem 3.3. We summarize the polynomially parameter-dependent approach to robust H_2 filter design in Algorithm 15.

Algorithm 15 Polynomially Parameter-Dependent Approach to Robust H_2 Filter Design

---

- Prescribe an integer $g \in \mathbb{N}_0^+$;
- Solve the minimization problem:

$$
\min_{\substack{\tilde{P}_{K_j(s,g)} = \tilde{P}^{\mathrm{T}}_{K_j(s,g)},\ Z_{K_j(s,g)} = Z^{\mathrm{T}}_{K_j(s,g)},\\ \tilde{W}_{1,K_j(s,g)},\ \tilde{W}_{2,K_j(s,g)},\ \tilde{F}_{1,K_j(s,g)},\ \tilde{F}_{2,K_j(s,g)},\\ \tilde{G}_{1,K_j(s,g)},\ j = 1,2,\ldots,J(s,g)\\ \tilde{W}_3,\ \tilde{A}_F,\ \tilde{B}_F,\ \tilde{C}_F,\ \gamma}} \gamma \quad \text{s.t. } (3.37) - (3.39)
$$

- Compute filter in (2.4) by (3.18) or (3.19).

We use a specific example to help reader to better understand the polynomially parameter-dependent approach and the expressions in Theorem 3.3. Let $s = 2$ and $g = 3$. The sets of $K(2, 3)$ and $I(2, 3)$ are given in (3.35). Moreover, we have

$$
\begin{aligned}
K(2,4) &= \{(0,4), (1,3), (2,2), (3,1), (4,0)\},\\
I(2,4) &= \{(2), (1,2), (1,2), (1,2), (1)\},
\end{aligned}
$$

and the possible combination $K_j^i(2,4)$, $j = 1, \ldots, 5$ and the associated $\beta_j^i(2,4)$ are given by

$$
\begin{aligned}
K_1^2(2,4) &= (0,3),\ \beta_1^2(2,4) = \frac{3!}{0!\,3!} = 1,\\
K_2^1(2,4) &= (0,3),\ \beta_2^1(2,4) = \frac{3!}{0!\,3!} = 1,\\
K_2^2(2,4) &= (1,2),\ \beta_2^2(2,4) = \frac{3!}{1!\,2!} = 3,\\
K_3^1(2,4) &= (1,2),\ \beta_3^1(2,4) = \frac{3!}{1!\,2!} = 3,\\
K_3^2(2,4) &= (2,1),\ \beta_3^2(2,4) = \frac{3!}{2!\,1!} = 3,\\
K_4^1(2,4) &= (2,1),\ \beta_4^1(2,4) = \frac{3!}{2!\,1!} = 3,\\
K_4^2(2,4) &= (3,0),\ \beta_4^2(2,4) = \frac{3!}{3!\,0!} = 1,\\
K_5^1(2,4) &= (3,0),\ \beta_5^1(2,4) = \frac{3!}{3!\,0!} = 1.
\end{aligned}
$$

Accordingly, the polynomially parameter-dependent matrix functions in (3.34) can be explicitly written as

$$
[\cdot](\lambda) = \lambda_2^3[\cdot]_{(0,3)} + \lambda_1\lambda_2^2[\cdot]_{(1,2)} + \lambda_1^2\lambda_2[\cdot]_{(1,2)} + \lambda_1^3[\cdot]_{(3,0)}
$$

where $[\cdot]$ are replaced by $\tilde{P}$, Z, $\tilde{W}_1$, $\tilde{W}_2$, $\tilde{F}_1$, $\tilde{F}_2$ and $\tilde{G}_1$, respectively. To guarantee the positive-definiteness of $\tilde{P}(\lambda)$ and $Z(\lambda)$ for all $\lambda \in \Gamma$, Theorem 3.3 requires

$$
\begin{aligned}
&\tilde{P}_{(0,3)} > 0,\ \tilde{P}_{(1,2)} > 0,\ \tilde{P}_{(2,1)} > 0,\ \tilde{P}_{(3,0)} > 0,\\
&Z_{(0,3)} > 0,\ Z_{(1,2)} > 0,\ Z_{(2,1)} > 0,\ Z_{(3,0)} > 0.
\end{aligned}
$$

Since $\lambda_1 + \lambda_2 = 1$, the inequality in (3.17) can be explicitly written as

$$
\begin{aligned}
\mathrm{Tr}[Z(\lambda)] - \gamma &= \mathrm{Tr}[\lambda_2^3 Z_{(0,3)} + \lambda_1\lambda_2^2 Z_{(1,2)} + \lambda_1^2\lambda_2 Z_{(2,1)} + \lambda_1^3 Z_{(3,0)}] - \gamma\\
&= (\lambda_1 + \lambda_2)\,\mathrm{Tr}[\lambda_2^3 Z_{(0,3)} + \lambda_1\lambda_2^2 Z_{(1,2)} + \lambda_1^2\lambda_2 Z_{(2,1)} + \lambda_1^3 Z_{(3,0)}]\\
&\quad - (\lambda_1 + \lambda_2)^4\,\gamma\\
&= \lambda_2^4\mathrm{Tr}[Z_{(0,3)} - \gamma] + \lambda_1^4\mathrm{Tr}[Z_{(3,0)} - \gamma]\\
&\quad + \lambda_1\lambda_2^3\mathrm{Tr}[\big(Z_{(0,3)} - \gamma\big) + \big(Z_{(1,2)} - 3\gamma\big)]\\
&\quad + \lambda_1^2\lambda_2^2\mathrm{Tr}[\big(Z_{(1,2)} - 3\gamma\big) + \big(Z_{(2,1)} - 3\gamma\big)]\\
&\quad + \lambda_1^3\lambda_2^1\mathrm{Tr}[\big(Z_{(2,1)} - 3\gamma\big) + \big(Z_{(3,0)} - \gamma\big)]\\
&= \lambda_2^4\mathrm{Tr}[Z_{(0,3)} - \beta_1^2(2,4)\gamma] + \lambda_1^4\mathrm{Tr}[Z_{(3,0)} - \beta_5^1(2,4)\gamma]\\
&\quad + \lambda_1\lambda_2^3\mathrm{Tr}[\Big(Z_{(0,3)} - \beta_2^1(2,4)\gamma\Big) + \Big(Z_{(1,2)} - \beta_2^2(2,4)\gamma\Big)]
\end{aligned}
$$

$$
\begin{aligned}
&+\lambda_1^2\lambda_2^2\mathrm{Tr}[\Big(Z_{(1,2)}-\beta_3^1(2,4)\gamma\Big)+\Big(Z_{(2,1)}-\beta_3^2(2,4)\gamma\Big)]\\
&+\lambda_1^3\lambda_2^1\mathrm{Tr}[\Big(Z_{(2,1)}-\beta_4^1(2,4)\gamma\Big)+\Big(Z_{(3,0)}-\beta_4^2(2,4)\gamma\Big)]\\
&<0.
\end{aligned}
$$

Hence, the inequality in (3.17) holds for all $(\lambda_1, \lambda_2) \in \Gamma$, if the inequalities in (3.37) are satisfied. Similarly, the inequalities in (3.15) and (3.16) can be explicitly related to those in (3.38) and (3.39), respectively.

The special case of Theorem 3.3 when $g = 1$ deserves some comments. As mentioned in Sect. 3.4.1, when $g = 1$, we have $J(s, 1) = s$, $J(s, 2) = s(s+1)/2$ and each parameter-dependent matrix defined in (3.34) is associated to s constant matrices. For convenience, the identified subscript of each constant matrix is directly specified to be the one of the coefficient, that is,

$$
\begin{aligned}
\tilde{P}(\lambda) &= \sum_{j=1}^{s}\lambda_j\tilde{P}_j,\ Z(\lambda) = \sum_{j=1}^{s}\lambda_j Z_j,\ \tilde{G}_1(\lambda) = \sum_{j=1}^{s}\lambda_j\tilde{G}_{1,j},\\
\tilde{W}_i(\lambda) &= \sum_{j=1}^{s}\lambda_j\tilde{W}_{i,j},\ \tilde{F}_i(\lambda) = \sum_{j=1}^{s}\lambda_j\tilde{F}_{i,j},\ i = 1, 2.
\end{aligned}
$$

With these notations, each of (3.37) and (3.38) is simplified into s LMIs as in (3.29) and (3.30), respectively. It should be pointed out that the reason why the number of LMIs reduces from $s(s+1)/2$ for (3.37) and (3.38) to s for (3.29) and (3.30) is that the remaining $s(s-1)/2$ LMIs all are the combinations of the sum of any two LMIs of the other s LMIs in (3.37) or (3.38).

For (3.39), take the continuous-time case for example. It can be re-written as

$$
\tilde{\Phi}_i + \tilde{\Phi}_j + \mathrm{sym}\left\{\tilde{\Psi}_{i,j} + \tilde{\Psi}_{j,i}\right\} < 0\,(i,\ j \in \mathbb{N}^+;\ 1 \le i \le j \le s), \qquad (3.40)
$$

where $\tilde{\Phi}_i$ is defined in (3.31) and $\tilde{\Psi}_{i,j}$ is re-defined as

$$
\tilde{\Psi}_{i,j} \triangleq \left[\begin{array}{cc|cc|c}
-\tilde{W}_{1,i} & -\tilde{W}_3 & \tilde{W}_{1,i}A_j + \tilde{B}_F C_j & \tilde{A}_F & \tilde{W}_{1,i}B_j + \tilde{B}_F D_j\\
-\tilde{W}_{2,i} & -\tilde{W}_3 & \tilde{W}_{2,i}A_j + \tilde{B}_F C_j & \tilde{A}_F & \tilde{W}_{2,i}B_j + \tilde{B}_F D_j\\
\hline
-\tilde{F}_{1,i} & -\tilde{W}_3 & \tilde{F}_{1,i}A_j + \tilde{B}_F C_j & \tilde{A}_F & \tilde{F}_{1,i}B_j + \tilde{B}_F D_j\\
-\tilde{F}_{2,i} & -\tilde{W}_3 & \tilde{F}_{2,i}A_j + \tilde{B}_F C_j & \tilde{A}_F & \tilde{F}_{2,i}B_j + \tilde{B}_F D_j\\
\hline
-\tilde{G}_{1,i} & 0 & \tilde{G}_{1,i}A_j & 0 & \tilde{G}_{1,i}B_j
\end{array}\right].
$$

The combination of (3.29), (3.30) and (3.40) gives another existence condition for robust H_2 filters in the linearly parameter-dependent framework, which can be used as a new linearly parameter-dependent approach to robust H_2 filter design. However, this result is different from the one in Lemma 3.1 in Sect. 3.3.1. Due to the linear parameter-dependence assumption for $\tilde{W}_i(\lambda)$, $\tilde{F}_i(\lambda)$, $i = 1, 2$, and $\tilde{G}_1(\lambda)$, this new result is generally less conservative than that in Lemma 3.1. Indeed, when $\tilde{W}_{1,i} = \tilde{W}_1$, $\tilde{W}_{2,i} = \tilde{W}_2$, $\tilde{G}_{1,i} = \tilde{G}_1$, $\tilde{F}_{1,i} = \tilde{F}_1$ and $\tilde{F}_{2,i} = \tilde{F}_2$, we have

$$\tilde{\Phi}_i + \tilde{\Phi}_j + \text{sym}\left\{\tilde{\Psi}_{i,j} + \tilde{\Psi}_{j,i}\right\} = \tilde{\Phi}_i + \tilde{\Phi}_j + \text{sym}\left\{\tilde{\Psi}_{j,j} + \tilde{\Psi}_{i,i}\right\},$$

which reveals that the $s(s+1)/2$ LMIs can reduce to s LMIs as

$$\tilde{\Phi}_i + \tilde{\Psi}_{i,i} + \tilde{\Psi}_{i,i}^{\mathrm{T}} < 0\,(i = 1, 2, \ldots, s),$$

This is exactly the one in (3.31). Thus, Theorem 3.1 recovers from Theorem 3.3 ($g = 1$) when $\tilde{W}_{1,i} = \tilde{W}_1$, $\tilde{W}_{2,i} = \tilde{W}_2$, $\tilde{G}_{1,i} = \tilde{G}_1$, $\tilde{F}_{1,i} = \tilde{F}_1$, $\tilde{F}_{2,i} = \tilde{F}_2$.

Although it has been shown that Theorem 3.1 is only a special case of Theorem 3.3 ($g = 1$), it is unnecessary to formulate the result in Theorem 3.1 into the general form of the conditions in Theorem 3.3. As the most basic application of the parameter-dependent method, an obvious merit of Theorem 3.1, compared with Theorem 3.3, is that much fewer LMIs and variables are involved, which generally means relatively lighter computational burden.

3.4.3 H_∞ Filter Design II

In this section, we investigate a polynomially parameter-dependent approach to robust H_∞ filter design. Applying the homogeneous parameter-dependence form defined in (3.34) to Lemma 3.6, we have the following result.

Theorem 3.4 *If there exist real matrices* $\tilde{P}_{K_j(s,g)} = \begin{bmatrix} \tilde{P}_{1,K_j(s,g)} & \tilde{P}_{2,K_j(s,g)} \\ * & \tilde{P}_{3,K_j(s,g)} \end{bmatrix} = \tilde{P}^{\mathrm{T}}_{K_j(s,g)}$, $\tilde{W}_{1,K_j(s,g)}$, $\tilde{W}_{2,K_j(s,g)}$, $\tilde{F}_{1,K_j(s,g)}$, $\tilde{F}_{2,K_j(s,g)}$, $\tilde{G}_{1,K_j(s,g)}$, $\tilde{H}_{1,K_j(s,g)}$, $j = 1, 2, \ldots, J(s,g)$, $\tilde{W}_3$, $\tilde{A}_F$, $\tilde{B}_F$ *and* $\tilde{C}_F$ *such that the following LMIs hold:*

$$\tilde{P}_{K_j(s,g)} > 0\,(j = 1, 2, \ldots, J(s,g)), \tag{3.41}$$

$$\sum_{i \in I_j(s,g+1)} \left\{\tilde{\Xi}_{i,j} + \tilde{\Upsilon}_{i,j} + \tilde{\Upsilon}_{i,j}^{\mathrm{T}}\right\} < 0\,(j = 1, 2, \ldots, J(s,g+1)), \tag{3.42}$$

where

- *for the continuous-time case*

$$\tilde{\Xi}_{i,j} \triangleq \begin{bmatrix} 0 & \tilde{P}_{K_j^i(s,g+1)} & 0 & 0 \\ \tilde{P}_{K_j^i(s,g+1)} & 0 & 0 & \beta_j^i(s,g+1)\tilde{C}_i^{\mathrm{T}} \\ 0 & 0 & -\beta_j^i(s,g+1)\gamma^2 I & \beta_j^i(s,g+1)\tilde{D}_i^{\mathrm{T}} \\ 0 & \beta_j^i(s,g+1)\tilde{C}_i & \beta_j^i(s,g+1)\tilde{D}_i & -\beta_j^i(s,g+1)I \end{bmatrix},$$

$$\tilde{\Upsilon}_{i,j} \triangleq \left[\begin{array}{cc|cc|c|c} -\tilde{W}_{1,K_j^i(s,g+1)} & -\beta_j^i(s,g+1)\tilde{W}_3 & \tilde{\Psi}_{1,i,j} & \beta_j^i(s,g+1)\tilde{A}_F & \tilde{\Psi}_{2,i,j} & 0 \\ -\tilde{W}_{2,K_j^i(s,g+1)} & -\beta_j^i(s,g+1)\tilde{W}_3 & \tilde{\Psi}_{3,i,j} & \beta_j^i(s,g+1)\tilde{A}_F & \tilde{\Psi}_{4,i,j} & 0 \\ \hline -\tilde{F}_{1,K_j^i(s,g+1)} & -\beta_j^i(s,g+1)\tilde{W}_3 & \tilde{\Psi}_{5,i,j} & \beta_j^i(s,g+1)\tilde{A}_F & \tilde{\Psi}_{6,i,j} & 0 \\ -\tilde{F}_{2,K_j^i(s,g+1)} & -\beta_j^i(s,g+1)\tilde{W}_3 & \tilde{\Psi}_{7,i,j} & \beta_j^i(s,g+1)\tilde{A}_F & \tilde{\Psi}_{8,i,j} & 0 \\ \hline -\tilde{G}_{1,K_j^i(s,g+1)} & 0 & \tilde{\Psi}_{9,i,j} & 0 & \tilde{\Psi}_{10,i,j} & 0 \\ \hline 0 & 0 & 0 & 0 & 0 & 0 \end{array}\right],$$

- *for the discrete-time case*

$$\tilde{\Xi}_{i,j} \triangleq \begin{bmatrix} \tilde{P}_{K_j^i(s,g+1)} & 0 & 0 & 0 \\ 0 & -\tilde{P}_{K_j^i(s,g+1)} & 0 & \beta_j^i(s,g+1)\tilde{C}_i^{\mathrm{T}} \\ 0 & 0 & -\beta_j^i(s,g+1)\gamma^2 I & \beta_j^i(s,g+1)\tilde{D}_i^{\mathrm{T}} \\ 0 & \beta_j^i(s,g+1)\tilde{C}_i & \beta_j^i(s,g+1)\tilde{D}_i & -\beta_j^i(s,g+1)I \end{bmatrix},$$

$$\tilde{\Upsilon}_{i,j} \triangleq \left[\begin{array}{cc|cc|c|c} -\tilde{W}_{1,K_j^i(s,g+1)} & -\beta_j^i(s,g+1)\tilde{W}_3 & \tilde{\Psi}_{1,i,j} & \beta_j^i(s,g+1)\tilde{A}_F & \tilde{\Psi}_{2,i,j} & 0 \\ -\tilde{W}_{2,K_j^i(s,g+1)} & -\beta_j^i(s,g+1)\tilde{W}_3 & \tilde{\Psi}_{3,i,j} & \beta_j^i(s,g+1)\tilde{A}_F & \tilde{\Psi}_{4,i,j} & 0 \\ \hline -\tilde{F}_{1,K_j^i(s,g+1)} & 0 & \tilde{\Psi}_{5,i,j} & 0 & \tilde{\Psi}_{6,i,j} & 0 \\ -\tilde{F}_{2,K_j^i(s,g+1)} & 0 & \tilde{\Psi}_{7,i,j} & 0 & \tilde{\Psi}_{8,i,j} & 0 \\ \hline -\tilde{G}_{1,K_j^i(s,g+1)} & 0 & \tilde{\Psi}_{9,i,j} & 0 & \tilde{\Psi}_{10,i,j} & 0 \\ \hline 0 & 0 & 0 & 0 & 0 & 0 \end{array}\right],$$

with $\tilde{C}_i$, $\tilde{D}_i$ *defined in* (3.33), $\tilde{\Psi}_{h,i,j}$, $h = 1, 2, 3, 4, 9, 10$, *defined in* (3.36) *and* $\tilde{\Psi}_{h,i,j}$, $h = 5, 6, 7, 8$, *defined in* (3.39). *Then, the homogeneous polynomial matrices given by* (3.34) *ensure* (3.22) *and* (3.23) *for all* $\lambda \in \Gamma$.

Moreover, if the LMIs of (3.41) *and* (3.42) *are satisfied for a given degree* $\tilde{g}$, *then the LMIs corresponding to any degree* $g > \tilde{g}$ *are also satisfied.*

Proof The proof can be completed by following similar lines to that of the proof of Theorem 3.3 and thus is omitted. □

The comments for Theorem 3.3 in Sect. 3.4.2 can be similarly made for the H_∞ filter design result, Theorem 3.4. Thus it is inferred that Theorem 3.2 is a special case of Theorem 3.4 ($g = 1$) and can recover from the case that each of the slack matrices $\tilde{W}_{1,K_j(s,g)}$, $\tilde{W}_{2,K_j(s,g)}$, $\tilde{F}_{1,K_j(s,g)}$, $\tilde{F}_{2,K_j(s,g)}$ and $\tilde{G}_{1,K_j(s,g)}$ in Theorem 3.4 ($g = 1$) are fixed the same for all $K_j(s, g) \in K(s, g)$.

Based on Theorem 3.4, a polynomially parameter-dependent approach to robust H_∞ filter design is summarized in Algorithm 16.

Algorithm 16 Polynomially Parameter-Dependent Approach to Robust H_∞ Filter Design

- Prescribe an integer $g \in \mathbb{N}_0^+$;
- Solve the minimization problem:

$$\min_{\substack{\tilde{P}_{K_j(s,g)} = \tilde{P}^{\mathrm{T}}_{K_j(s,g)},\ \tilde{W}_{1,K_j(s,g)},\\ \tilde{W}_{2,K_j(s,g)},\ \tilde{F}_{1,K_j(s,g)},\ \tilde{F}_{2,K_j(s,g)},\\ \tilde{G}_{1,K_j(s,g)},\ j=1,2,\ldots,J(s,g),\\ \tilde{W}_3,\ \tilde{A}_F,\ \tilde{B}_F,\ \tilde{C}_F,\ \tilde{D}_F,\ \mu}} \mu = \gamma^2 \quad \text{s.t. (3.41) and (3.42)}$$

- Compute filter in (2.66) by (3.24) or (3.25).

Table 3.1 Calculated results by different methods in Example 3.1-(1)

Methods	γ^*	Number of variables	Filter realization $\left[\begin{array}{cc} A_F & B_F \\ C_F & 0 \end{array}\right]$
Algorithm 13	2.1750	77	$\left[\begin{array}{cc\|c} -12.4875 & 11.7680 & 0.1279 \\ 10.8185 & -10.4980 & -0.0992 \\ \hline -0.7952 & -0.0387 & 0 \end{array}\right]$
Algorithm 15 ($g=1$)	2.1721	137	$\left[\begin{array}{cc\|c} -13.8111 & 13.0890 & 0.1411 \\ 11.7812 & -11.4672 & -0.1089 \\ \hline -0.7915 & -0.0431 & 0 \end{array}\right]$
Algorithm 15 ($g=2$)	2.1691	323	$\left[\begin{array}{cc\|c} -15.3990 & 14.6582 & 0.1559 \\ 13.4703 & -13.1295 & -0.1250 \\ \hline -0.8041 & -0.0451 & 0 \end{array}\right]$

3.5 Illustrative Examples

In the section, we provide several numerical examples to illustrate the effectiveness and advantage of the parameter-dependent methods.

Example 3.1 (Robust H_2 filter design, continuous-time)

(1) Consider the uncertain continuous-time system in (2.97) that consists of four vertices. Parameters α and β are also assumed to be unknown but satisfy $|\alpha| \le 1$ and $|\beta| \le 1$. We will design robust H_2 filters by Algorithms 13 and 15 (continuous-time case).

Table 3.1 presents the results on the obtained minimum upper bound γ^* of the squared H_2 filtering performance, the involved number of variables and the resulting state space realizations of the filters for the continuous-time case of Algorithms 13 and 15. It is shown that the parameter-dependent approaches in this chapter obtain much smaller γ^* than the one $\gamma^* = 5.7281$ obtained by the quadratic approaches in Chap. 2, which verifies that the parameter-dependent approaches to H_2 filter design are generally less conservative than the quadratic approaches. Moreover, it can still

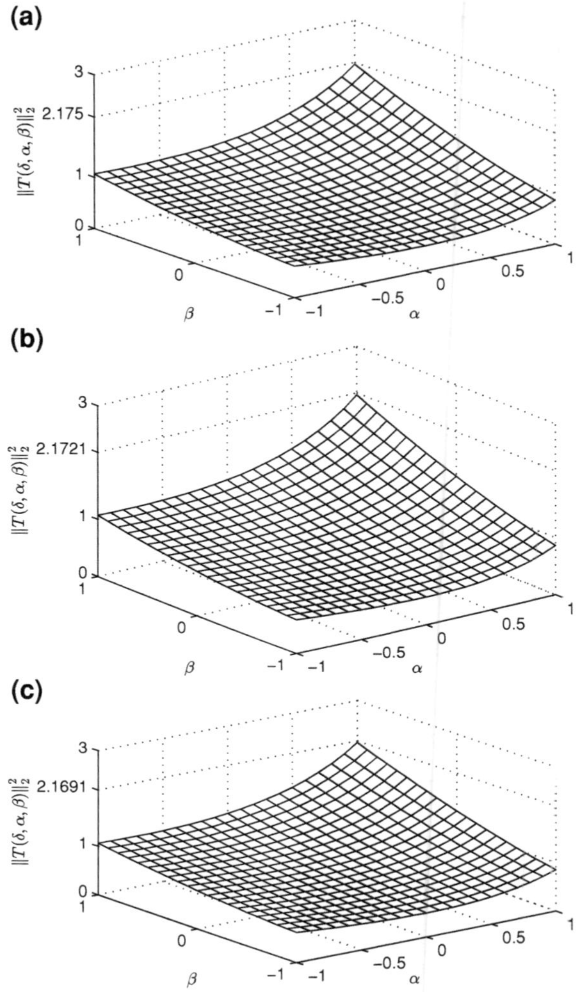

Fig. 3.1 Actual performance $\|T(\delta, \alpha, \beta)\|_2^2$ with different α and β in Example 3.1. **a** Algorithm 13 (continuous-time); **b** Algorithm 15 ($g = 1$, continuous-time); **c** Algorithm 15 ($g = 2$, continuous-time)

be seen that the growth of the degree of the polynomial employed for the parameter-dependent conditions can further improve the results to some extent. On the other hand, it should be noted from the third column of Table 3.1 that the design algorithms based on the parameter-dependent approaches inevitably involve more decision variables than that the quadratic approaches have, which usually leads to more computational cost. However, as pointed out in [3], a robust filter is usually designed off-line, where the filter performance is the main concern.

Figure 3.1 displays the actual values of the H_2 norm square of the filtering error systems under different parameters α and β, where the filters correspond to the ones in Table 3.1. The effectiveness of the designed filters by the parameter-dependent approaches is clearly demonstrated.

Table 3.2 Comparison between different methods in Example 3.1-(2)

Methods	γ^*			
	Case (a)	Case (b)	Case (c)	Case (d)
[7, 8]	5.7281	4.8188	Infeasible	Infeasible
[5]	4.876	4.373	Infeasible	Infeasible
[9]	2.382	2.382	93.365	100.963
[4]	2.187	2.186	15.8000	14.6200
[6] ($\lambda_1 = 0,\ \lambda_2 = 0$)	2.1750	2.1687	15.7966	14.2271
Algorithm 15 ($g = 2$)	2.1691	2.1687	15.7740	14.0583

(2) Still consider the uncertain continuous-time system in (2.97), where the following four cases for parameters α and β will be discussed:

- Case (a): $|\alpha| \le 1$ and $|\beta| \le 1$;
- Case (b): $|\alpha| \le 1$ and $\alpha = \beta$;
- Case (c): $|\alpha| \le 3$ and $|\beta| \le 1$;
- Case (d): $|\alpha| \le 3$ and $\alpha = \beta$.

Case (a) and Case (c) each can be modeled by uncertain system $\mathcal{S}$ in (2.1) with four vertices, and Case (b) and Case (d) each can be modeled by that with two vertices. Table 3.2 collects the results on the minimum upper bound (γ^*) of the squared H_2 performance obtained by the methods proposed in [4–9] and the polynomially parameter-dependent approach, Algorithm 15 ($g = 2$) in this chapter. From the table, it can be seen that our approach generally produces less conservative results than those in the literature.

(3) Consider the uncertain continuous system $\mathcal{S}$ in (2.1), where the vertices of the polytope $\mathcal{M}$ are given by

$$A_1 = \begin{bmatrix} 4 & 3 \\ -8.6 & -4.2 \end{bmatrix},\ A_2 = \begin{bmatrix} -12.1 & 2.3 \\ -3.6 & 0.6 \end{bmatrix},\ A_3 = \begin{bmatrix} -6.1 & 4.7 \\ -20.9 & 5.9 \end{bmatrix},$$
$$B_i = \begin{bmatrix} -2 & 1 \end{bmatrix}^{\mathrm{T}},\ C_i = \begin{bmatrix} 0.1 & 0.2 \end{bmatrix},\ D_i = 0,\ L_i = \begin{bmatrix} 1 & 0 \end{bmatrix},\ i = 1, 2, 3$$

For this example, none of the methods in [4–9], the quadratic approaches in Chap. 2, or the linearly parameter-dependent approach, Algorithm 13 in this chapter, can find feasible solutions. However, by applying Algorithm 15, it is found that $\gamma^* = 34.0380$ for $g = 1$ and $\gamma^* = 26.6108$ for $g = 2$, respectively. This again shows the improvement of the polynomially parameter-dependent approach to robust H_2 filter design.

Example 3.2 (*Robust H_2 filter design, discrete-time*) Consider the uncertain discrete-time system given by Eq. (2.105). Parameters α and β are also assumed to satisfy $|\alpha| \le 1$ and $|\beta| \le 1$.

Table 3.3 Calculated results by different methods in Example 3.2

Methods	γ^*	Number of variables	Filter realization $\begin{bmatrix} A_F & B_F \\ C_F & 0 \end{bmatrix}$
Algorithm 13	16.0898	79	$\begin{bmatrix} 0.0242 & -0.0756 & -1.0230 \\ 1.0542 & 1.0100 & 1.0603 \\ -1.2907 & -0.3281 & 0 \end{bmatrix}$
Algorithm 15 ($g = 1$)	11.9756	145	$\begin{bmatrix} 0.3193 & 0.0362 & -0.6599 \\ 0.0713 & 0.8529 & -0.0203 \\ -1.4126 & -0.1865 & 0 \end{bmatrix}$
Algorithm 15 ($g = 2$)	11.5095	343	$\begin{bmatrix} 0.3290 & 0.0332 & -0.6524 \\ 0.0480 & 0.8652 & -0.0368 \\ -1.3975 & -0.1638 & 0 \end{bmatrix}$

By applying the discrete-time case of Algorithms 13 and 15, the calculated results are listed in Table 3.3. Note that $\gamma^* = 100.0276$ by the quadratic approaches in Example 2.2. Table 3.3 shows that the parameter-dependent approaches in this chapter yield much smaller γ^*. On the other hand, similar to the continuous-time case, the parameter-dependent approaches usually involve more variables. However, the fact that the filters obtained by the parameter-dependent approaches can guarantee a lower H_2 filtering performance level, as clearly shown in Table 3.3, justifies their more computational costs. Figure 3.2 depicts the actual H_2 filtering performance of the filters in Table 3.3 for different parameters α and β. The effectiveness of the designed filters is apparent.

In addition, for this example, the obtained minimum squared H_2 costs are $\gamma^* = 44.0039$ by [10], $\gamma^* = 19.4682$ by [11] and $\gamma^* = 16.0925$ by [6] ($\lambda_1 = \lambda_2 = 0$), respectively. Comparing these results with those in Table 3.3, one sees that the parameter-dependent approaches in this chapter outperform the ones in [6, 10, 11]. Note that two scalars are tunable for the method in [6] and they are also shown to be effective in further refining the designed filters. For instance, [6] exposes that the searched $\lambda_1 = -0.89$ and $\lambda_2 = -0.921$ provide the best result $\gamma^* = 13.4561$, better than $\gamma^* = 16.0898$ given by Algorithm 13. Although this exceptional case shows that the method in [6] may produce less conservative results than the linear parameter-dependent approach Algorithm 13 could, no systematic and general procedure can be found in the literature for tuning these two parameters. However, all the design approaches in this chapter are given in terms of LMI without tuning scalars, which can be solved directly by the existing numerical algorithms. Moreover, it is obvious from Table 3.3 that the polynomially parameter-dependent approach can yield better results than the method in [6] under searched tuning parameters.

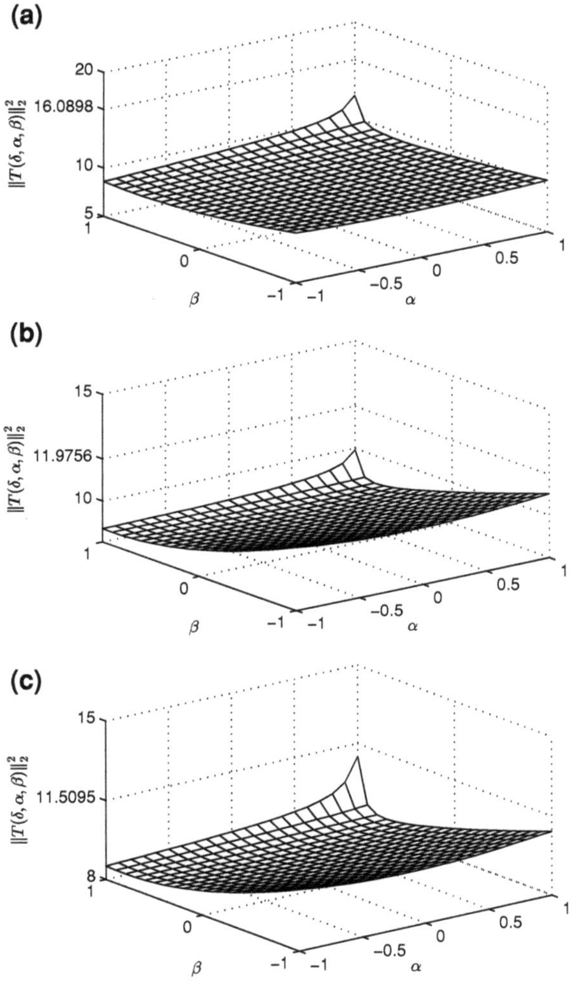

Fig. 3.2 Actual performance $\|T(\delta, \alpha, \beta)\|_2^2$ with different α and β in Example 3.2. **a** Algorithm 13 (discrete-time); **b** Algorithm 15 ($g = 1$, discrete-time); **c** Algorithm 15 ($g = 2$, discrete-time)

Example 3.3 (Robust H_∞ filter design, continuous-time) Consider an uncertain continuous-time system, borrowed from [12], given by

$$\begin{aligned}
\dot{x}(t) &= \begin{bmatrix} -0.6 & 4+\alpha \\ -4 & -0.6 \end{bmatrix} x(t) + \begin{bmatrix} 0 & 0 \\ 1.5 & 0 \end{bmatrix} w(t), \\
y(t) &= \begin{bmatrix} 0 & -1.2 \end{bmatrix} x(t) + \begin{bmatrix} 0 & 1 \end{bmatrix} w(t), \\
z(t) &= \begin{bmatrix} 0 & 1 \end{bmatrix} x(t),
\end{aligned} \tag{3.43}$$

where α is an uncertain parameter satisfying $\alpha \leq \bar{\alpha}$.

Table 3.4 Comparison between different methods in Example 3.3

Methods	γ^* $\alpha = 0.5$	$\alpha = 1.0$	$\alpha = 1.5$	$\alpha = 2.0$
[8, 12, 13], Algorithm 8	0.7773	0.8283	Infeasible	Infeasible
[6] ($\lambda_1 = \lambda_2 = 1$)	0.7170	0.7496	0.7976	0.8653
[14]	0.7086	0.7234	0.7421	0.7643
Algorithm 14	0.6977	0.7017	0.7094	0.7208
Algorithm 16 ($g = 1$)	0.6972	0.7017	0.7094	0.7208
Algorithm 16 ($g = 2$)	0.6960	0.6964	0.6970	0.6987

By applying some robust H_∞ filter design methods in the literature as well as Algorithms 14 and 16 in this chapter to design filters for (3.43), the calculated minimum guaranteed H_∞ costs γ^* under different parameter $\bar{\alpha}$ are summarized in Table 3.4. From the table, it can be seen that the parameter-dependent approaches in this chapter, Algorithms 14 and 16, are able to yield less conservative results than the methods in [6, 8, 12–14] and the quadratic approach in Chap. 2.

For $\bar{\alpha} = 1$, the filters obtained from (3.24) in Algorithms 14, 16 ($g = 1$) and Algorithm 16 ($g = 2$) are respectively given by

$$\begin{bmatrix} A_F & B_F \\ C_F & D_F \end{bmatrix} = \left[\begin{array}{cc|c} -2.8650 & 2.3277 & -0.2540 \times 10^{-5} \\ -10.2609 & -1.4844 & 0.4718 \times 10^{-5} \\ \hline -0.0085 & -0.0043 & -0.5908 \end{array}\right], \tag{3.44}$$

$$\begin{bmatrix} A_F & B_F \\ C_F & D_F \end{bmatrix} = \left[\begin{array}{cc|c} -3.0153 & 2.3281 & 0.0015 \\ -9.8806 & -1.5905 & 0.0037 \\ \hline -0.0080 & -0.0045 & -0.5908 \end{array}\right], \tag{3.45}$$

$$\begin{bmatrix} A_F & B_F \\ C_F & D_F \end{bmatrix} = \left[\begin{array}{cc|c} -5.4818 & 6.5166 & 0.4732 \\ -3.8604 & 0.5134 & 0.3322 \\ \hline -0.0073 & -0.0105 & -0.5806 \end{array}\right]. \tag{3.46}$$

The singular value curves of the filtering error systems under the filters in (3.44)–(3.46) are respectively depicted in Fig. 3.3, which clearly demonstrates the effectiveness of the designed filters.

Example 3.4 (H_∞ filter design, discrete-time) Consider the uncertain discrete-time system given by (2.105) and parameters α and β are also assumed to satisfy $|\alpha| \leq 1$ and $|\beta| \leq 1$. In Examples 2.2 and 3.2, the problem of robust H_2 filtering for (2.105) has been investigated. Here, we consider robust H_∞ filtering for this example.

By applying the methods in [6, 16–19], the quadratic approach Algorithm 8, and the parameter-dependent approaches, Algorithms 14 and 16 ($g = 1$), the corresponding minimum H_∞ filtering performance upper bound γ^* is presented in Table 3.5, from which, it can be seen that the parameter-dependent approach, Algorithm 16,

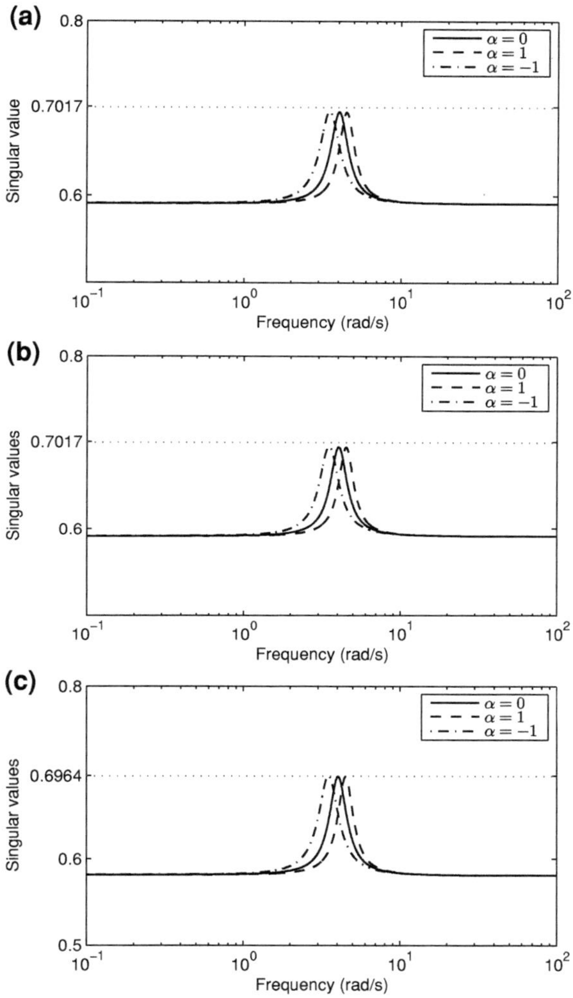

Fig. 3.3 Singular value *curves* under $\alpha = 0$, 1 and -1 in Example 3.3, **a** the filter in Eq. (3.44); **b** the filter in Eq. (3.45); **c** the filter in Eq. (3.46)

Table 3.5 Comparison between different methods in Example 3.4

Methods	γ^*
[15, 16]	21.1368
[17], Algorithm 8	19.9627
[18]	19.3411
[6] ($\lambda_1 = \lambda_2 = 0$)	19.1057
Algorithm 14	19.1057
Algorithm 16 ($g = 1$)	11.0517

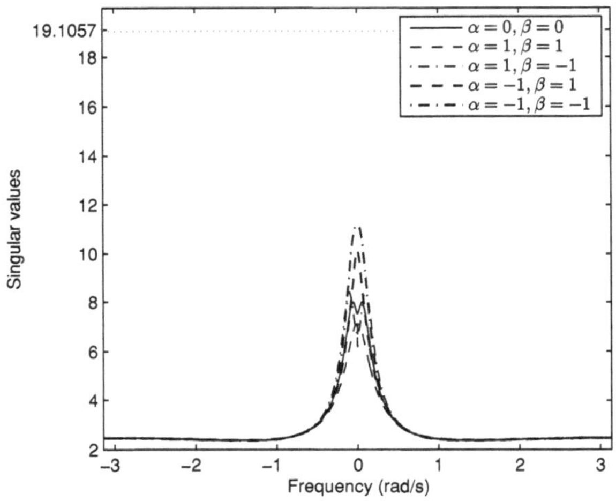

Fig. 3.4 Singular value *curves* for the filter in Eq. (3.47) in Example 3.4

produces the best guaranteed H_∞ filtering performance level. The filters designed by Algorithms 14 and 16 ($g = 1$) are respectively given by

$$\begin{bmatrix} A_F & B_F \\ C_F & D_F \end{bmatrix} = \left[\begin{array}{cc|c} 0.0887 & 0.0058 & -0.9077 \\ -0.3183 & 0.8715 & -0.3983 \\ \hline 0.5799 & -1.0618 & 1.5414 \end{array}\right], \tag{3.47}$$

$$\begin{bmatrix} A_F & B_F \\ C_F & D_F \end{bmatrix} = \left[\begin{array}{cc|c} 0.1753 & 0.0401 & -0.8007 \\ -0.7457 & 0.8054 & -0.8620 \\ \hline 0.7883 & -1.0407 & 1.7640 \end{array}\right]. \tag{3.48}$$

Connecting the filters in (3.47) and (3.48) to the original system in (2.105), we plot the singular value curves of the corresponding filtering error system, which are displayed in Figs. 3.4 and 3.5, respectively. All the singular values are shown to be smaller than the corresponding H_∞ performance upper bound, which verifies the effectiveness of the designed filters.

3.6 Summary and Notes

3.6.1 Summary

This chapter focuses on deriving parameter-dependent approaches to robust H_2 and H_∞ filter design for linear polytopic uncertain systems. The developed parameter-dependent approaches are based on the new H_2 and H_∞ performance criteria with the product terms between the Lyapunov matrices and system matrices eliminated.

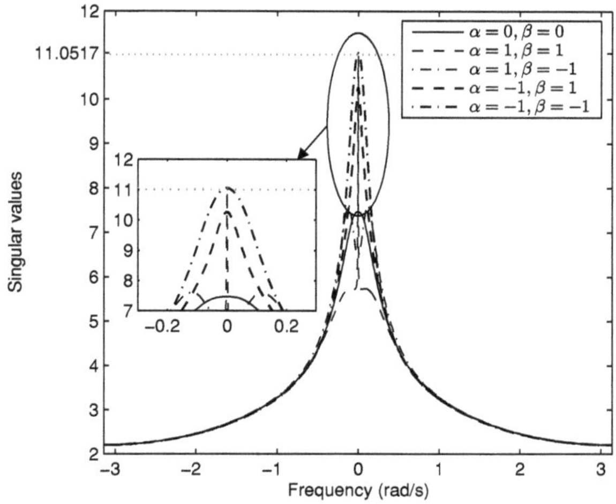

Fig. 3.5 Singular value curves for the filter in Eq. (3.48) in Example 3.4

The linearly parameter-dependent approaches are obtained by restricting the slack matrices to be independent of the uncertain parameter and specifying the Lyapunov matrices to be linearly dependent on the parameter over the entire uncertainty domain. To make full use of the parameter-dependent method, polynomially parameter-dependent approaches are also presented by choosing the Lyapunov matrices and part of slack matrices to be of the general homogeneous polynomial parameter-dependence form with a prescribed degree. Numerical examples have clearly demonstrated the effectiveness and the advantages of the presented parameter-dependent approaches.

3.6.2 Notes

Slack matrix Projection Lemma plays a fundamental role in eliminating the product terms between Lyapunov matrix and system matrices through introducing slack matrix. Oliveira et al. [19] presented new parameter-dependent conditions for robust stability of uncertain discrete-time systems, which may be the most basic and concerned result in this framework during the past years; [20–22] introduced more general slack matrices for robust stability analysis. The authors in [23] developed a variation of Projection Lemma, called Reciprocal Projection Lemma, to deal with the robust performance analysis for uncertain continuous-time systems. Other developments for robust performance analysis can be found in [24–26]. In particular, [27] provided some elaborate discussions on how to use Projection Lemma to introduce different degree of slack matrix, such that most of the existing results can be regarded in the same framework. The conditions with slack matrices for H_2 and H_∞ performances in Sect. 3.1 are also based on the method in [27].

Linearly parameter-dependent approaches to robust filtering Linearly parameter-dependent approaches are still the main method in the literature for designing robust filters in the parameter-dependent Lyapunov function framework, because of its simpleness. Following [19, 23], parameter-dependent approaches to robust H_2 filter design were proposed in [3, 10] for discrete-time systems and in [9] for continuous-time systems, respectively. Counterparts of H_∞ filtering for discrete-time systems and energy-to-peak filtering for both continuous- and discrete-time systems can be found in [18, 28]. [6, 11] developed improved robust H_2 and H_∞ filter designs through introducing more general slack matrices as in [20], where [6] studied both continuous- and discrete-time systems. Other results on continuous-time systems can be found in [14, 29] for H_∞ filtering and [4] for H_2 filtering. Notes on more complicated systems will be given in Chaps. 4–6.

Polynomially parameter-dependent approaches to robust filtering The polynomial approaches in Sect. 3.4 are based mainly on the parameter-dependent analysis method in [30]. Some related results on filtering can be found in [1, 2]. Gao et al. [2] combined the method of introducing slack matrices in [20] and the polynomially parameter-dependent approach in [30] to deal with robust H_2 filter design, while [1] introduced full slack matrices based on Projection Lemma to deal with robust H_2 and H_∞ filter design. Both continuous- and discrete-time systems were addressed in the two references. Meng et al. [31, 32] investigated robust energy-to-peak filtering for a special case where both Lyapunov matrices and slack matrices take the linearly parameter-dependent form. Notes on more complicated systems will be given in Chaps. 4–6.

References

1. Lacerda, M.J., Oliveira, R.C.L.F., Peres, P.L.D.: Robust H_2 and H_∞ filter design for uncertain linear systems via LMIs and polynomial matrices. Signal Process. **91**(5), 1115–1122 (2011)
2. Gao, H., Meng, X., Chen, T.: A new design of robust H_2 filters for uncertain systems. Syst. Control Lett. **57**(7), 585–593 (2008)
3. Shaked, U., Xie, L., Soh, Y.C.: New approaches to robust minimum variance filter design. IEEE Trans. Signal Process. **49**(11), 2620–2629 (2001)
4. Barbosa, K.A., de Souza, C.E., Trofino, A.: Robust H_2 filtering for uncertain linear systems: LMI based methods with parametric Lyapunov functions. Syst. Control Lett. **54**(3), 251–262 (2005)
5. de Souza, C.E., Trofino, A.: An LMI approach to the design of robust H_2 filters. In: Ghaoui, L.E., Niculescu, S. (eds.) Recent Advances on Linear Matrix Inequality Methods in Control. SIAM, Philadelphia (1999)
6. Duan, Z., Zhang, J.C.Z., Mosca, E.: Robust H_2 and H_∞ filtering for uncertain linear systems. Automatica **42**(11), 1919–1926 (2006)
7. Geromel, J.C.: Optimal linear filtering under parameter uncertainty. IEEE Trans. Signal Process. **47**(1), 168–175 (1999)
8. Geromel, J.C., de Oliveira, M.C.: H_2 and H_∞ robust filtering for convex bounded uncertain systems. IEEE Trans. Autom. Control **46**(1), 100–107 (2001)
9. Tuan, H.D., Apkarian, P., Nguyen, T.Q.: Robust and reduced-order filtering: new LMI-based characterizations and methods. IEEE Trans. Signal Process. **49**(12), 2975–2984 (2001)

10. Geromel, J.C., De Oliveira, M.C., Bernussou, J.: Robust filtering of discrete-time linear systems with parameter dependent Lyapunov functions. SIAM J. Control Optim. **41**(3), 700–711 (2002)
11. Xie, L., Lu, L., Zhang, H.: Improved robust H_2 and H_∞ filtering for uncertain discrete-time systerms. Automatica **40**, 873–880 (2004)
12. Jin, S.H., Park, J.B.: Robust H_∞ filter for polytopic uncertain systems via convex optimization. IEE Proc. Control Theory Appl. **148**(1), 55–59 (2001)
13. Palhares, R.M., Peres, P.L.D.: Robust H_∞-filtering design with pole placement constraint via linear matrix inequalities. J. Optim. Theory Appl. **102**(2), 239–261 (1999)
14. Wu, A., Duan, G.: Robust H-infinity estimation for continuous-time polytopic uncertain systems. J. Control Theory Appl. **4**, 393–398 (2005)
15. Geromel, J.C., Bernussou, J., Garcia, G., de Oliveira, M.C.: H_2 and H_∞ robust filtering for discrete-time linear systems. In: Proceedings of the 37th IEEE Conference on Decision and Control, pp. 632–637. Tampa, Florida USA (1998)
16. Geromel, J.C., Bernussou, J., Garcia, G., de Oliveira, M.C.: H_2 and H_∞ robust filtering for discrete-time linear systems. SIAM J. Control Optim. **38**, 1353–1368 (2000)
17. Palhares, R.M., Peres, P.L.D.: Robust H_∞ filter design with pole constraints for discrete-time systems. J. Frankl. Inst. **337**(6), 713–723 (2000)
18. Gao, H., Wang, C.: New approaches to robust l_2-l_∞ and H_∞ filtering for uncertain discrete-time systems. Sci. China (Ser. F) **46**(5):355–370 (2003)
19. de Oliveira, M.C., Bernussou, J., Geromel, J.C.: A new discrete-time robust stability condition. Syst. Control Lett. **37**, 261–265 (1999)
20. Peaucelle, D., Arzelier, D., Bachelier, J., Bernussou, J.: A new robust D-stability condition for real convex polytopic uncertainty. Syst. Control Lett. **40**, 21–30 (2000)
21. Geromel, J.C., De Oliveira, M.C., Hsu, L.: LMI characterization of structural and robust stability. Linear Algebra Its Appl. **285**, 69–80 (1998)
22. de Oliveira, M.C., Geromel, J.C., Hsu, L.: LMI characterization of structural and robust stability: the discrete-time case. Linear Algebra Its Appl. **296**, 27–38 (1999)
23. Apkarian, P., Tuan, H.D., Bernussou, J.: Continuous-time analysis, eigenstructure assignment, and H_2 synthesis with enhanced linear matrix inequalities (LMI) characterizations. IEEE Trans. Autom. Control **46**(12), 1941–1946 (2001)
24. Shaked, U.: Improved LMI representations for the analysis and the design of continuous-time systems with polytopic type uncertainty. IEEE Trans. Autom. Control **46**(4), 652–656 (2001)
25. de Oliveira, M.C., Geromel, J.C., Bernussou, J.: Extended H_2 and H_∞ norm characterizations and controller parametrizations for discrete-time systems. Int. J. Control **75**(9), 666–679 (2002)
26. Ebihara, Y., Hagiwara, T.: New dilated LMI characterizations for continuous-time multiobjective controller synthesis. Automatica **40**, 2003–2009 (2004)
27. Pipeleer, G., Demeulenaere, B., Swevers, J., Vandenberghe, L.: Extended LMI characterizations for stability and performance of linear systems. Syst. Control Lett. **58**(7), 510–518 (2009)
28. Gao, H., Wang, C.: Robust energy-to-peak filtering with improved LMI representations. IEE Proc. Vis. Image Signal Process. **150**(2), 82–89 (2003)
29. Gao, H.J., Wang, C.H.: Robust H_∞ state estimation for systems with uncertain parameters. Control Decis. **19**(2), 147–152 (2004)
30. Oliveira, R.C.L.F., Peres, P.L.D.: LMI conditions for robust stability analysis based on polynomially parameter-dependent Lyapunov functions. Syst. Control Lett. **55**(1), 52–61 (2006)
31. Meng, X., Gao, H., Mou, S.: A new parameter-dependent approach to robust energy-to-peak filter design. Circuits Syst. Signal Process. **26**(4), 451–471 (2007)
32. Zhang, H., Mehr, A.S., Shi, Y.: Improved robust energy-to-peak filtering for uncertain linear systems. Signal Process. **90**(9), 2667–2675 (2010)

Part II
Robust Filtering for Time-Delay Systems and 2-D Systems

Chapter 4
Robust Filtering for Continuous Time-Delay Systems

4.1 Introduction

As is well known, time delay commonly exists in many processes due to the after-effect phenomena in internal dynamics and/or signal transmission, which has been recognized to be an important source of instability and performance degradation. Therefore, time-delay systems have drawn much attention in the last few decades. H_∞ filtering was introduced in 1989, in which the input signal is assumed to be energy-bounded and the main objective is to minimize the H_∞ norm of the filtering error system. In recent years, an advanced research topic in this area is to investigate the problem of robust filtering via parameter-dependent Lyapunov functions, the aim of which is to reduce the overdesign in the quadratic framework. So far, there have been several LMI conditions for robust filter design based on parameter-dependent Lyapunov functions, such as [1–4] and the results in Chap. 3, most of which have been shown either theoretically or via numerical examples to be less conservative than those in the quadratic framework, thanks to the utilization of the parameter-dependent idea.

In this chapter, we apply the parameter-dependent idea to study the problem of robust H_∞ filtering for uncertain linear continuous-time systems with a time-varying delay in the state. The problem of robust H_∞ filtering has been addressed in [5, 6] for continuous time-delay systems and discrete time-delay systems, respectively. These results are developed through a delay-independent approach and thus provide admissible filters for arbitrarily large delays. To reduce the conservatism, delay-dependent filtering strategies have been reported in [7, 8]. It is worth noting that most of these filtering results on time-delay systems are obtained in the quadratic framework, and they are thus inevitably conservative. It is, therefore, our intention to further reduce the conservatism in those existing results, by employing the idea of parameter-dependence.

This chapter develops a polynomially parameter-dependent approach to solving the problem of robust H_∞ filtering for linear continuous-time systems with polytopic parameter uncertainties and time-varying delay in the state. New LMI conditions

H. Gao and X. Li, *Robust Filtering for Uncertain Systems*,
Communications and Control Engineering, DOI: 10.1007/978-3-319-05903-7_4,

are obtained for the existence of admissible filters. These conditions are developed based on homogeneously polynomially parameter-dependent matrices of arbitrary degree. The proposed approach is different from the existing results in the quadratic framework that entail fixed matrices for the entire uncertainty domain, or those in the linearly parameter-dependent framework that use linear convex combinations of matrices. Moreover, the obtained conditions are dependent on the delay information, which is achieved without performing any model transformation to the original system. The delay-dependence and polynomial parameter-dependence nature guarantees the approach presented in this chapter to be potentially less conservative, as shown via two numerical examples.

4.2 Problem Formulation

Consider a class of linear uncertain continuous time-delay systems:

$$\begin{aligned}\dot{x}(t) &= A(\lambda)x(t) + A_d(\lambda)x(t-d(t)) + B(\lambda)v(t),\\ y(t) &= C(\lambda)x(t) + C_d(\lambda)x(t-d(t)) + D(\lambda)v(t),\\ z(t) &= L(\lambda)x(t) + L_d(\lambda)x(t-d(t)) + G(\lambda)v(t),\\ x(t) &= \phi(t),\ \ \forall t \in \left[-\bar{d}, 0\right].\end{aligned} \tag{4.1}$$

Here $x(t) \in \mathbb{R}^n$ is the state vector; $y(t) \in \mathbb{R}^m$ is the measurement output; $z(t) \in \mathbb{R}^p$ is the signal to be estimated; $\phi(t)$ is the initial vector function that is continuous on the segment $\left[-\bar{d}, 0\right]$; and $v(t) \in \mathbb{R}^q$ is an exogenous noise input in $L_2[0, -\infty)$. The time delay $d(t)$ is assumed to be time-varying and satisfies $0 < d(t) \le \bar{d}$, $\dot{d}(t) \le \tau < \infty$, where $\bar{d}$ and τ are constant positive scalars. Suppose that the system matrices are uncertain but belong to a given polyhedral domain:

$$\Omega(\lambda) \triangleq \begin{pmatrix} A(\lambda), A_d(\lambda), B(\lambda),\\ C(\lambda), C_d(\lambda), D(\lambda),\\ L(\lambda), L_d(\lambda), G(\lambda)\end{pmatrix} \in \mathcal{R}, \tag{4.2}$$

where $\mathcal{R}$ denotes a polytope defined as $\mathcal{R} \triangleq \left\{\Omega(\lambda)\,\middle|\,\Omega(\lambda) = \sum_{i=1}^{s} \lambda_i\Omega_i;\ \lambda \in \Gamma\right\}$ with $\Omega_i \triangleq (A_i, A_{di}, B_i, C_i, C_{di}, D_i, L_i, L_{di}, G_i)$ denoting the vertices of the polytope and $\Gamma \triangleq \left\{(\lambda_1, \lambda_2, \ldots, \lambda_s) : \sum_{i=1}^{s} \lambda_i = 1,\ \lambda_i \ge 0\right\}$ denoting the unit simplex.

The purpose of this chapter is to design a filter of the form

$$\begin{aligned}\dot{x}_F(t) &= A_F x_F(t) + B_F y(t),\\ z_F(t) &= C_F x_F(t) + D_F y(t),\end{aligned} \tag{4.3}$$

where $x_F(t) \in \mathbb{R}^n$ is the state vector of the filter, $z_F(t)$ is the estimate of $z(t)$ and A_F, B_F, C_F and D_F are appropriately dimensioned real matrices to be designed.

Defining $\xi(t) \triangleq \left[x(t)^{\mathrm{T}}\ x_F(t)^{\mathrm{T}}\right]^{\mathrm{T}}$ and $e(t) \triangleq z(t) - z_F(t)$, the filtering error system is given by

$$\begin{aligned} \dot{\xi}(t) &= \bar{A}(\lambda)\xi(t) + \bar{A}_d(\lambda)K\xi(t - d(t)) + \bar{B}(\lambda)v(t), \\ e(t) &= \bar{C}(\lambda)\xi(t) + \bar{C}_d(\lambda)K\xi(t - d(t)) + \bar{D}(\lambda)v(t), \end{aligned} \tag{4.4}$$

where $K = \left[I\ 0\right]$ and

$$\begin{aligned} \bar{A}(\lambda) &= \begin{bmatrix} A(\lambda) & \mathbf{0} \\ B_F C(\lambda) & A_F \end{bmatrix}, \quad \bar{A}_d(\lambda) = \begin{bmatrix} A_d(\lambda) \\ B_F C_d(\lambda) \end{bmatrix}, \\ \bar{B}(\lambda) &= \begin{bmatrix} B(\lambda) \\ B_F D(\lambda) \end{bmatrix}, \quad \bar{C}_d(\lambda) = L_d(\lambda) - D_F C_d(\lambda), \\ \bar{C}(\lambda) &= \left[L(\lambda) - D_F C(\lambda)\ -C_F\right], \quad \bar{D}(\lambda) = G(\lambda) - D_F D(\lambda). \end{aligned} \tag{4.5}$$

Then, the robust H_∞ filtering problem for continuous time-delay systems to be addressed in this chapter is expressed as follows.

Robust H_∞ Filtering for Continuous Time-Delay Systems: Given the system in (4.1), design a filter of the form in (4.3) such that for all admissible uncertainty and time delay:

(i) the filtering error system in (4.4) is robustly asymptotically stable;
(ii) the filtering error system in (4.4) satisfies

$$\|e\|_{L_2}^2 < \gamma^2 \|v\|_{L_2}^2, \tag{4.6}$$

for all $0 \neq v(t) \in L_2[0, +\infty)$ under zero initial conditions, where γ is a given positive constant.

Remark 4.1 Very recently, some researchers consider the delay case where both upper and lower bounds of the delay are assumed. The results in this chapter can be further extended to this case. In addition, addressing the multiple delay case would make the notations much more complicated and overshadow the main idea; hence, the system under consideration only contains one single delay in the state. However, it is not difficult to further extend the results in the chapter to the case of multiple delays, which has been considered in [5, 7].

4.3 Filter Analysis and Design

4.3.1 Filter Analysis

We first introduce the following BRL, a delay-dependent H_∞ performance criterion for continuous time-delay systems, which shows that the H_∞ performance of the filtering error system in (4.4) can be guaranteed if there exist some matrices satisfying some parameter-dependent LMIs.

Lemma 4.1 *Consider the system in* (4.1) *and the filter in* (4.3). *For any fixed* $\lambda \in \Gamma$, *the filtering error system in* (4.4) *is asymptotically stable with a guaranteed* H_∞ *performance upper bound* γ *if there exist real matrices* $P(\lambda) = P(\lambda)^{\mathrm{T}} > 0$, $Q(\lambda) = Q(\lambda)^{\mathrm{T}} > 0$, $M(\lambda) = M(\lambda)^{\mathrm{T}} > 0$, $X(\lambda)$ *and* $Y(\lambda)$ *satisfying*

$$\begin{bmatrix} -I & \bar{C}(\lambda) & \bar{C}_d(\lambda) & \bar{D}(\lambda) & 0 & 0 \\ * & \Gamma_1(\lambda) & \Gamma_2(\lambda) & P(\lambda)\bar{B}(\lambda) & -\bar{d}K^{\mathrm{T}}X(\lambda) & \bar{d}\bar{A}(\lambda)^{\mathrm{T}}K^{\mathrm{T}}M(\lambda) \\ * & * & \Gamma_3(\lambda) & 0 & -\bar{d}Y(\lambda) & \bar{d}\bar{A}_d(\lambda)^{\mathrm{T}}K^{\mathrm{T}}M(\lambda) \\ * & * & * & -\gamma^2 I & 0 & \bar{d}\bar{B}(\lambda)^{\mathrm{T}}K^{\mathrm{T}}M(\lambda) \\ * & * & * & * & -\bar{d}M(\lambda) & 0 \\ * & * & * & * & * & -\bar{d}M(\lambda) \end{bmatrix} < 0, \quad (4.7)$$

where

$$\begin{aligned} \Gamma_1(\lambda) &\triangleq \bar{A}(\lambda)^{\mathrm{T}}P(\lambda) + P(\lambda)\bar{A}(\lambda) + K^{\mathrm{T}}\left(X(\lambda) + X(\lambda)^{\mathrm{T}} + Q(\lambda)\right)K, \\ \Gamma_2(\lambda) &\triangleq P(\lambda)\bar{A}_d(\lambda) - K^{\mathrm{T}}X(\lambda) + K^{\mathrm{T}}Y(\lambda)^{\mathrm{T}}, \\ \Gamma_3(\lambda) &\triangleq -(1-\tau)Q(\lambda) - Y(\lambda) - Y(\lambda)^{\mathrm{T}}. \end{aligned} \quad (4.8)$$

Proof Choose the following Lyapunov–Krasovskii functional (LKF) candidate:

$$V(t) \triangleq V_1(t) + V_2(t) + V_3(t)$$

where

$$\begin{aligned} V_1(t) &\triangleq \xi(t)^{\mathrm{T}} P(\lambda)\xi(t), \\ V_2(t) &\triangleq \int_{t-d(t)}^{t} \xi(s)^{\mathrm{T}} K^{\mathrm{T}}Q(\lambda)K\xi(s)\,\mathrm{d}s, \\ V_3(t) &\triangleq \int_{-\bar{d}}^{0}\int_{t+\beta}^{t} \dot{\xi}(\alpha)^{\mathrm{T}} K^{\mathrm{T}}M(\lambda)K\dot{\xi}(\alpha)\,\mathrm{d}\alpha\mathrm{d}\beta, \end{aligned}$$

and $P(\lambda) > 0$, $Q(\lambda) > 0$ and $M(\lambda) > 0$. Then, by taking the derivative of $V(\cdot)$ with respect to t along the trajectory of the filtering error system in (4.4), we have

$$\dot{V}_1(t) = \dot{\xi}(t)^{\mathrm{T}} P(\lambda)\xi(t) + \xi(t)^{\mathrm{T}} P(\lambda)\dot{\xi}(t), \quad (4.9)$$

$$\begin{aligned} \dot{V}_2(t) &= \xi(t)^{\mathrm{T}} K^{\mathrm{T}}Q(\lambda)K\xi(t) - \left(1-\dot{d}(t)\right)\xi(t-d(t))^{\mathrm{T}} \\ &\quad \times K^{\mathrm{T}}Q(\lambda)K\xi(t-d(t)) \\ &\le \xi(t)^{\mathrm{T}} K^{\mathrm{T}}Q(\lambda)K\xi(t) - (1-\tau)\,\xi(t-d(t))^{\mathrm{T}} \\ &\quad \times K^{\mathrm{T}}Q(\lambda)K\xi(t-d(t)), \end{aligned} \quad (4.10)$$

$$\begin{aligned}\dot{V}_3(t) &= \bar{d}\dot{\xi}(t)^{\mathrm{T}} K^{\mathrm{T}} M(\lambda) K \dot{\xi}(t) \\ &\quad - \int_{t-\bar{d}}^{t} \dot{\xi}(\beta)^{\mathrm{T}} K^{\mathrm{T}} M(\lambda) K \dot{\xi}(\beta)\,\mathrm{d}\beta \\ &\le \bar{d}\dot{\xi}(t)^{\mathrm{T}} K^{\mathrm{T}} M(\lambda) K \dot{\xi}(t) \\ &\quad - \int_{t-d(t)}^{t} \dot{\xi}(s)^{\mathrm{T}} K^{\mathrm{T}} M(\lambda) K \dot{\xi}(s)\,\mathrm{d}s. \end{aligned} \tag{4.11}$$

Applying the free weighting matrix idea [9–11] and the Newton-Leibniz formula, the following relation can be easily verified:

$$\begin{aligned} U &\triangleq 2\left(\xi(t)^{\mathrm{T}} K^{\mathrm{T}} X(\lambda) K + \xi(t-d(t))^{\mathrm{T}} K^{\mathrm{T}} Y(\lambda) K\right) \\ &\quad \times \left(\xi(t) - \xi(t-d(t)) - \int_{t-d(t)}^{t} \dot{\xi}(s)\,\mathrm{d}s\right) \\ &\equiv 0. \end{aligned} \tag{4.12}$$

Combining the relations in (4.9)–(4.12), we have

$$\begin{aligned}\dot{V}(t) &= \dot{V}_1(t) + \dot{V}_2(t) + \dot{V}_3(t) + U \\ &\le \frac{1}{d(t)} \int_{t-d(t)}^{t} \eta(t,s)^{\mathrm{T}} \left(\Gamma + \bar{d}\Gamma_4^{\mathrm{T}} K^{\mathrm{T}} M(\lambda) K \Gamma_4\right) \eta(t,s)\,\mathrm{d}s, \end{aligned} \tag{4.13}$$

where

$$\Gamma \triangleq \begin{bmatrix} \Gamma_1(\lambda) & \Gamma_2(\lambda) & P(\lambda)\bar{B}(\lambda) & -d(t)K^{\mathrm{T}}X(\lambda) \\ * & \Gamma_3(\lambda) & 0 & -d(t)Y(\lambda) \\ * & * & 0 & 0 \\ * & * & * & -d(t)M(\lambda) \end{bmatrix},$$

$$\Gamma_4 \triangleq \left[\, \bar{A}(\lambda)\ \bar{A}_d(\lambda)\ \bar{B}(\lambda)\ 0 \,\right],$$

$$\eta(t,s) \triangleq \left[\, \xi(t)^{\mathrm{T}}\ \xi(t-d(t))^{\mathrm{T}} K^{\mathrm{T}}\ v(t)^{\mathrm{T}}\ \dot{\xi}(s)\,K^{\mathrm{T}} \,\right]^{\mathrm{T}}$$

and $\Gamma_i(\lambda)$, $i = 1, 2, 3$, are defined in (4.7).

When $v(t) = 0$, $t \ge 0$, we obtain from (4.13) that

$$\dot{V}(t) \le \frac{1}{d(t)} \int_{t-d(t)}^{t} \tilde{\eta}(t,s)^{\mathrm{T}} \left(\tilde{\Gamma} + \bar{d}\tilde{\Gamma}_4^{\mathrm{T}} K^{\mathrm{T}} M(\lambda) K \tilde{\Gamma}_4\right) \tilde{\eta}(t,s)\,\mathrm{d}s,$$

where

$$\tilde{\Gamma} \triangleq \begin{bmatrix} \Gamma_1(\lambda) & \Gamma_2(\lambda) & -d(t)K^{\mathrm{T}}X(\lambda) \\ * & \Gamma_3(\lambda) & -d(t)Y(\lambda) \\ * & * & -d(t)M(\lambda) \end{bmatrix},$$

$$\tilde{\Gamma}_4 \triangleq \left[\bar{A}(\lambda) \; \bar{A}_d(\lambda) \; 0 \right],$$
$$\tilde{\eta}(t,s) \triangleq \left[\xi(t)^{\mathrm{T}} \; \xi(t-d(t))^{\mathrm{T}} K^{\mathrm{T}} \; \dot{\xi}(s) K^{\mathrm{T}} \right]^{\mathrm{T}}.$$

Using the Schur complement, one sees that (4.7) implies $\tilde{\Gamma} + \bar{d}\tilde{\Gamma}_4^{\mathrm{T}} K^{\mathrm{T}} M(\lambda) K \tilde{\Gamma}_4 < 0$, which further implies that $\dot{V}(t) < -\epsilon |\xi(t)|^2$, where ϵ is a positive scalar. Then from the Lyapunov stability theory, we can conclude that the filtering error system (4) is asymptotically stable, if the conditions in Lemma 4.1 are satisfied.

To establish the H_∞ performance for the filtering error system in (4.4), consider the index:

$$J \triangleq \int_0^\infty \left(e(t)^{\mathrm{T}} e(t) - \gamma^2 v(t)^{\mathrm{T}} v(t) \right) \mathrm{d}t. \tag{4.14}$$

Noting that $V(\infty) \geq 0$ and considering (4.13), we have, under zero initial conditions,

$$\begin{aligned} J &\leq \int_0^\infty \left(e(t)^{\mathrm{T}} e(t) - \gamma^2 v(t)^{\mathrm{T}} v(t) \right) dt + V(\infty) - V(0) \\ &= \int_0^\infty \left(e(t)^{\mathrm{T}} e(t) - \gamma^2 v(t)^{\mathrm{T}} v(t) + \dot{V}(t) \right) \mathrm{d}t \\ &\leq \int_0^\infty \frac{1}{d(t)} \left[\int_{t-d(t)}^t \eta(t,s)^{\mathrm{T}} \left(\hat{\Gamma} + \bar{d}\Gamma_4^{\mathrm{T}} K^{\mathrm{T}} M(\lambda) K \Gamma_4 + \Gamma_5^{\mathrm{T}} \Gamma_5 \right) \eta(t,s) \, \mathrm{d}s \right] \mathrm{d}t \end{aligned}$$

where

$$\hat{\Gamma} \triangleq \begin{bmatrix} \Gamma_1(\lambda) & \Gamma_2(\lambda) & P(\lambda)\bar{B}(\lambda) & -d(t)K^{\mathrm{T}} X(\lambda) \\ * & \Gamma_3(\lambda) & 0 & -d(t)Y(\lambda) \\ * & * & -\gamma^2 I & 0 \\ * & * & * & -d(t)M(\lambda) \end{bmatrix},$$
$$\Gamma_5 \triangleq \left[\bar{C}(\lambda) \; \bar{C}_d(\lambda) \; \bar{D}(\lambda) \; 0 \right],$$

$\Gamma_i(\lambda)$, $i = 1, 2, 3$, are defined in (4.7) and $\Gamma_4(\lambda)$ is in (4.13). Again using the Schur complement, (4.7) implies that $\hat{\Gamma} + \bar{d}\Gamma_4^{\mathrm{T}} K^{\mathrm{T}} M(\lambda) K \Gamma_4 + \Gamma_5^{\mathrm{T}} \Gamma_5 < 0$. Then for any nonzero $v \in L_2[0, +\infty)$, there holds $J < 0$, that is, $\|e\|_{L_2}^2 < \gamma^2 \|v\|_{L_2}^2$, which completes the proof. □

4.3.2 Filter Realization

In the following, we will focus on the design of robust H_∞ filters based on Lemma 4.1, that is, to determine the filter matrices (A_F, B_F, C_F, D_F), which will guarantee the filtering error system to be robustly asymptotically stable with a robust H_∞ performance upper bound. We have the following parameter-dependent LMI condition for the existence of such a filter realization.

Lemma 4.2 *Consider the system in* (4.1) *and the filter in* (4.3). *For any fixed* $\lambda \in \Gamma$, *there exist real matrices* $P(\lambda) = \begin{bmatrix} R(\lambda) & V \\ V^{\mathrm{T}} & W \end{bmatrix} = P(\lambda)^{\mathrm{T}} > 0$, $Q(\lambda) = Q(\lambda)^{\mathrm{T}} > 0$, $M(\lambda) = M(\lambda)^{\mathrm{T}} > 0$, $X(\lambda)$ *and* $Y(\lambda)$ *and filter matrices* A_F, B_F, C_F *and* D_F *satisfying* (4.7) *if and only if there exist matrices* $R(\lambda) = R(\lambda)^{\mathrm{T}} > 0$, $F = F^{\mathrm{T}} > 0$, $Q(\lambda) = Q(\lambda)^{\mathrm{T}} > 0$, $M(\lambda) = M(\lambda)^{\mathrm{T}} > 0$, $X(\lambda)$, $Y(\lambda)$, $U(\lambda)$ *and matrices* $\bar{A}_F$, $\bar{B}_F$, $\bar{C}_F$ *and* $\bar{D}_F$ *satisfying*

$$\Delta(\lambda) \triangleq \begin{bmatrix} \Upsilon(\lambda) & W(\lambda)U(\lambda) \\ * & \bar{d}M(\lambda) - U(\lambda) - U(\lambda)^{\mathrm{T}} \end{bmatrix} < 0, \tag{4.15}$$

$$\Theta(\lambda) \triangleq \begin{bmatrix} R(\lambda) & F \\ F & F \end{bmatrix} > 0, \tag{4.16}$$

where $W(\lambda) = \begin{bmatrix} 0 & A(\lambda) & 0 & A_d(\lambda) & B(\lambda) & 0 \end{bmatrix}^{\mathrm{T}}$ *and*

$$\Upsilon(\lambda) \triangleq \begin{bmatrix} -I & \Pi_1(\lambda) & -\bar{C}_F & \Pi_4(\lambda) & \Pi_8(\lambda) & 0 \\ * & \Pi_2(\lambda) & \Pi_3(\lambda) & \Pi_5(\lambda) & \Pi_9(\lambda) & -\bar{d}X(\lambda) \\ * & * & sym\{\bar{A}_F\} & \Pi_6(\lambda) & \Pi_{10}(\lambda) & 0 \\ * & * & * & \Pi_7(\lambda) & 0 & -\bar{d}Y(\lambda) \\ * & * & * & * & -\gamma^2 I & 0 \\ * & * & * & * & * & -\bar{d}M(\lambda) \end{bmatrix},$$

$$\begin{aligned}
\Pi_1(\lambda) &\triangleq L(\lambda) - \bar{D}_F C(\lambda),\\
\Pi_2(\lambda) &\triangleq sym\left\{R(\lambda)A(\lambda) + X(\lambda) + \bar{B}_F C(\lambda)\right\} + Q(\lambda),\\
\Pi_3(\lambda) &\triangleq \bar{A}_F + A(\lambda)^{\mathrm{T}} F + C(\lambda)^{\mathrm{T}} \bar{B}_F^{\mathrm{T}},\\
\Pi_4(\lambda) &\triangleq L_d(\lambda) - \bar{D}_F C_d(\lambda),\\
\Pi_5(\lambda) &\triangleq R(\lambda)A_d(\lambda) + \bar{B}_F C_{d\lambda}(\lambda) - X(\lambda) + Y(\lambda)^{\mathrm{T}},\\
\Pi_6(\lambda) &\triangleq FA_d(\lambda) + \bar{B}_F C_d(\lambda),\\
\Pi_7(\lambda) &\triangleq -(1-\tau)Q(\lambda) - sym\{Y(\lambda)\},\\
\Pi_8(\lambda) &\triangleq G(\lambda) - \bar{D}_F D(\lambda),\\
\Pi_9(\lambda) &\triangleq R(\lambda)B(\lambda) + \bar{B}_F D(\lambda),\\
\Pi_{10}(\lambda) &\triangleq FB(\lambda) + \bar{B}_F D(\lambda).
\end{aligned}$$

Furthermore, under the above conditions, the matrices for an admissible H_∞ *filter in the form of* (4.3) *are given by*

$$\begin{bmatrix} A_F & B_F \\ C_F & D_F \end{bmatrix} = \begin{bmatrix} F^{-1} & 0 \\ 0 & 0 \end{bmatrix} \begin{bmatrix} \bar{A}_F & \bar{B}_F \\ \bar{C}_F & \bar{D}_F \end{bmatrix}. \tag{4.17}$$

Proof First, it is not difficult to show that $\Delta(\lambda) < 0$ is equivalent to

$$\begin{bmatrix} \Upsilon(\lambda) & \bar{d}W(\lambda)M(\lambda) \\ * & -\bar{d}M(\lambda) \end{bmatrix} < 0. \tag{4.18}$$

Indeed, by letting $U(\lambda) = M(\lambda)$, one can recover (4.15) from (4.18). On the other hand, according to the inequality [12]

$$\bar{d}M(\lambda) - U(\lambda) - U(\lambda)^{\mathrm{T}} \geq -U(\lambda)^{\mathrm{T}} \left(\bar{d}M(\lambda)\right)^{-1} U(\lambda),$$

we can conclude from (4.15) that the following inequality holds:

$$\begin{bmatrix} \Upsilon(\lambda) & W(\lambda)U(\lambda) \\ * & -U(\lambda)^{\mathrm{T}} \left(\bar{d}M(\lambda)\right)^{-1} U(\lambda) \end{bmatrix} < 0,$$

which is equivalent to (4.18) by performing a congruence transformation to it by $\mathrm{diag}\left\{I, U(\lambda)^{-1}\left(\bar{d}M(\lambda)\right)\right\}$.

Necessity: Suppose there exist matrices $P(\lambda) = \begin{bmatrix} R(\lambda) & V \\ V^{\mathrm{T}} & W \end{bmatrix} > 0$, $Q(\lambda) > 0$ and $M(\lambda) > 0$ and filter matrices A_F, B_F, C_F and D_F satisfying (4.7). Without loss of generality, we assume that V and W are invertible. Define matrix

$$J \triangleq \begin{bmatrix} I & 0 \\ 0 & W^{-1}V^{\mathrm{T}} \end{bmatrix}. \tag{4.19}$$

Applying a congruence transformation to (4.7) by $\mathrm{diag}\{I, J, I, I, I, I\}$ and taking into account (4.5), we obtain

$$\begin{bmatrix} -I & \Psi_1(\lambda) & \Psi_3(\lambda) & \Psi_6(\lambda) & \Psi_{10}(\lambda) & 0 & 0 \\ * & \Psi_2(\lambda) & \Psi_4(\lambda) & \Psi_7(\lambda) & \Psi_{11}(\lambda) & -\bar{d}X(\lambda) & \bar{d}A(\lambda)^{\mathrm{T}}M(\lambda) \\ * & * & \Psi_5(\lambda) & \Psi_8(\lambda) & \Psi_{12}(\lambda) & 0 & 0 \\ * & * & * & \Psi_9(\lambda) & 0 & -\bar{d}Y(\lambda) & \bar{d}A_d(\lambda)^{\mathrm{T}}M(\lambda) \\ * & * & * & * & -\gamma^2 I & 0 & \bar{d}B(\lambda)^{\mathrm{T}}M(\lambda) \\ * & * & * & * & * & -\bar{d}M(\lambda) & 0 \\ * & * & * & * & * & * & -\bar{d}M(\lambda) \end{bmatrix} < 0, \tag{4.20}$$

where

$$\begin{aligned} \Psi_1(\lambda) &\triangleq L(\lambda) - D_F C(\lambda), \\ \Psi_2(\lambda) &\triangleq \mathrm{sym}\{R(\lambda)A(\lambda) + X(\lambda)\} + VB_F C(\lambda) \\ &\quad + C(\lambda)^{\mathrm{T}} B_F^{\mathrm{T}} V^{\mathrm{T}} + Q(\lambda), \\ \Psi_3(\lambda) &\triangleq -C_F W^{-1} V^{\mathrm{T}}, \\ \Psi_4(\lambda) &\triangleq VA_F W^{-1} V^{\mathrm{T}} + A(\lambda)^{\mathrm{T}} V W^{-1} V^{\mathrm{T}} + C(\lambda)^{\mathrm{T}} B_F^{\mathrm{T}} V^{\mathrm{T}}, \\ \Psi_5(\lambda) &\triangleq \mathrm{sym}\{V W^{-1} A_F^{\mathrm{T}} V^{\mathrm{T}}\}, \end{aligned}$$

$$\Psi_6(\lambda) \triangleq L_d(\lambda) - D_F C_d(\lambda),$$
$$\Psi_7(\lambda) \triangleq R(\lambda)A_d(\lambda) + V B_F C_d(\lambda) - X(\lambda) + Y(\lambda)^{\mathrm{T}},$$
$$\Psi_8(\lambda) \triangleq V W^{-1} V^{\mathrm{T}} A_d(\lambda) + V B_F C_d(\lambda),$$
$$\Psi_9(\lambda) \triangleq -(1-\tau)\, Q(\lambda) - \left(Y(\lambda) + Y(\lambda)^{\mathrm{T}}\right),$$
$$\Psi_{11}(\lambda) \triangleq R(\lambda)B(\lambda) + V B_F D(\lambda),$$
$$\Psi_{10}(\lambda) \triangleq G(\lambda) - D_F D(\lambda),$$
$$\Psi_{12}(\lambda) \triangleq V W^{-1} V^{\mathrm{T}} B(\lambda) + V B_F D(\lambda).$$

Define

$$F \triangleq V W^{-1} V^{\mathrm{T}},$$
$$\begin{bmatrix} \bar{A}_F & \bar{B}_F \\ \bar{C}_F & \bar{D}_F \end{bmatrix} \triangleq \begin{bmatrix} V & 0 \\ 0 & I \end{bmatrix} \begin{bmatrix} A_F & B_F \\ C_F & D_F \end{bmatrix} \begin{bmatrix} W^{-1} V^{\mathrm{T}} & 0 \\ 0 & I \end{bmatrix}. \tag{4.21}$$

Substituting the above matrices into (4.20), we obtain (4.18) and thus (4.15). In addition, noting that $P(\lambda) > 0$ and $F = V W^{-1} V^{\mathrm{T}}$, we have $\Theta(\lambda) = J^{\mathrm{T}} P(\lambda) J > 0$, which is exactly (4.16). The proof of necessity is completed.

Sufficiency: Suppose there exist matrices $R(\lambda) > 0$, $F > 0$, $Q(\lambda) > 0$, $M(\lambda) > 0$, $X(\lambda)$, $Y(\lambda)$ and $U(\lambda)$ and matrices $\bar{A}_F$, $\bar{B}_F$, $\bar{C}_F$ and $\bar{D}_F$ satisfying (4.15) and (4.16). Note that (4.15) is equivalent to (4.18). Since $F > 0$, we can always find square and nonsingular matrices V and W satisfying $F = V W^{-1} V^{\mathrm{T}}$. Now introduce matrices $P(\lambda) = \begin{bmatrix} R(\lambda) & V \\ V^{\mathrm{T}} & W \end{bmatrix}$, J as in (4.19) and

$$\begin{bmatrix} A_F & B_F \\ C_F & D_F \end{bmatrix} \triangleq \begin{bmatrix} V^{-1} & 0 \\ 0 & I \end{bmatrix} \begin{bmatrix} \bar{A}_F & \bar{B}_F \\ \bar{C}_F & \bar{D}_F \end{bmatrix} \begin{bmatrix} V^{-T} W & 0 \\ 0 & I \end{bmatrix}. \tag{4.22}$$

By Schur complement, (4.16) guarantees $R(\lambda) - F > 0$. Then, from $W > 0$ and $R(\lambda) - V W^{-1} V^{\mathrm{T}} = R(\lambda) - F > 0$, we have that matrix $P(\lambda)$ defined above is positive definite.

By some algebraic matrix manipulations, (4.18) is equivalent to

$$\mathrm{diag}\left\{I, J^{\mathrm{T}}, I, I, I, I\right\} \begin{bmatrix} \Upsilon(\lambda) & \bar{d} W(\lambda) M(\lambda) \\ * & -\bar{d} M(\lambda) \end{bmatrix} \mathrm{diag}\{I, J, I, I, I, I\} < 0. \tag{4.23}$$

Applying a congruence transformation to (4.23) by $\mathrm{diag}\left\{I, J^{-1}, I, I, I, I\right\}$, we obtain (4.7). The proof of sufficiency is completed.

Let us denote the transfer function of filter (4.3) from $y(t)$ to $z_F(t)$ by $T_{z_F y}(s) = C_F (sI - A_F)^{-1} B_F + D_F$. Substituting the filter matrices in (4.22) and considering $F = V W^{-1} V^{\mathrm{T}}$ together with some transformations, we obtain

$$T_{z_F y}(s) = \bar{C}_F \left(sI - F^{-1}\bar{A}_F\right)^{-1} F^{-1}\bar{B}_F + \bar{D}_F.$$

Therefore, an admissible filter in (4.4) can be obtained from (4.17). The proof is completed. □

Lemma 4.2 provides a method for designing H_∞ filters for fixed λ, which casts the NLMI in Lemma 4.1 into a set of LMIs. It is noted that the conditions in Lemma 4.2 are dependent on the parameter λ, therefore it cannot be used to design robust H_∞ filters due to the infinite dimension nature of the parameter λ. In the sequel, based on Lemma 4.2, we focus on the design of robust H_∞ filters via a polynomially parameter-dependent approach. Also, note that the existence condition for filters, Lemma 4.2, is also delay-dependent.

4.3.3 Polynomially Parameter-Dependent Approach to Filter Design

Define the notations $K(s,g)$, $K_j(s,g)$, $J(s,g)$, $I(s,g)$, $K_j^i(s,g)$ and $\beta_j^i(s,g+1)$ as in Sect. 3.4.1. Now let the λ-dependent matrices in Lemma 4.2 take the homogeneous polynomial form of arbitrary degree g and depend polynomially on the uncertain parameters λ_i, $i = 1, 2, \ldots, s$. That is,

$$\begin{aligned}
R_g(\lambda) &= \sum_{j=1}^{J(s,g)} \lambda_1^{k_1}\lambda_2^{k_2}\cdots\lambda_s^{k_s} R_{K_j(s,g)},\\
Q_g(\lambda) &= \sum_{j=1}^{J(s,g)} \lambda_1^{k_1}\lambda_2^{k_2}\cdots\lambda_s^{k_s} Q_{K_j(s,g)},\\
M_g(\lambda) &= \sum_{j=1}^{J(s,g)} \lambda_1^{k_1}\lambda_2^{k_2}\cdots\lambda_s^{k_s} M_{K_j(s,g)},\\
X_g(\lambda) &= \sum_{j=1}^{J(s,g)} \lambda_1^{k_1}\lambda_2^{k_2}\cdots\lambda_s^{k_s} X_{K_j(s,g)},\\
Y_g(\lambda) &= \sum_{j=1}^{J(s,g)} \lambda_1^{k_1}\lambda_2^{k_2}\cdots\lambda_s^{k_s} Y_{K_j(s,g)},\\
U_g(\lambda) &= \sum_{j=1}^{J(s,g)} \lambda_1^{k_1}\lambda_2^{k_2}\cdots\lambda_s^{k_s} U_{K_j(s,g)},\\
K_j(s,g) &= (k_1, k_2, \ldots, k_s) \in K(s,g).
\end{aligned} \tag{4.24}$$

For more detailed explanation upon these notations, please refer to Sect. 3.4.1. With these notations, we provide the following polynomially parameter-dependent approach to robust H_∞ filter design for continuous time-delay systems, which is also a delay-dependent design approach.

Theorem 4.1 *Consider the system in* (4.1). *If there exist real matrices* $F = F^{\mathrm{T}} > 0$, $R_{K_j(s,g)} = R^{\mathrm{T}}_{K_j(s,g)} > 0$, $Q_{K_j(s,g)} = Q^{\mathrm{T}}_{K_j(s,g)} > 0$, $M_{K_j(s,g)} = M^{\mathrm{T}}_{K_j(s,g)} > 0$, $X_{K_j(s,g)}$, $Y_{K_j(s,g)}$ *and* $U_{K_j(s,g)}$, $K_j(s,g) \in K(s,g)$, $j = 1, \ldots, J(s,g)$, *and matrices* $\bar{A}_F$, $\bar{B}_F$, $\bar{C}_F$ *and* $\bar{D}_F$ *such that the following LMIs hold,*

$$\Lambda_j \triangleq \sum_{i \in I_j(s,g+1)} \begin{bmatrix} \Lambda_1 & \Lambda_2 & \Lambda_4 & \Lambda_7 & \Lambda_{11} & 0 & 0 \\ * & \Lambda_3 & \Lambda_5 & \Lambda_8 & \Lambda_{12} & \Lambda_{15} & \Lambda_{18} \\ * & * & \Lambda_6 & \Lambda_9 & \Lambda_{13} & 0 & 0 \\ * & * & * & \Lambda_{10} & 0 & \Lambda_{16} & \Lambda_{19} \\ * & * & * & * & \Lambda_{14} & 0 & \Lambda_{20} \\ * & * & * & * & * & \Lambda_{17} & 0 \\ * & * & * & * & * & * & \Lambda_{21} \end{bmatrix} < 0, \tag{4.25}$$

$$\Xi_j \triangleq \sum_{i \in I_j(s,g+1)} \begin{bmatrix} R_{K^i_j(s,g+1)} & \beta^i_j(s,g+1)F \\ \beta^i_j(s,g+1)F & \beta^i_j(s,g+1)F \end{bmatrix} > 0, \tag{4.26}$$

for all $K_j(s,g+1) \in K(s,g+1)$, $j = 1, \ldots, J(s,g+1)$, *where*

$$\begin{aligned}
\Lambda_1 &\triangleq -\beta^i_j(s,g+1)I, \\
\Lambda_2 &\triangleq \beta^i_j(s,g+1)L_i - \beta^i_j(s,g+1)\bar{D}_F C_i, \\
\Lambda_3 &\triangleq \mathrm{sym}\left\{R_{K^i_j(s,g+1)}A_i + \beta^i_j(s,g+1)\bar{B}_F C_i + X_{K^i_j(s,g+1)}\right\} \\
&\quad + Q_{K^i_j(s,g+1)}, \\
\Lambda_4 &\triangleq -\beta^i_j(s,g+1)\bar{C}_F, \\
\Lambda_5 &\triangleq \beta^i_j(s,g+1)\left(\bar{A}_F F + A_i^{\mathrm{T}} F + C_i^{\mathrm{T}} \bar{B}_F^{\mathrm{T}}\right), \\
\Lambda_6 &\triangleq \beta^i_j(s,g+1)\mathrm{sym}\left\{\bar{A}_F\right\}, \\
\Lambda_7 &\triangleq \beta^i_j(s,g+1)\left(L_{di} - \bar{D}_F C_{di}\right), \\
\Lambda_8 &\triangleq R_{K^i_j(s,g+1)}A_{di} + \beta^i_j(s,g+1)\bar{B}_F C_{di} - X_{K^i_j(s,g+1)} \\
&\quad + Y^{\mathrm{T}}_{K^i_j(s,g+1)}, \\
\Lambda_9 &\triangleq \beta^i_j(s,g+1)\left(F A_{di} + \bar{B}_F C_{di}\right), \\
\Lambda_{10} &\triangleq -(1-\tau)Q_{K^i_l(s,g+1)} - \mathrm{sym}\left\{Y_{K^i_j(s,g+1)}\right\}, \\
\Lambda_{11} &\triangleq \beta^i_j(s,g+1)\left(G_i - \bar{D}_F D_i\right), \\
\Lambda_{12} &\triangleq R_{K^i_j(s,g+1)}B_i + \beta^i_j(s,g+1)\bar{B}_F D_i,
\end{aligned}$$

$$
\begin{aligned}
\Lambda_{13} &\triangleq \beta_j^i(s, g+1)\left(FB_i + \bar{B}_F D_i\right),\\
\Lambda_{14} &\triangleq -\beta_j^i(s, g+1)\gamma^2 I,\\
\Lambda_{15} &\triangleq -\bar{d}X_{K_j^i(s,g+1)},\\
\Lambda_{16} &\triangleq -\bar{d}Y_{K_j^i(s,g+1)},\\
\Lambda_{17} &\triangleq -\bar{d}M_{K_j^i(s,g+1)},\\
\Lambda_{18} &\triangleq A_i^{\mathrm{T}} U_{K_l^i(s,g+1)},\\
\Lambda_{19} &\triangleq A_{di}^{\mathrm{T}} U_{K_l^i(s,g+1)},\\
\Lambda_{20} &\triangleq B_i^{\mathrm{T}} U_{K_l^i(s,g+1)},\\
\Lambda_{21} &\triangleq \bar{d}M_{K_j^i(s,g+1)} - \mathrm{sym}\left\{U_{K_j^i(s,g+1)}\right\},
\end{aligned}
$$

then the homogeneous polynomial matrices given by (4.24) *assure* $\Delta(\lambda) < 0$ *and* $\Theta(\lambda) < 0$ *for all* $\lambda \in \Gamma$.

Moreover, if the LMIs of (4.25) *and* (4.26) *are fulfilled for a given degree* $\hat{g}$, *then the LMIs corresponding to any degree* $g > \hat{g}$ *are also satisfied and the matrices for an admissible robust* H_∞ *filter in the form of* (4.3) *are given by* (4.17).

Proof Since $R_{K_j(s,g)} > 0$, $Q_{K_j(s,g)} > 0$ and $M_{K_j(s,g)} > 0$, $K_j(s,g) \in K(s,g)$, $j = 1, \ldots, J(s,g)$, we know that $R_g(\lambda)$, $Q_g(\lambda)$ and $M_g(\lambda)$ defined in (4.24) are all positive definite for all $\lambda \in \Gamma$. Now, note that $\Delta(\lambda)$ in (4.15), and $\Theta(\lambda)$ in (4.16) for $(A(\lambda), A_d(\lambda), B(\lambda), C(\lambda), C_d(\lambda), D(\lambda), L(\lambda), L_d(\lambda), G(\lambda)) \in \mathcal{R}$ and $R_g(\lambda)$, $Q_g(\lambda)$, $M_g(\lambda)$, $X_g(\lambda)$, $Y_g(\lambda)$ and $U_g(\lambda)$ given by (4.24) are homogeneous polynomials of degree $g+1$ that can be written as

$$
\begin{aligned}
\Delta(\lambda) &= \sum_{j=1}^{J(g+1)} \lambda_1^{k_1}\lambda_2^{k_2}\cdots\lambda_s^{k_s}\Lambda_j,\\
\Theta(\lambda) &= \sum_{j=1}^{J(g+1)} \lambda_1^{k_1}\lambda_2^{k_2}\cdots\lambda_s^{k_s}\Xi_j,\\
k_1 k_2 \cdots k_s &= K_j(s, g+1).
\end{aligned}
$$

Conditions (4.25) and (4.26) imposed for all j, $j = 1, 2, \ldots, J(s, g+1)$ assure that $\Delta(\lambda) < 0$ and $\Theta(\lambda) < 0$ for all $\lambda \in \Gamma$, and thus the first part is proven.

Suppose that LMIs in (4.25) and (4.26) are fulfilled for a certain degree $\hat{g}$, that is, there exist $J(s,\hat{g})$ symmetric positive definite matrices $R_{K_j(s,\hat{g})}$, $Q_{K_j(s,\hat{g})}$, $M_{K_j(s,\hat{g})}$ and matrices $X_{K_j(s,\hat{g})}$, $Y_{K_j(s,\hat{g})}$, $U_{K_j(s,\hat{g})}$, $j = 1, 2, \ldots, J(s,\hat{g})$ such that $R_{\hat{g}}(\lambda)$, $Q_{\hat{g}}(\lambda)$, $M_{\hat{g}}(\lambda)$, $X_{\hat{g}}(\lambda)$, $Y_{\hat{g}}(\lambda)$ and $U_{\hat{g}}(\lambda)$ are homogeneously polynomially parameter-dependent Lyapunov matrices assuring $\Delta(\lambda) < 0$ and $\Theta(\lambda) < 0$. Then, the terms of the polynomial matrices

Algorithm 17 Robust H_∞ Filter Design for Continuous Time-Delay Systems

- Prescribe an integer $g \in \mathbb{N}_0^+$;
- Solve the minimization problem:

$$\min_{\substack{R_{K_j(s,g)} = R^{\mathrm{T}}_{K_j(s,g)},\\ Q_{K_j(s,g)} = Q^{\mathrm{T}}_{K_j(s,g)} > 0,\\ M_{K_j(s,g)} = M^{\mathrm{T}}_{K_j(s,g)} > 0,\\ X_{K_j(s,g)},\ Y_{K_j(s,g)},\ U_{K_j(s,g)},\\ K_j(s,g) \in K(s,g),\ j=1,2,\ldots,J(s,g),\\ F = F^{\mathrm{T}},\ \bar{A}_F,\ \bar{B}_F,\ \bar{C}_F,\ \bar{D}_F,\ \mu}} \mu = \gamma^2 \quad \text{s.t. (4.25) and (4.26)}$$

- Compute the filter matrices from (4.17).

$$\begin{aligned}
R_{\hat{g}+1}(\lambda) &= (\lambda_1 + \cdots + \lambda_s) R_{\hat{g}}(\lambda),\\
Q_{\hat{g}+1}(\lambda) &= (\lambda_1 + \cdots + \lambda_s) Q_{\hat{g}}(\lambda),\\
M_{\hat{g}+1}(\lambda) &= (\lambda_1 + \cdots + \lambda_s) M_{\hat{g}}(\lambda),\\
X_{\hat{g}+1}(\lambda) &= (\lambda_1 + \cdots + \lambda_s) X_{\hat{g}}(\lambda),\\
Y_{\hat{g}+1}(\lambda) &= (\lambda_1 + \cdots + \lambda_s) Y_{\hat{g}}(\lambda),\\
U_{\hat{g}+1}(\lambda) &= (\lambda_1 + \cdots + \lambda_s) U_{\hat{g}}(\lambda),
\end{aligned}$$

also satisfy the LMIs in (4.25) and (4.26) corresponding to the degree $\hat{g}+1$, which can be obtained by the linear combination of these LMIs for $\hat{g}$. □

Based on Theorem 4.1, an algorithm for designing suboptimal robust H_∞ filters for uncertain continuous time-delay systems can be obtained in Algorithm 17.

4.4 Illustrative Examples

In this section, we present two numerical examples to demonstrate the effectiveness and merits of the robust H_∞ filter design method presented in this chapter.

Example 4.1 Consider the system given in (4.1) where

$$\begin{aligned}
A_1 &= \begin{bmatrix} -0.05 & 0.97 \\ -1.00 & 0.01 \end{bmatrix},\ A_2 = \begin{bmatrix} -1.68 & -1.44 \\ 0.94 & 0.22 \end{bmatrix},\\
A_{di} &= \begin{bmatrix} -1 & -1 \\ -1 & -1 \end{bmatrix},\ B_i = \begin{bmatrix} 1 \\ 0 \end{bmatrix},\ C_{di} = \begin{bmatrix} 1 & 2 \end{bmatrix},\\
C_i &= \begin{bmatrix} 0 & 1 \end{bmatrix},\ L_i = \begin{bmatrix} 2 & 1 \end{bmatrix},\ D_i = G_i = 0,\ i = 1, 2.
\end{aligned}$$

Table 4.1 The minimum γ obtained for $\bar{d} = 0.44$ by Algorithm 17 in Example 4.1

g	γ					
	$\tau = 0$	$\tau = 0.2$	$\tau = 0.4$	$\tau = 0.6$	$\tau = 0.8$	$\tau = 1.0$
1	6.1566	6.4461	6.7444	7.1525	7.9190	8.7940
2	4.3153	4.3474	4.3678	4.3964	4.4321	4.4731
3	3.6242	3.6644	3.6900	3.7095	3.7338	3.7588
4	3.3617	3.3905	3.4118	3.4262	3.4437	3.4576

Table 4.2 The minimum γ obtained for $\bar{d} = 0.44$ by different methods in Example 4.2

Method	γ					
	$\tau = 0$	$\tau = 0.2$	$\tau = 0.4$	$\tau = 0.6$	$\tau = 0.8$	$\tau = 1.0$
[13]	1.5107	1.6331	1.7819	2.0104	2.4557	4.2334
Algorithm 17 ($g = 1$)	1.2463	1.3391	1.4678	1.6567	1.9913	3.3241

It is assumed that the filter in (4.3) is strictly proper, that is, $D_F = 0$. For this example, the existing results on robust filtering for time-delay systems in [7, 13] and the method presented in this chapter, Algorithm 17, for $g = 0$ fail to find feasible solutions, even for a very small value of delay bound $\bar{d}$. However, when $\bar{d} = 0.44$ and $\tau = 0.4$, by applying Algorithm 17 in this chapter, we can find feasible solutions with a guaranteed H_∞ performance bound $\gamma = 6.7444$ for $g = 1$, and $\gamma = 4.3474$ for $g = 2$, respectively. Table 4.1 gives more results on the minimum H_∞ filtering performance upper bound γ achieved by Algorithm 17 for different polynomial degree g and time derivative bound τ. From Table 4.1, we can see that the guaranteed H_∞ filtering performance bound decreases with g increasing.

In addition, we notice, by calculation, that when $\bar{d} = 0.4894$ and $\tau = 1$, Algorithm 17 for $g = 1$ fails to generate a feasible solution, but Algorithm 17 for $g = 2$ reaches feasibility with a guaranteed performance upper bound $\gamma = 8.2231$.

This example illustrates the fact that if there exist solutions of degree $g > 1$, then there may exist a solution (possibly with larger values of γ) for $g = 1$. But it is not always the case, which has been illustrated in the above. For some cases, an affine parameter-dependent LKF cannot be used to find an admissible robust H_∞ filter, but a high-degree parameter-dependent LKF might be able to provide feasible solutions.

Example 4.2 Consider another example for the system given in (4.1) where

$$
\begin{aligned}
A_1 &= \begin{bmatrix} -2 & 0 \\ 0 & -0.7+\alpha \end{bmatrix}, \; A_d = \begin{bmatrix} -1 & -1+\beta \\ -1 & -1 \end{bmatrix}, \; B = \begin{bmatrix} -0.5 \\ 2 \end{bmatrix}, \\
C &= \begin{bmatrix} 0 & 1 \end{bmatrix}, \; C_d = \begin{bmatrix} 1 & 2 \end{bmatrix}, \; D = 1, \\
L &= \begin{bmatrix} 2 & 1 \end{bmatrix}, \; L_d = \begin{bmatrix} 0 & 0 \end{bmatrix}, \; G = 0,
\end{aligned}
$$

where the uncertain parameters (α, β) satisfy $|\alpha| \le 0.2$ and $|\beta| \le 0.5$.

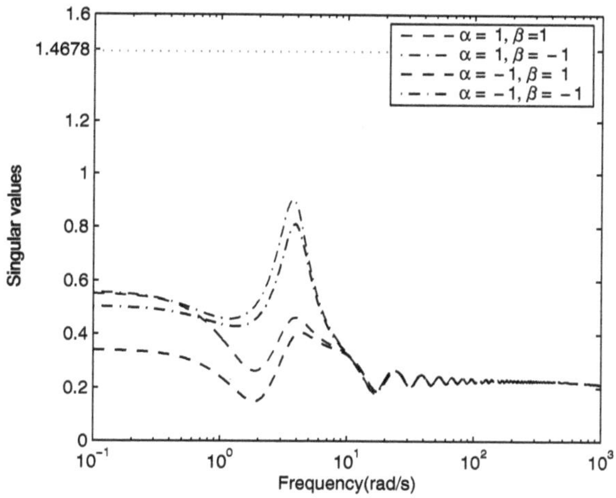

Fig. 4.1 Singular value curves under different (α, β) with the filter in Eq. (4.27) in Example 4.2

For this example when $\bar{d} = 0.44$ and $\tau = 0.4$, the minimum H_∞ performance upper bound obtained by the method in [13] is $\gamma = 1.7819$. By applying Algorithm 17 ($g = 1$) in this chapter, we can find the minimum H_∞ performance bound $\gamma = 1.4678$, which is much smaller than that for [13]. More detailed results on the obtained γ for different τ when $\bar{d} = 0.44$ are summarized in Table 4.2, which clearly shows that Algorithm 17 can yield less conservative results than the method in [13].

When $\bar{d} = 0.44$ and $\tau = 0.4$, the filter of the strictly proper form in (4.3) obtained by Algorithm 17 ($g = 1$) is given by

$$\begin{bmatrix} A_F & B_F \\ C_F & D_F \end{bmatrix} = \left[\begin{array}{cc|c} -441.5 & -530.7 & -97.569 \\ -1812.5 & -2191.8 & -404.38 \\ \hline -2.0000 & -1.0000 & 0 \end{array}\right]. \tag{4.27}$$

The singular value curves of the filtering error system for different (α, β) under the filter in (4.27) are depicted in Fig. 4.1. Obviously, all the singular values at the four vertices of the uncertainty polytope are below the guaranteed H_∞ performance level $\gamma = 1.4678$, which demonstrates the effectiveness of the designed filter.

4.5 Summary and Notes

4.5.1 Summary

This chapter has addressed the problem of robust H_∞ filtering for uncertain linear continuous-time systems with parameter uncertainty residing in a polytope and time-varying delay in the state. Different from those results in the quadratic framework that entails fixed matrices for the entire uncertainty domain, or the linearly parameter-dependent framework that uses linear convex combinations of the uncertain parameters, the method in this chapter utilizes a polynomially parameter-dependent idea to solve the robust H_∞ filtering problem for continuous time-delay systems. The result presented in this chapter is both parameter-dependent and delay-dependent, which aims at reducing the design conservatism. Two numerical examples have shown that the filter design method presented in this chapter is much less conservative than some existing ones.

4.5.2 Notes

[5, 7, 13–17] have investigated the problem of robust H_∞ filtering for uncertain continuous-time systems with time-varying delay. Except [5] that proposes a delay-independent approach to robust filter design, other references all were concerned with delay-dependent design methods so as to reduce conservatism of filter design methods. To handle polytopic uncertainty, [5, 7, 13–15, 17] made use of parameter-independent LKFs to analyze the robust performance of the error system, the notion of which is similar to the quadratic stability. Different from these quadratic results, [16] utilized Projection Lemma to eliminate the product terms between Lyapunov matrices and system matrices, and proposed a parameter-dependent design approach that employs linearly parameter-dependent LKF and slack matrices.

Robust filtering for uncertain continuous-time delayed systems also has been investigated under the energy-to-peak performance [4, 7, 18]. Gao and Wang [7] proposed a parameter-independent design method while [4, 18] both considered linearly parameter-dependent approaches. All these results are delay-dependent.

References

1. Gonçalves, E.N., Palhares, R.M., Takahashi, R.H.C.: H_2/H_∞ filter design for systems with polytopic-bounded uncertainty. IEEE Trans. Signal Process. **54**(9), 3620–3626 (2006)
2. Gao, H., Lam, J., Shi, P., Wang, C.: Parameter-dependent filter design with guaranteed H_∞ performance. IEE Proc. Control Theory Appl. **152**(5), 531–537 (2005)
3. Sato, M.: Filter design for LPV systems using quadratically parameter-dependent Lyapunov functions. Automatica **42**(11), 2017–2023 (2006)
4. Zhang, W.A., Yu, L., Jiang, X.: Delay-dependent generalized H_2 filtering for uncertain systems with multiple time-varying state delays. Signal Process. **87**(4), 709–724 (2007)
5. de Souza, C.E., Palhares, R.M., Peres, P.L.D.: Robust H_∞ filter design for uncertain linear systems with multiple time-varying state delays. IEEE Trans. Signal Process. **49**(3), 569–576 (2001)
6. Palhares, R.M., de Souza, C.E., Peres, P.L.D.: Robust H_∞ filtering for uncertain discrete-time state-delayed systems. IEEE Trans. Signal Process. **49**(8), 1696–1703 (2001)
7. Gao, H., Wang, C.: Delay-dependent robust H_∞ and L_2-L_∞ filtering for a class of uncertain nonlinear time-delay systems. IEEE Trans. Autom. Control **48**(9), 1661–1666 (2003)

8. Gao, H., Wang, C.: A delay-dependent approach to robust H_∞ filtering for uncertain discrete-time state-delayed systems. IEEE Trans. Signal Process. **52**(6), 1631–1640 (2004)
9. He, Y., Wang, Q., Lin, Q., Wu, M.: Delay-range-dependent stability for systems with time-varying delay. Automatica **43**, 371–376 (2007)
10. He, Y., Wu, M., She, J.H.: G.-P., L.: Parameter-dependent Lyapunov functional for stability of time-delay systems with polytopic-type uncertainties. IEEE Trans. Autom. Control **49**(5), 828–832 (2004)
11. Xu, S., Lam, J.: Improved delay-dependent stability criteria for time-delay systems. IEEE Trans. Autom. Control **50**(3), 384–387 (2005)
12. de Oliveira, M.C., Bernussou, J., Geromel, J.C.: A new discrete-time robust stability condition. Syst. Control Lett. **37**, 261–265 (1999)
13. Zhang, X.M., Han, Q.L.: Robust H_∞ filtering for a class of uncertain linear systems with time-varying delay. Automatica **44**(1), 157–166 (2008)
14. Fridman, E., Shaked, U., Xie, L.: Robust H_∞ filtering of linear systems with time-varying delay. IEEE Trans. Autom. Control **48**(1), 159–165 (2003)
15. He, Y., Liu, G.P., Rees, D., Wu, M.: Improved H_∞ filtering for systems with a time-varying delay. Circuits, Syst. Signal Process. **29**(3), 377–389 (2010)
16. Qiu, J., Feng, G., Yang, J.: A new design of delay-dependent robust H_∞ filtering for continuous-time polytopic systems with time-varying delay. Int. J. Robust Nonlinear Control **20**(3), 346–365 (2010)
17. Dong, X., Qiu, J., Ma, Y., Gao, H.: A new approach to H_∞ filter design for systems with time-varying state delay. IEEE Trans. Circuits Syst. II Express Briefs **59**(11), 825–839 (2012)
18. Li, X., Gao, H.: A delay-dependent approach to robust generalized H_2 filtering for uncertain continuous-time systems with interval delay. Signal Process. **91**(10), 2371–2378 (2011)

Chapter 5
Robust Filtering for Discrete Time-Delay Systems

In Chap. 4, H_∞ filtering of time-delay systems for the continuous-time case has been considered. In this chapter, we turn our attention to the discrete-time case. The problem of robust H_∞ filtering for uncertain discrete time-delay systems will be investigated through a polynomially parameter-dependent approach. A new BRL is first obtained by using advanced techniques for achieving delay dependence. Instead of directly extending the performance condition to polytopic uncertain systems, we further modify the obtained criterion by introducing slack matrix variables to eliminate the products between the positive definite matrices and system matrices. Then, by applying the polynomially parameter-dependent method, a new filter design procedure is proposed, which formulates the existence of admissible robust H_∞ filters into a set of LMIs. These conditions are developed based on homogeneously polynomially parameter-dependent matrices of an arbitrary degree. As the degree grows, less conservative filter designs can be obtained. Based on these conditions, the robust H_∞ filter designs are cast into convex optimizations. If these conditions are satisfied, a desired robust H_∞ filter can be readily constructed. Several numerical examples are given to illustrate the effectiveness and advantages of the proposed filter design methods.

5.1 Problem Formulation

Consider a class of uncertain linear discrete time-delay systems with a time-varying delay in the state:

$$\begin{aligned} x(k+1) &= A(\lambda)x(k) + A_d(\lambda)x(k-d(k)) + B(\lambda)v(k), \\ y(k) &= C(\lambda)x(k) + C_d(\lambda)x(k-d(k)) + D(\lambda)v(k), \\ z(k) &= L(\lambda)x(k) + L_d(\lambda)x(k-d(k)) + G(\lambda)v(k), \\ x(k) &= \phi(k),\ k = -\bar{d}, -\bar{d}+1, \ldots, 0. \end{aligned} \tag{5.1}$$

H. Gao and X. Li, *Robust Filtering for Uncertain Systems*,
Communications and Control Engineering, DOI: 10.1007/978-3-319-05903-7_5,

Here $x(k) \in \mathbb{R}^n$ is the system state vector; $y(k) \in \mathbb{R}^m$ is the measurement output; $z(k) \in \mathbb{R}^p$ is the signal to be estimated; and $v(k) \in \mathbb{R}^q$ is the noise input belonging to $l_2 [0, +\infty)$. Time delay $d(k)$ is a positive integer and assumed to be constant or time-varying in the whole dynamic process and satisfies $1 \leq \underline{d} \leq d(k) \leq \bar{d}$, where $\underline{d}$ and $\bar{d}$ are constant positive scalars representing the minimum and maximum delays, respectively. $\{\phi(k),\ k = -\bar{d}, -\bar{d}+1, \ldots, 0\}$ is a known given initial condition sequence. The system matrices are supposed to be uncertain but belong to a given convex bounded polyhedral domain described by s vertices:

$$\Omega(\lambda) \triangleq \begin{pmatrix} A(\lambda), A_d(\lambda), B(\lambda), \\ C(\lambda), C_d(\lambda), D(\lambda), \\ L(\lambda), L_d(\lambda), G(\lambda) \end{pmatrix} \in \mathcal{R},$$

$$\mathcal{R} \triangleq \left\{ \Omega(\lambda) \,\middle|\, \Omega(\lambda) = \sum_{i=1}^{s} \lambda_i \Omega_i;\ \lambda \in \Gamma \right\}, \tag{5.2}$$

with $\Omega_i \triangleq (A_i, A_{di}, B_i, C_i, C_{di}, D_i, L_i, L_{di}, G_i)$ denoting the i-th vertex of the polytope, and Γ denoting the unit simplex as

$$\Gamma \triangleq \left\{ (\lambda_1, \lambda_2, \ldots, \lambda_s) : \sum_{i=1}^{s} \lambda_i = 1, \lambda_i \geq 0 \right\}.$$

Suppose that system (5.1) is robustly asymptotically stable over the entire polytopic domain $\mathcal{R}$. Here we are interested in estimating the signal $z(k)$ by a linear filter of general structure described by

$$\begin{aligned} x_F(k+1) &= A_F x_F(k) + B_F y(k), \\ z_F(k) &= C_F x_F(k) + D_F y(k), \end{aligned} \tag{5.3}$$

where $x_F(k) \in \mathbb{R}^n$ is the filter state vector and $z_F(k)$ is the estimate of $z(k)$. The constant matrices $A_F \in \mathbb{R}^{n \times n}$, $B_F \in \mathbb{R}^{n \times m}$, $C_F \in \mathbb{R}^{p \times n}$ and $D_F \in \mathbb{R}^{p \times m}$ are filter parameters to be determined.

Letting

$$\xi(k) \triangleq \left[x(k)^{\mathrm{T}}\ x_F(k)^{\mathrm{T}} \right]^{\mathrm{T}},\ e(k) \triangleq z(k) - z_F(k),$$

the filtering error system can be described by

$$\begin{aligned} \xi(k+1) &= \bar{A}(\lambda)\xi(k) + \bar{A}_d(\lambda)K\xi(k - d(k)) + \bar{B}(\lambda)v(k), \\ e(k) &= \bar{C}(\lambda)\xi(k) + \bar{C}_d(\lambda)K\xi(k - d(k)) + \bar{D}(\lambda)v(k), \\ \xi(k) &= \left[\phi(k)^{\mathrm{T}}\ 0 \right]^{\mathrm{T}},\ k = -\bar{d}, -\bar{d}+1, \ldots, 0, \end{aligned} \tag{5.4}$$

where $K = \left[I\ 0 \right]$ and

$$\bar{A}(\lambda) = \begin{bmatrix} A(\lambda) & 0 \\ B_F C(\lambda) & A_F \end{bmatrix}, \; \bar{A}_d(\lambda) = \begin{bmatrix} A_d(\lambda) \\ B_F C_d(\lambda) \end{bmatrix},$$
$$\bar{B}(\lambda) = \begin{bmatrix} B(\lambda) \\ B_F D(\lambda) \end{bmatrix}, \; \bar{C}_d(\lambda) = \begin{bmatrix} L_d(\lambda) - D_F C_d(\lambda) \end{bmatrix},$$
$$\bar{C}(\lambda) = \begin{bmatrix} L(\lambda) - D_F C(\lambda) & -C_F \end{bmatrix}, \; \bar{D}(\lambda) = \begin{bmatrix} G(\lambda) - D_F D(\lambda) \end{bmatrix}. \quad (5.5)$$

Denote the transfer function from the noise signal $v(k)$ to the error signal $e(k)$ by $T_{ev}(z, \lambda)$. Then, the robust H_∞ filtering problem for uncertain discrete time-delay systems to be addressed in this chapter can be formulated as follows.

Robust H_∞ Filtering for Discrete Time-Delay Systems: Given the system in (5.1), design a robust filter of the form (5.3) such that for all admissible uncertainty and time delay:

(i) the filtering error system (5.4) is robustly asymptotically stable;
(ii) the filtering error system (5.4) satisfies

$$\|e\|_{l_2}^2 < \gamma^2 \|v\|_{l_2}^2, \quad (5.6)$$

for $0 \neq v(t) \in l_2[0, +\infty)$ under zero initial conditions, where γ is a given positive constant.

Filters satisfying the above conditions are called robust H_∞ filters. Denote $\tau \triangleq \bar{d} - \underline{d} + 1$. Clearly, $\tau = 1$ means that the time-delay $d(k)$ is time-invariant.

Remark 5.1 The problem considered in this chapter is basically the same as that in [1]. To make our idea more lucid, here we only consider the single delay case. However, it is not difficult to extend the results in the chapter to the case where multiple time delays appear. In addition, it is worth mentioning that in this chapter, the delay considered is time-varying, which is more general than the constant delay considered in [1]. This setting is more realistic than constant delay in practical situations, especially in NCSs.

5.2 H_∞ Filtering Analysis

In the sequel, sufficient conditions assuring a guaranteed γ level of noise attenuation for the filtering error system are established. The following BRL in terms of parameter-dependent LMI will play an instrumental role in the filter design problem of discrete time-delay systems.

Lemma 5.1 *Consider the system in* (5.1)*. Given a filter in* (5.3) *and a scalar* $\gamma > 0$*, the filtering error system in* (5.4) *for any fixed* $\lambda \in \Gamma$ *is asymptotically stable with* (5.6) *guaranteed if there exist real matrices* $P(\lambda) = P(\lambda)^{\mathrm{T}} > 0$, $Q(\lambda) = Q(\lambda)^{\mathrm{T}} > 0$, $M(\lambda) = M(\lambda)^{\mathrm{T}} > 0$, $X(\lambda)$ *and* $Y(\lambda)$ *with appropriate dimensions such that the following matrix inequality is satisfied*

$$\begin{bmatrix} \Xi_0(\lambda) & \Xi_1(\lambda)^{\mathrm{T}} & \Xi_2(\lambda)^{\mathrm{T}} P(\lambda) & \bar{d}\Xi_3(\lambda)^{\mathrm{T}} M(\lambda) \\ * & -I & 0 & 0 \\ * & * & -P(\lambda) & 0 \\ * & * & * & -\bar{d}M(\lambda) \end{bmatrix} < 0, \tag{5.7}$$

where

$$\Xi_0(\lambda) \triangleq \begin{bmatrix} \Xi_{0,1}(\lambda) & -K^{\mathrm{T}} X(\lambda) + K^{\mathrm{T}} Y(\lambda)^{\mathrm{T}} & 0 & -\bar{d}K^{\mathrm{T}} X(\lambda) \\ * & -Q(\lambda) - \mathrm{sym}\{Y(\lambda)\} & 0 & -\bar{d}Y(\lambda) \\ * & * & -\gamma^2 I & 0 \\ * & * & * & -\bar{d}M(\lambda) \end{bmatrix},$$

$$\begin{aligned}
\Xi_1(\lambda) &\triangleq \left[\bar{C}(\lambda) \; \bar{C}_d(\lambda) \; \bar{D} \; 0 \right], \\
\Xi_2(\lambda) &\triangleq \left[\bar{A}(\lambda) \; \bar{A}_d(\lambda) \; \bar{B}(\lambda) \; 0 \right], \\
\Xi_3(\lambda) &\triangleq \left[K\left(\bar{A}(\lambda) - I\right) \; K\bar{A}_d(\lambda) \; K\bar{B}(\lambda) \; 0 \right], \\
\Xi_{0,1}(\lambda) &\triangleq -P(\lambda) + K^{\mathrm{T}}\left(\tau Q(\lambda) + X(\lambda) + X(\lambda)^{\mathrm{T}}\right) K.
\end{aligned} \tag{5.8}$$

Proof For any uncertain parameter $\lambda \in \Gamma$, we construct an LKF candidate as:

$$V(k) \triangleq V_1(k) + V_2(k) + V_3(k) + V_4(k), \tag{5.9}$$

with

$$\begin{aligned}
V_1(k) &\triangleq \xi(k)^{\mathrm{T}} P(\lambda)\xi(k), \;\; \delta(k) \triangleq \xi(k+1) - \xi(k), \\
V_2(k) &\triangleq \sum_{i=k-d(k)}^{k-1} \xi(i)^{\mathrm{T}} K^{\mathrm{T}} Q(\lambda) K\xi(i), \\
V_3(k) &\triangleq \sum_{j=-\bar{d}+2}^{-\underline{d}+1} \sum_{i=k+j-1}^{k-1} \xi(i)^{\mathrm{T}} K^{\mathrm{T}} Q(\lambda) K\xi(i), \\
V_4(k) &\triangleq \sum_{i=\bar{d}}^{-1} \sum_{m=k+i}^{k-1} \delta(m)^{\mathrm{T}} K^{\mathrm{T}} M(\lambda) K\delta(m),
\end{aligned}$$

where $P(\lambda) > 0$, $Q(\lambda) > 0$ and $M(\lambda) > 0$ are appropriately dimensioned matrices to be determined. Taking the forward difference of the functional (5.9) along the solution of system (5.4), we have

$$\begin{aligned}
\Delta V_1(k) &= \xi(k+1)^{\mathrm{T}} P(\lambda)\xi(k+1) - \xi(k)^{\mathrm{T}} P(\lambda)\xi(k), \\
\Delta V_2(k) &\leq \xi(k)^{\mathrm{T}} K^{\mathrm{T}} Q(\lambda) K\xi(k) - \xi(k - d(k))^{\mathrm{T}} K^{\mathrm{T}} Q(\lambda) \\
&\quad \times K\xi(k - d(k)) + \sum_{i=k-\bar{d}+1}^{k-\underline{d}} \xi(i)^{\mathrm{T}} K^{\mathrm{T}} Q(\lambda) K\xi(i),
\end{aligned}$$

$$\begin{aligned}\triangle V_3(k) &= \left(\bar{d} - \underline{d}\right)\xi(k)^{\mathrm{T}} K^{\mathrm{T}} Q(\lambda) K \xi(k) \\ &\quad - \sum_{i=k-\bar{d}+1}^{k-\underline{d}} \xi(k)^{\mathrm{T}} K^{\mathrm{T}} Q(\lambda) K \xi(i), \\ \triangle V_4(k) &\leq \bar{d}\delta(k)^{\mathrm{T}} K^{\mathrm{T}} M(\lambda) K \delta(k) \\ &\quad - \sum_{m=k-d(k)}^{k-1} \delta(m)^{\mathrm{T}} K^{\mathrm{T}} M(\lambda) K \delta(m).\end{aligned}$$

Then, we can obtain

$$\begin{aligned}\triangle V(k) &= \triangle V_1(k) + \triangle V_2(k) + \triangle V_3(k) + \triangle V_4(k) \\ &\leq \frac{1}{d(k)} \sum_{m=k-d(k)}^{k-1} \varXi(k, m),\end{aligned} \tag{5.10}$$

where

$$\begin{aligned}\varXi(k, m) \triangleq\ & \xi(k+1)^{\mathrm{T}} P(\lambda) \xi(k+1) - \xi(k)^{\mathrm{T}} P(\lambda) \xi(k) + \tau \xi(k)^{\mathrm{T}} K^{\mathrm{T}} \\ & \times Q(\lambda) K \xi(k) - \xi(k - d(k))^{\mathrm{T}} K^{\mathrm{T}} Q(\lambda) K \xi(k - d(k)) \\ & + \bar{d}\left[\xi(k+1) - \xi(k)\right]^{\mathrm{T}} K^{\mathrm{T}} M(\lambda) K \left[\xi(k+1) - \xi(k)\right] \\ & - d(k)\, \delta(m)^{\mathrm{T}} K^{\mathrm{T}} M(\lambda) K \delta(m).\end{aligned}$$

According to $\xi(k - d(k)) = \xi(k) - \sum_{m=k-d(k)}^{k-1} \delta(m)$, for any appropriately dimensioned matrices $X(\lambda)$ and $Y(\lambda)$, the following equation always holds

$$\begin{aligned}\Lambda &\triangleq \frac{1}{d(k)} \sum_{m=k-d(k)}^{k-1} \left[\xi(k)^{\mathrm{T}}\ \xi(k - d(k))^{\mathrm{T}}\right] \begin{bmatrix} K^{\mathrm{T}} X(\lambda) K \\ K^{\mathrm{T}} Y(\lambda) K \end{bmatrix} \\ &\quad \times \left[\xi(k) - \xi(k - d(k)) - d(k)\delta(m)\right] \\ &\equiv 0,\end{aligned} \tag{5.11}$$

which is an extension of equalities introduced in [2].

Combining (5.10), (5.11) and

$$\xi(k+1) = \bar{A}(\lambda)\xi(k) + \bar{A}_d(\lambda) K \xi(k - d(k)) + \bar{B}(\lambda) v(k),$$

one obtains

$$\triangle V(k) = V(k) + 2\Lambda \leq \frac{1}{d(k)} \sum_{m=k-d(k)}^{k-1} \eta(t, m)^{\mathrm{T}} \tilde{\varXi}(\lambda) \eta(t, m) \tag{5.12}$$

with

$$\begin{aligned}
\eta(t,m) &\triangleq \left[\xi(k)^{\mathrm{T}}\ \xi(k-d(k))^{\mathrm{T}}K^{\mathrm{T}}\ v(k)^{\mathrm{T}}\ \delta(m)^{\mathrm{T}}K^{\mathrm{T}}\right]^{\mathrm{T}},\\
\tilde{\Xi}(\lambda) &\triangleq \tilde{\Xi}_0(\lambda)+\Xi_2(\lambda)^{\mathrm{T}}P(\lambda)\Xi_2(\lambda)+\Xi_3(\lambda)^{\mathrm{T}}K^{\mathrm{T}}M(\lambda)K\Xi_3(\lambda),\\
\tilde{\Xi}_0(\lambda) &\triangleq \begin{bmatrix} \Xi_{0,1} & -K^{\mathrm{T}}X(\lambda)+K^{\mathrm{T}}Y(\lambda)^{\mathrm{T}} & 0 & -\bar{d}K^{\mathrm{T}}X(\lambda)\\ * & -Q(\lambda)-\mathrm{sym}\{Y(\lambda)\} & 0 & -\bar{d}Y(\lambda)\\ * & * & 0 & 0\\ * & * & * & -\bar{d}M(\lambda)\end{bmatrix}
\end{aligned}$$

and $\Xi_{0,1}$, Ξ_2 and Ξ_3 defined in (5.8).

Assuming $v=0$, we can obtain that

$$\triangle V(k)\leq \frac{1}{d(k)}\sum_{m=k-d(k)}^{k-1}\bar{\eta}(k,m)^{\mathrm{T}}\bar{\Xi}(\lambda)\bar{\eta}(k,m),$$

where

$$\begin{aligned}
\bar{\eta}(t,m) &\triangleq \left[\xi(k)^{\mathrm{T}}\ \xi(k-d(k))^{\mathrm{T}}K^{\mathrm{T}}\ \delta(m)^{\mathrm{T}}K^{\mathrm{T}}\right]^{\mathrm{T}},\\
\bar{\Xi}(\lambda) &\triangleq \bar{\Xi}_0(\lambda)+\bar{\Xi}_2(\lambda)^{\mathrm{T}}P(\lambda)\bar{\Xi}_2(\lambda)+\bar{\Xi}_3(\lambda)^{\mathrm{T}}K^{\mathrm{T}}M(\lambda)K\bar{\Xi}_3(\lambda),\\
\bar{\Xi}_0(\lambda) &\triangleq \begin{bmatrix} \Xi_{0,1} & -K^{\mathrm{T}}X(\lambda)+K^{\mathrm{T}}Y(\lambda)^{\mathrm{T}} & -\bar{d}K^{\mathrm{T}}X(\lambda)\\ * & -Q(\lambda)-\mathrm{sym}\{Y(\lambda)\} & -\bar{d}Y(\lambda)\\ * & * & -\bar{d}M(\lambda)\end{bmatrix},\\
\bar{\Xi}_2(\lambda) &\triangleq \left[\bar{A}(\lambda)\ \bar{A}_d(\lambda)\ 0\right],\\
\bar{\Xi}_3(\lambda) &\triangleq \left[K\left(\bar{A}(\lambda)-I\right)\ K\bar{A}_d(\lambda)\ 0\right].
\end{aligned}$$

Using the Schur complement, (5.7) implies $\bar{\Xi}(\lambda)<0$. That is $\triangle V(k)<-\varepsilon\left|\xi(k)\right|^2$, where ε is a positive scalar. Then from the Lyapunov stability theory, we can conclude that the filtering error system (5.4) is asymptotically stable.

The next task is to establish the H_∞ performance for the filtering error system. Consider the index:

$$J\triangleq\sum_{k=0}^{\infty}\left[e(k)^{\mathrm{T}}e(k)-\gamma^2v(k)^{\mathrm{T}}v(k)\right]. \tag{5.13}$$

Under zero initial conditions, that is, $\xi(k)=0, k=-2\bar{d},-2\bar{d}+1,\ldots,0$, $V(0)=0$. It follows from $V(\infty)\geq 0$ that

$$J\leq\sum_{k=0}^{\infty}\left[e(k)^{\mathrm{T}}e(k)-\gamma^2v(k)^{\mathrm{T}}v(k)+\triangle V(k)\right]. \tag{5.14}$$

Substituting (5.12) into (5.14) and considering (5.4), we have

$$e(k)^{\mathrm{T}}e(k) - \gamma^2 v(k)^{\mathrm{T}}v(k) + \Delta V(k) \leq \frac{1}{d(k)} \sum_{m=k-d(k)}^{k-1} \eta(t,m)^{\mathrm{T}} \Xi(\lambda)\eta(t,m),$$

where

$$\begin{aligned}\Xi(\lambda) \triangleq\ & \Xi_0(\lambda) + \Xi_1(\lambda)^{\mathrm{T}}\Xi_1(\lambda) + \Xi_2(\lambda)^{\mathrm{T}}P(\lambda)\Xi_2(\lambda) \\ & + \Xi_3(\lambda)^{\mathrm{T}}K^{\mathrm{T}}M(\lambda)K\Xi_3(\lambda).\end{aligned}$$

Using the Schur complement, inequality (5.7) implies $\Xi(\lambda) < 0$. Then for any nonzero $v \in l_2[0, +\infty)$, there holds $J < 0$. That is, $\|e\|_{l_2}^2 < \gamma^2 \|v\|_{l_2}^2$ is true for all nonzero $v \in l_2[0, +\infty)$, which completes the proof. □

5.3 H_∞ Filtering Synthesis

5.3.1 Slack Matrices

Before moving on to derive the filter synthesis conditions, we first present the following alternative result in terms of parameter-dependent LMI for testing the robust H_∞ filtering performance, where the idea of slack matrix in [3] is applied.

Lemma 5.2 *There exist real matrices $P(\lambda) = P(\lambda)^{\mathrm{T}} > 0$, $Q(\lambda) = Q(\lambda)^{\mathrm{T}} > 0$, $M(\lambda) = M(\lambda)^{\mathrm{T}} > 0$, $X(\lambda)$ and $Y(\lambda)$ with appropriate dimensions such that the matrix inequality in* (5.7) *is satisfied if and only if there exist real matrices $P(\lambda) = P(\lambda)^{\mathrm{T}} > 0$, $Q(\lambda) = Q(\lambda)^{\mathrm{T}} > 0$, $M(\lambda) = M(\lambda)^{\mathrm{T}} > 0$, $X(\lambda)$, $Y(\lambda)$, $V(\lambda)$ and $U(\lambda)$ such that the following matrix inequality is satisfied*

$$\begin{bmatrix} \Xi_0(\lambda) & \Xi_1(\lambda)^{\mathrm{T}} & \Xi_2(\lambda)^{\mathrm{T}}V(\lambda) & \Xi_3(\lambda)^{\mathrm{T}}U(\lambda) \\ * & -I & 0 & 0 \\ * & * & P(\lambda) - \mathrm{sym}\{V(\lambda)\} & 0 \\ * & * & * & \bar{d}M(\lambda) - \mathrm{sym}\{U(\lambda)\} \end{bmatrix} < 0, \quad (5.15)$$

where $\Xi_i(\lambda)$, $i = 0, 1, 2, 3$, are defined in (5.7).

Proof Necessity: Suppose that there exist real matrices $P(\lambda) > 0$, $Q(\lambda) > 0$, $M(\lambda) > 0$, $X(\lambda)$, and $Y(\lambda)$ with appropriate dimensions such that (5.7) is satisfied. Letting $V(\lambda) = P(\lambda)$ and $U(\lambda) = \bar{d}M(\lambda)$, one can obtain (5.15).

Sufficiency: Suppose that there exist real matrices $P(\lambda) > 0$, $Q(\lambda) > 0$, $M(\lambda) > 0$, $X(\lambda)$, $Y(\lambda)$, $V(\lambda)$, and $U(\lambda)$ such that (5.15) is satisfied. From [3], there always hold

$$\begin{aligned} P(\lambda) - V(\lambda) - V(\lambda)^{\mathrm{T}} &\geq -V(\lambda)^{\mathrm{T}}P(\lambda)^{-1}V(\lambda), \\ \bar{d}M(\lambda) - U(\lambda) - U(\lambda)^{\mathrm{T}} &\geq -U(\lambda)^{\mathrm{T}}\left(\bar{d}M(\lambda)\right)^{-1}U(\lambda). \end{aligned}$$

Thus, the following inequality is satisfied

$$\begin{bmatrix} \Xi_0(\lambda) & \Xi_1(\lambda)^{\mathrm{T}} & \Xi_2(\lambda)^{\mathrm{T}} V(\lambda) & \Xi_3(\lambda)^{\mathrm{T}} U(\lambda) \\ * & -I & 0 & 0 \\ * & * & -V(\lambda)^{\mathrm{T}} P(\lambda)^{-1} V(\lambda) & 0 \\ * & * & * & -U(\lambda)^{\mathrm{T}} \left(\bar{d} M(\lambda)\right)^{-1} U(\lambda) \end{bmatrix} < 0,$$

which is equivalent to the inequality in (5.7) by performing a congruence transformation to it by diag $\{I, I, V(\lambda)^{-1} P(\lambda), \bar{d} U(\lambda)^{-1} M(\lambda)\}$. It completes the proof. □

5.3.2 Filter Realization

In this section, we will focus on the design of robust H_∞ filters based on Lemma 5.2, that is, to determine the filter matrices (A_F, B_F, C_F, D_F) which will guarantee the filtering error system to be asymptotically stable with a guaranteed H_∞ performance level. We have the following delay-dependent and parameter-dependent LMI condition for the existence of such a filter.

Lemma 5.3 *Consider the system in* (5.1) *and the filter in* (5.3). *For any fixed* $\lambda \in \Gamma$, *there exist real matrices* $P(\lambda) = P(\lambda)^{\mathrm{T}} > 0$, $Q(\lambda) = Q(\lambda)^{\mathrm{T}} > 0$, $M(\lambda) = M(\lambda)^{\mathrm{T}} > 0$, $X(\lambda)$, $Y(\lambda)$, $V(\lambda) = \begin{bmatrix} V_1(\lambda) & V_2(\lambda) \\ V_4 & V_3 \end{bmatrix}$ *and* $U(\lambda)$ *and filter matrices* A_F, B_F, C_F *and* D_F *satisfying* (5.7) *if and only if there exist real matrices* $\bar{P}(\lambda) \triangleq \begin{bmatrix} \bar{P}_{1\lambda} & \bar{P}_{2\lambda} \\ * & \bar{P}_{3\lambda} \end{bmatrix} = \bar{P}(\lambda)^{\mathrm{T}} > 0$, $Q(\lambda) = Q(\lambda)^{\mathrm{T}} > 0$, $M(\lambda) = M(\lambda)^{\mathrm{T}} > 0$, $X(\lambda)$, $Y(\lambda)$, $R(\lambda)$, $S(\lambda)$, T *and* $U(\lambda)$ *and matrices* $\bar{A}_F$, $\bar{B}_F$, $\bar{C}_F$ *and* $\bar{D}_F$ *satisfying*

$$\Upsilon(\lambda) \triangleq \begin{bmatrix} \Upsilon_0(\lambda) & \Upsilon_1(\lambda)^{\mathrm{T}} & \Upsilon_2(\lambda)^{\mathrm{T}} & \Upsilon_3(\lambda)^{\mathrm{T}} \\ * & -I & 0 & 0 \\ * & * & \Upsilon_4(\lambda) & 0 \\ * & * & * & \Upsilon_5(\lambda) \end{bmatrix} < 0, \tag{5.16}$$

where

$$\Upsilon_0(\lambda) \triangleq \begin{bmatrix} \Upsilon_{0,1}(\lambda) & -\bar{P}_2(\lambda) & -X(\lambda) + Y(\lambda)^{\mathrm{T}} & 0 & -\bar{d} X(\lambda) \\ * & -\bar{P}_3(\lambda) & 0 & 0 & 0 \\ * & * & -Q(\lambda) - \mathrm{sym}\{Y(\lambda)\} & 0 & -\bar{d} Y(\lambda) \\ * & * & * & -\gamma^2 I & 0 \\ * & * & * & * & -\bar{d} M(\lambda) \end{bmatrix},$$

$$\Upsilon_1(\lambda) \triangleq \begin{bmatrix} \Upsilon_{1,1}(\lambda) & -\bar{C}_F & \Upsilon_{1,2}(\lambda) & \Upsilon_{1,3}(\lambda) & 0 \end{bmatrix},$$

$$\Upsilon_2(\lambda) \triangleq \begin{bmatrix} \Upsilon_{2,1}(\lambda) & \bar{A}_F & \Upsilon_{2,2}(\lambda) & \Upsilon_{2,3}(\lambda) & 0 \\ \Upsilon_{2,4}(\lambda) & \bar{A}_F & \Upsilon_{2,5}(\lambda) & \Upsilon_{2,6}(\lambda) & 0 \end{bmatrix},$$

$$\begin{aligned}
\Upsilon_3(\lambda) &\triangleq \begin{bmatrix} U(\lambda)^{\mathrm{T}}(A(\lambda)-I) & 0 & U(\lambda)^{\mathrm{T}}A_d(\lambda) & U(\lambda)^{\mathrm{T}}B(\lambda) & 0 \\ 0 & 0 & 0 & 0 & 0 \end{bmatrix},\\
\Upsilon_4(\lambda) &\triangleq \begin{bmatrix} \bar{P}_1(\lambda)-\mathrm{sym}\{R(\lambda)\} & \bar{P}_2(\lambda)-S(\lambda)-T \\ \bar{P}_2(\lambda)^{\mathrm{T}}-S(\lambda)^{\mathrm{T}}-T^{\mathrm{T}} & \bar{P}_3(\lambda)-\mathrm{sym}\{T\} \end{bmatrix},\\
\Upsilon_5(\lambda) &\triangleq \bar{d}M(\lambda)-\mathrm{sym}\{U(\lambda)\},\\
\Upsilon_{0,1}(\lambda) &\triangleq -\bar{P}_1(\lambda)+\tau Q(\lambda)+X(\lambda)+X(\lambda)^{\mathrm{T}},\\
\Upsilon_{1,1}(\lambda) &\triangleq L(\lambda)-\bar{D}_F C(\lambda),\\
\Upsilon_{1,2}(\lambda) &\triangleq L_d(\lambda)-\bar{D}_F C_d(\lambda),\\
\Upsilon_{1,3}(\lambda) &\triangleq G(\lambda)-\bar{D}_F D(\lambda),\\
\Upsilon_{2,1}(\lambda) &\triangleq R(\lambda)^{\mathrm{T}}A(\lambda)+\bar{B}_F C(\lambda),\\
\Upsilon_{2,2}(\lambda) &\triangleq R(\lambda)^{\mathrm{T}}A_d(\lambda)+\bar{B}_F C_d(\lambda),\\
\Upsilon_{2,3}(\lambda) &\triangleq R(\lambda)^{\mathrm{T}}B(\lambda)+\bar{B}_F D(\lambda),\\
\Upsilon_{2,4}(\lambda) &\triangleq S(\lambda)^{\mathrm{T}}A(\lambda)+\bar{B}_F C(\lambda),\\
\Upsilon_{2,5}(\lambda) &\triangleq S(\lambda)^{\mathrm{T}}A_d(\lambda)+\bar{B}_F C_d(\lambda),\\
\Upsilon_{2,6}(\lambda) &\triangleq S(\lambda)^{\mathrm{T}}B(\lambda)+\bar{B}_F D(\lambda).
\end{aligned}$$

Moreover, if (5.16) *is feasible, the desired filter parameters can be given by*

$$\begin{bmatrix} A_F & B_F \\ C_F & D_F \end{bmatrix} = \begin{bmatrix} T^{-1} & 0 \\ 0 & I \end{bmatrix} \begin{bmatrix} \bar{A}_F & \bar{B}_F \\ \bar{C}_F & \bar{D}_F \end{bmatrix}. \tag{5.17}$$

Proof Necessity: First, partition the matrix variable $P(\lambda)$ in (5.15) as

$$P(\lambda) = \begin{bmatrix} P_1(\lambda) & P_2(\lambda) \\ * & P_3(\lambda) \end{bmatrix} > 0. \tag{5.18}$$

Without loss of generality, assume that the blocks V_4 and V_3 in $V(\lambda)$ are invertible. Let

$$J \triangleq \begin{bmatrix} I & 0 \\ 0 & V_3^{-1}V_4 \end{bmatrix},\ \phi_1 \triangleq \mathrm{diag}\{J, I, I, I\},\ \phi \triangleq \mathrm{diag}\{\phi_1, I, J, I\}. \tag{5.19}$$

Performing a congruence transformation to (5.15) by ϕ and considering (5.5), we have

$$\begin{bmatrix} \Phi_0(\lambda) & \Phi_1(\lambda)^{\mathrm{T}} & \Phi_2(\lambda)^{\mathrm{T}} & \Upsilon_3(\lambda)^{\mathrm{T}} \\ * & -I & 0 & 0 \\ * & * & \Phi_4(\lambda) & 0 \\ * & * & * & \Upsilon_5(\lambda) \end{bmatrix} < 0, \tag{5.20}$$

where

$$
\begin{aligned}
\Phi_0(\lambda) &\triangleq \phi_1^{\mathrm{T}} \Xi_0(\lambda)\phi_1 \\
&= \begin{bmatrix} \Phi_{0,1}(\lambda) & -P_2(\lambda)V_3^{-1}V_4 & -X(\lambda)+Y(\lambda)^{\mathrm{T}} & 0 & -\bar{d}X(\lambda) \\ * & -V_4^{\mathrm{T}}V_3^{-\mathrm{T}}P_3(\lambda)V_3^{-1}V_4 & 0 & 0 & 0 \\ * & * & -Q(\lambda)-\mathrm{sym}\{Y(\lambda)\} & 0 & -\bar{d}Y(\lambda) \\ * & * & * & -\gamma^2 I & 0 \\ * & * & * & * & -\bar{d}M(\lambda) \end{bmatrix}, \\
\Phi_1(\lambda) &\triangleq \Xi_1(\lambda)\phi_1 \\
&= \begin{bmatrix} L(\lambda)-D_F C(\lambda) & -C_F V_3^{-1}V_4 & L_d(\lambda)-D_F C_d(\lambda) & G(\lambda)-D_F D(\lambda) & 0 \end{bmatrix}, \\
\Phi_2(\lambda) &\triangleq J^{\mathrm{T}}V(\lambda)^{\mathrm{T}}\Xi_2(\lambda)\phi_1 = \begin{bmatrix} \Phi_{2,1}(\lambda) & V_4^{\mathrm{T}}A_F V_3^{-1}V_4 & \Phi_{2,2}(\lambda) & \Phi_{2,3}(\lambda) & 0 \\ \Phi_{2,4}(\lambda) & V_4^{\mathrm{T}}A_F V_3^{-1}V_4 & \Phi_{2,5}(\lambda) & \Phi_{2,6}(\lambda) & 0 \end{bmatrix}, \\
\Phi_4(\lambda) &\triangleq J^{\mathrm{T}}\left[P(\lambda)-\mathrm{sym}\{V(\lambda)\}\right]J \\
&= \begin{bmatrix} \bar{P}_1(\lambda)-\mathrm{sym}\{V_1(\lambda)\} & P_2(\lambda)V_3^{-1}V_4 - V_2(\lambda)V_3^{-1}G_4 - V_4^{\mathrm{T}}V_3^{-1}V_4 \\ * & V_4^{\mathrm{T}}V_3^{-\mathrm{T}}P_3(\lambda)V_3^{-1}V_4 - \mathrm{sym}\left\{V_4^{\mathrm{T}}V_3^{-1}V_4\right\} \end{bmatrix},
\end{aligned}
$$

$$
\begin{aligned}
\Phi_{0,1}(\lambda) &\triangleq -P_1(\lambda)+\tau Q(\lambda)+X(\lambda)+X(\lambda)^{\mathrm{T}}, \\
\Phi_{2,1}(\lambda) &\triangleq V_1(\lambda)^{\mathrm{T}}A(\lambda)+V_4^{\mathrm{T}}B_F C(\lambda), \\
\Phi_{2,2}(\lambda) &\triangleq V_1(\lambda)^{\mathrm{T}}A_d(\lambda)+V_4^{\mathrm{T}}B_F C_d(\lambda), \\
\Phi_{2,3}(\lambda) &\triangleq V_1(\lambda)^{\mathrm{T}}B(\lambda)+V_4^{\mathrm{T}}B_F D(\lambda), \\
\Phi_{2,4}(\lambda) &\triangleq V_4^{\mathrm{T}}V_3^{-\mathrm{T}}V_2(\lambda)^{\mathrm{T}}A(\lambda)+V_4^{\mathrm{T}}B_F C(\lambda), \\
\Phi_{2,5}(\lambda) &\triangleq V_4^{\mathrm{T}}V_3^{-\mathrm{T}}V_2(\lambda)^{\mathrm{T}}A_d(\lambda)+V_4^{\mathrm{T}}B_F C_d(\lambda), \\
\Phi_{2,6}(\lambda) &\triangleq V_4^{\mathrm{T}}V_3^{-\mathrm{T}}V_2(\lambda)^{\mathrm{T}}B(\lambda)+V_4^{\mathrm{T}}B_F D(\lambda).
\end{aligned}
$$

and $\Upsilon_3(\lambda)$ and $\Upsilon_5(\lambda)$ are in (5.16). Defining the following matrix variables

$$
\bar{P}(\lambda) \triangleq \begin{bmatrix} \bar{P}_{1\lambda} & \bar{P}_{2\lambda} \\ * & \bar{P}_{3\lambda} \end{bmatrix} = J^{\mathrm{T}}P(\lambda)J, \tag{5.21}
$$

$$
\begin{aligned}
R(\lambda) &\triangleq V_1(\lambda), \\
S(\lambda) &\triangleq V_2(\lambda)V_3^{-1}V_4, \\
T &\triangleq V_4^{\mathrm{T}}V_3^{-1}V_4, \\
\begin{bmatrix} \bar{A}_F & \bar{B}_F \\ \bar{C}_F & \bar{D}_F \end{bmatrix} &\triangleq \begin{bmatrix} V_4^{\mathrm{T}} & 0 \\ 0 & I \end{bmatrix}\begin{bmatrix} A_F & B_F \\ C_F & D_F \end{bmatrix}\begin{bmatrix} V_3^{-1}V_4 & 0 \\ 0 & I \end{bmatrix}.
\end{aligned}
\tag{5.22}
$$

Substituting the above matrices into (5.20), we have (5.16). The proof of necessity is completed.

Sufficiency: If the matrix inequality (5.16) is true, then $T + T^{\mathrm{T}} > 0$, therefore T is nonsingular. Then we can always find square and nonsingular matrices V_3 and V_4 satisfying $T = V_4^{\mathrm{T}}V_3^{-1}V_4$. Introduce matrices J and ϕ as in (5.19) and the following matrix variables:

$$\begin{aligned}
V(\lambda) &\triangleq \begin{bmatrix} R(\lambda) & S(\lambda)V_4^{-1}V_3 \\ V_4 & V_3 \end{bmatrix}, \\
P(\lambda) &\triangleq J^{-\mathrm{T}}\bar{P}_\lambda J^{-1}, \\
\begin{bmatrix} A_F & B_F \\ C_F & D_F \end{bmatrix} &\triangleq \begin{bmatrix} V_4^{-\mathrm{T}} & 0 \\ 0 & I \end{bmatrix} \begin{bmatrix} \bar{A}_F & \bar{B}_F \\ \bar{C}_F & \bar{D}_F \end{bmatrix} \begin{bmatrix} V_4^{-1}V_3 & 0 \\ 0 & I \end{bmatrix}
\end{aligned} \tag{5.23}$$

which imply that $P(\lambda) > 0$. By some matrix manipulations, it can be established that (5.16) is equivalent to

$$\phi^{\mathrm{T}} \begin{bmatrix} \varXi_0(\lambda) & \varXi_1(\lambda)^{\mathrm{T}} & \varXi_2(\lambda)^{\mathrm{T}}V(\lambda) & \varXi_3(\lambda)^{\mathrm{T}}U(\lambda) \\ * & -I & 0 & 0 \\ * & * & P(\lambda) - \mathrm{sym}\{V(\lambda)\} & 0 \\ * & * & * & \bar{d}M(\lambda) - \mathrm{sym}\{U(\lambda)\} \end{bmatrix} \phi < 0, \tag{5.24}$$

Performing a congruence transformation to (5.24) by ϕ^{-1} can lead to (5.7). The proof of sufficiency is completed.

Now let us denote the filter by

$$T_{z_F y}(z) = C_F\,(zI - A_F)^{-1}\,B_F + D_F.$$

Substituting the filter matrices with (5.23) and considering the relationship $T = V_4^{\mathrm{T}}V_3^{-1}V_4$ result in

$$T_{z_F y} = \bar{C}_F\left(zI - T^{-1}\bar{A}_F\right)^{-1} T^{-1}\bar{B}_F + \bar{D}_F.$$

Therefore, an admissible filter can be given by (5.17), and the proof is completed. □

Remark 5.2 Lemma 5.3 provides a method for designing H_∞ filters for fixed λ, which casts the nonlinear matrix inequality in Lemma 5.1 into a linear matrix inequality. It is noted that the conditions in Lemma 5.3 are dependent on the parameter λ, therefore it still cannot be used to design robust H_∞ filters due to the infinite dimension nature of the parameter λ. In the sequel, based on Lemma 5.3, we devote ourselves to the design of robust H_∞ filters via a structured polynomially parameter-dependent approach.

5.3.3 Polynomially Parameter-Dependent Approaches to Filter Design

To apply the polynomial parameter-dependence idea, define the notations $K(s,g)$, $K_j(s,g)$, $J(s,g)$, $I(s,g)$, $K_j^i(s,g)$, and $\beta_j^i(s,g+1)$ as in Sect. 3.4.1. Now let the matrices $\bar{P}(\lambda)$, $Q(\lambda)$, $M(\lambda)$, $X(\lambda)$, $Y(\lambda)$, $R(\lambda)$, $S(\lambda)$, and $U(\lambda)$ in Lemma 5.3 take the homogeneous polynomial of arbitrary degree g and depend polynomially on the uncertain parameters λ_i, $i = 1, \ldots, s$. That is,

$$
\begin{aligned}
\bar{P}_g(\lambda) &= \sum_{j=1}^{J(s,g)} \lambda_1^{k_1} \lambda_2^{k_2} \cdots \lambda_s^{k_s} \bar{P}_{K_j(s,g)}, \\
Q_g(\lambda) &= \sum_{j=1}^{J(s,g)} \lambda_1^{k_1} \lambda_2^{k_2} \cdots \lambda_s^{k_s} Q_{K_j(s,g)}, \\
M_g(\lambda) &= \sum_{j=1}^{J(s,g)} \lambda_1^{k_1} \lambda_2^{k_2} \cdots \lambda_s^{k_s} M_{K_j(s,g)}, \\
X_g(\lambda) &= \sum_{j=1}^{J(s,g)} \lambda_1^{k_1} \lambda_2^{k_2} \cdots \lambda_s^{k_s} X_{K_j(s,g)}, \\
Y_g(\lambda) &= \sum_{j=1}^{J(s,g)} \lambda_1^{k_1} \lambda_2^{k_2} \cdots \lambda_s^{k_s} Y_{K_j(s,g)}, \\
R_g(\lambda) &= \sum_{j=1}^{J(s,g)} \lambda_1^{k_1} \lambda_2^{k_2} \cdots \lambda_s^{k_s} R_{K_j(s,g)}, \\
S_g(\lambda) &= \sum_{j=1}^{J(s,g)} \lambda_1^{k_1} \lambda_2^{k_2} \cdots \lambda_s^{k_s} S_{K_j(s,g)}, \\
U_g(\lambda) &= \sum_{j=1}^{J(s,g)} \lambda_1^{k_1} \lambda_2^{k_2} \cdots \lambda_s^{k_s} U_{K_j(s,g)}, \\
K_j(s,g) &= (k_1, k_2, \ldots, k_s) \in K(s,g).
\end{aligned} \tag{5.25}
$$

For more detailed explanations upon the above notations, one can refer to Sect. 3.4.1

Remark 5.3 Note that, when $g = 0$, we have $\bar{P}(\lambda) = \bar{P}_0(\lambda)$, $Q(\lambda) = Q_0(\lambda)$, $M(\lambda) = M_0(\lambda)$, $X(\lambda) = X_0(\lambda)$, $Y(\lambda) = Y_0(\lambda)$, $R(\lambda) = R_0(\lambda)$, $S(\lambda) = S_0(\lambda)$ and $U(\lambda) = U_0(\lambda)$, which lead to a filtering result in the quadratic framework. In addition, when $g = 1$, $\bar{P}(\lambda) = \bar{P}_1(\lambda)$, $Q(\lambda) = Q_1(\lambda)$, $M(\lambda) = M_1(\lambda)$, $X(\lambda) = X_1(\lambda)$, $Y(\lambda) = Y_1(\lambda)$, $R(\lambda) = R_1(\lambda)$, $S(\lambda) = S_1(\lambda)$ and $U(\lambda) = U_1(\lambda)$ are linearly dependent on the parameter λ. Notice that a simple way to ensure $\bar{P}_g(\lambda) > 0$, $Q_g(\lambda) > 0$, and $M_g(\lambda) > 0$ is to impose $\bar{P}_{K_j(s,g)} > 0$, $Q_{K_j(s,g)} > 0$ and $M_{K_j(s,g)} > 0$ for $j = 1, 2, \ldots, J(s,g)$.

Then, we are in a position to give the main result of this chapter, that is, a polynomially parameter-dependent approach to robust H_∞ filter design for discrete time-delay systems with polytopic uncertainty.

Theorem 5.1 *Consider the system in* (5.1). *If there exist real matrices* T, $\bar{P}_{K_j(s,g)} = \begin{bmatrix} \bar{P}_{1,K_j(s,g)} & \bar{P}_{2,K_j(s,g)} \\ * & \bar{P}_{3,K_j(s,g)} \end{bmatrix} = \bar{P}^{\mathrm{T}}_{K_j(s,g)} > 0$, $Q_{K_j(s,g)} = Q^{\mathrm{T}}_{K_j(s,g)} > 0$, $M_{K_j(s,g)} = M^{\mathrm{T}}_{K_j(s,g)} > 0$, $X_{K_j(s,g)}$, $Y_{K_j(s,g)}$, $R_{K_j(s,g)}$, $S_{K_j(s,g)}$ *and* $U_{K_j(s,g)}$, $K_j(s,g) \in K(s,g)$, $j = 1, 2, \ldots, J(s,g)$, *and matrices* $\bar{A}_F$, $\bar{B}_F$, $\bar{C}_F$ *and* $\bar{D}_F$ *such that the following LMIs hold for all* $K_j(s, g+1) \in K(s, g+1)$, $j = 1, 2, \ldots, J(s, g+1)$*:*

$$\Theta_j \triangleq \sum_{i \in I_j(s,g+1)} \begin{bmatrix} \Sigma_0 & \Sigma_1^{\mathrm{T}} & \Sigma_2^{\mathrm{T}} & \Sigma_3^{\mathrm{T}} \\ * & -\beta_j^i(s,g+1)I & 0 & 0 \\ * & * & \Sigma_4 & 0 \\ * & * & * & \Sigma_5 \end{bmatrix} < 0, \tag{5.26}$$

where

$$\Sigma_0 \triangleq \begin{bmatrix} \Sigma_{0,1} & -\bar{P}_{2,K_j(s,g)} & \Sigma_{0,2} & 0 & -\bar{d}X_{K_j^i(s,g+1)} \\ * & -\bar{P}_{3,K_j(s,g)} & 0 & 0 & 0 \\ * & * & \Sigma_{0,3} & 0 & -\bar{d}Y_{K_j^i(s,g+1)} \\ * & * & * & -\gamma^2\beta_j^i(s,g+1)I & 0 \\ * & * & * & * & -\bar{d}M_{K_j^i(s,g+1)} \end{bmatrix},$$

$$\Sigma_1 \triangleq \begin{bmatrix} \Sigma_{1,1} & -\beta_j^i(s,g+1)\bar{C}_F & \Sigma_{1,2} & \Sigma_{1,3} & 0 \end{bmatrix},$$

$$\Sigma_2 \triangleq \begin{bmatrix} \Sigma_{2,1} & \beta_j^i(s,g+1)\bar{A}_F & \Sigma_{2,2} & \Sigma_{2,3} & 0 \\ \Sigma_{2,4} & \beta_j^i(s,g+1)\bar{A}_F & \Sigma_{2,5} & \Sigma_{2,6} & 0 \end{bmatrix},$$

$$\Sigma_3 \triangleq \begin{bmatrix} U^{\mathrm{T}}_{K_j^i(s,g+1)}(A_i - I) & 0 & U^{\mathrm{T}}_{K_j^i(s,g+1)}A_{di} & U^{\mathrm{T}}_{K_j^i(s,g+1)}B_i & 0 \\ 0 & 0 & 0 & 0 & 0 \end{bmatrix},$$

$$\Sigma_4 \triangleq \begin{bmatrix} \Sigma_{4,1} & \Sigma_{4,2} \\ * & \Sigma_{4,3} \end{bmatrix},$$

$$\Sigma_5 \triangleq \bar{d}M_{K_j^i(s,g+1)} - \mathrm{sym}\left\{U_{K_j^i(s,g+1)}\right\},$$

$$\Sigma_{0,1} \triangleq -\bar{P}_{1,K_j^i(s,g+1)} + \tau Q_{K_j^i(s,g+1)} + X_{K_j^i(s,g+1)} + X^{\mathrm{T}}_{K_j^i(s,g+1)},$$

$$\Sigma_{0,2} \triangleq -X_{K_j^i(s,g+1)} + Y^{\mathrm{T}}_{K_j^i(s,g+1)},$$

$$\Sigma_{0,3} \triangleq -Q_{K_j^i(s,g+1)} - \mathrm{sym}\left\{Y_{K_j^i(s,g+1)}\right\},$$

$$\Sigma_{1,1} \triangleq \beta_j^i(s,g+1)L_i - \beta_j^i(s,g+1)\bar{D}_F C_i,$$

$$\Sigma_{1,2} \triangleq \beta_j^i(s,g+1)L_{di} - \beta_j^i(s,g+1)\bar{D}_F C_{di},$$

$$\Sigma_{1,3} \triangleq \beta_j^i(s,g+1)G_i - \beta_j^i(s,g+1)\bar{D}_F D_i,$$

$$\Sigma_{2,1} \triangleq R^{\mathrm{T}}_{K_j^i(s,g+1)}A_i + \beta_j^i(s,g+1)\bar{B}_F C_i,$$

$$\Sigma_{2,4} \triangleq S^{\mathrm{T}}_{K^i_j(s,g+1)} A_i + \beta^i_j(s,g+1)\bar{B}_F C_i,$$
$$\Sigma_{2,2} \triangleq R^{\mathrm{T}}_{K^i_j(s,g+1)} A_{di} + \beta^i_j(s,g+1)\bar{B}_F C_{di},$$
$$\Sigma_{2,5} \triangleq S^{\mathrm{T}}_{K^i_j(s,g+1)} A_{di} + \beta^i_j(s,g+1)\bar{B}_F C_{di},$$
$$\Sigma_{2,3} \triangleq R^{\mathrm{T}}_{K^i_j(s,g+1)} B_i + \beta^i_j(s,g+1)\bar{B}_F D_i,$$
$$\Sigma_{2,6} \triangleq S^{\mathrm{T}}_{K^i_j(s,g+1)} B_i + \beta^i_j(s,g+1)\bar{B}_F D_i,$$
$$\Sigma_{4,1} \triangleq \bar{P}_{1,K^i_j(s,g+1)} - \mathrm{sym}\left\{R_{K^i_j(s,g+1)}\right\},$$
$$\Sigma_{4,2} \triangleq \bar{P}_{2,K^i_j(s,g+1)} - S_{K^i_j(s,g+1)} - \beta^i_j(s,g+1)T,$$
$$\Sigma_{4,3} \triangleq \bar{P}_{3,K^i_j(s,g+1)} - \beta^i_j(s,g+1)\mathrm{sym}\{T\},$$

then, the homogeneous polynomial matrices given by (5.25) *assure* (5.16) *for all* $\lambda \in \Gamma$.

Moreover, if the LMIs of (5.26) *are fulfilled for a given degree* $\hat{g}$, *then the LMIs corresponding to any degree* $g > \hat{g}$ *are also satisfied.*

Proof Since $\bar{P}_{K_j(s,g)} > 0$, $Q_{K_j(s,g)} > 0$ and $M_{K_j(s,g)} > 0$, $K_j(s,g) \in K(s,g)$, $j = 1, 2, \ldots, J(s,g)$, we know that $\bar{P}_g(\lambda)$, $Q_g(\lambda)$ and $M_g(\lambda)$ defined in (5.25) are all positive definite for all $\lambda \in \Gamma$. Now, note that $\Upsilon(\lambda)$ in (5.16) for any $\Omega(\lambda) \in \mathcal{R}$ and $\bar{P}_g(\lambda)$, $Q_g(\lambda)$, $M_g(\lambda)$, $X_g(\lambda)$, $Y_g(\lambda)$, $R_g(\lambda)$, $S_g(\lambda)$ and $U_g(\lambda)$ given by (5.25) are homogeneous polynomials of degree $g+1$ that can be written as

$$\Upsilon(\lambda) = \sum_{j=1}^{J(s,g+1)} \lambda_1^{k_1}\lambda_2^{k_2}\cdots\lambda_s^{k_s}\Theta_j,\ (k_1, k_2, \ldots, k_s) = K_j(s,g+1).$$

Conditions (5.26) imposed for all $j = 1, 2, \ldots, J(s,g+1)$ assure that $\Upsilon(\lambda) < 0$ for all $\lambda \in \Gamma$, and thus the first part is proved.

Suppose that LMIs in (5.26) are fulfilled for a certain degree $\hat{g}$, that is, there exist $J(s,\hat{g})$ symmetric positive definite matrices $\bar{P}_{K_j(s,\hat{g})}$, $Q_{K_j(s,\hat{g})}$, $M_{K_j(s,\hat{g})}$ and $X_{K_j(s,\hat{g})}$, $Y_{K_j(s,\hat{g})}$, $R_{K_j(s,\hat{g})}$, $S_{K_j(s,\hat{g})}$, $U_{K_j(s,\hat{g})}$, $j = 1, 2, \cdots, J(s,\hat{g})$ such that $\bar{P}_{\hat{g}}(\lambda)$, $Q_{\hat{g}}(\lambda)$, $M_{\hat{g}}(\lambda)$, $X_{\hat{g}}(\lambda)$, $Y_{\hat{g}}(\lambda)$, $R_{\hat{g}}(\lambda)$, $S_{\hat{g}}(\lambda)$ and $U_{\hat{g}}(\lambda)$ are homogeneously polynomially parameter-dependent matrices assuring $\Upsilon(\lambda)$. Then, the terms of the polynomial matrices

$$\bar{P}_{\hat{g}+1}(\lambda) = (\lambda_1 + \cdots + \lambda_s)\bar{P}_{\hat{g}}(\lambda),$$
$$Q_{\hat{g}+1}(\lambda) = (\lambda_1 + \cdots + \lambda_s)Q_{\hat{g}}(\lambda),$$
$$M_{\hat{g}+1}(\lambda) = (\lambda_1 + \cdots + \lambda_s)M_{\hat{g}}(\lambda),$$
$$X_{\hat{g}+1}(\lambda) = (\lambda_1 + \cdots + \lambda_s)X_{\hat{g}}(\lambda),$$
$$Y_{\hat{g}+1}(\lambda) = (\lambda_1 + \cdots + \lambda_s)Y_{\hat{g}}(\lambda),$$

$$
\begin{aligned}
R_{\hat{g}+1}(\lambda) &= (\lambda_1 + \cdots + \lambda_s)R_{\hat{g}}(\lambda),\\
S_{\hat{g}+1}(\lambda) &= (\lambda_1 + \cdots + \lambda_s)S_{\hat{g}}(\lambda),\\
U_{\hat{g}+1}(\lambda) &= (\lambda_1 + \cdots + \lambda_s)U_{\hat{g}}(\lambda),
\end{aligned}
$$

also satisfy the LMIs of Theorem 5.1 corresponding to the degree $\hat{g}+1$, which can be obtained by linear combination of the LMIs of Theorem 5.1 for $\hat{g}$. □

Based on Theorem 5.1, we summarize this polynomially parameter-dependent approach in Algorithm 18 for designing suboptimal robust H_∞ filters for uncertain discrete time-delay systems. Note that this method is also delay-dependent.

Algorithm 18 Delay-Dependent Robust H_∞ Filter Design for Discrete Time-Delay Systems

- Prescribe an integer $g \in \mathbb{N}_0^+$;
- Solve the minimization problem:

$$
\begin{aligned}
&\min_{\substack{\bar{P}_{K_j(s,g)} = \begin{bmatrix} \bar{P}_{1,K_j(s,g)} & \bar{P}_{2,K_j(s,g)} \\ * & \bar{P}_{3,K_j(s,g)} \end{bmatrix} = \bar{P}^{\mathrm{T}}_{K_j(s,g)} > 0,\\ Q_{K_j(s,g)} = Q^{\mathrm{T}}_{K_j(s,g)} > 0,\ M_{K_j(s,g)} = M^{\mathrm{T}}_{K_j(s,g)} > 0,\\ X_{K_j(s,g)},\ Y_{K_j(s,g)},\ R_{K_j(s,g)},\ S_{K_j(s,g)},\ U_{K_j(s,g)},\\ K_j(s,g) \in K(s,g),\ j=1,2,\ldots,J(s,g),\\ T,\ \bar{A}_F,\ \bar{B}_F,\ \bar{C}_F,\ \bar{D}_F,\ \mu}} \quad \mu = \gamma^2 \ \text{ s.t. (5.26)}
\end{aligned}
$$

- Compute the filter matrices from (5.17)

It is noted that with $X_{K_j(s,g)} \equiv Y_{K_j(s,g)} \equiv 0$ and $U_{K_j(s,g)} \equiv M_{K_j(s,g)} \equiv \varepsilon I$, for a sufficiently small positive constant ε, the inequality (5.26) implies a delay-independent H_∞ filtering result for the system (5.4). In this case, one has the following corollary.

Corollary 5.1 *Consider the system in* (5.1) *with uncertainty in* (5.2). *An admissible robust H_∞ filter of the form* (5.3) *exists if the there exist real matrices T, $\bar{P}_{K_j(s,g)} = \begin{bmatrix} \bar{P}_{1,K_j(s,g)} & \bar{P}_{2,K_j(s,g)} \\ * & \bar{P}_{3,K_j(s,g)} \end{bmatrix} = \bar{P}^{\mathrm{T}}_{K_j(s,g)} > 0$, $Q_{K_j(s,g)} = Q^{\mathrm{T}}_{K_j(s,g)} > 0$, $R_{K_j(s,g)}$ and $S_{K_j(s,g)}$, $K_j(s,g) \in K(s,g)$, $j = 1,2,\ldots,J(s,g)$ and matrices $\bar{A}_F$, $\bar{B}_F$, $\bar{C}_F$ and $\bar{D}_F$ such that the following LMIs hold for all $K_j(s,g+1) \in K(s,g+1)$, $j = 1,2,\ldots,J(s,g+1)$:*

$$
\sum_{i \in I_j(s,g+1)} \begin{bmatrix} \bar{\Sigma}_0 & \bar{\Sigma}_1^{\mathrm{T}} & \bar{\Sigma}_2^{\mathrm{T}} \\ * & -\beta_j^i(s,g+1)I & 0 \\ * & * & \Sigma_4 \end{bmatrix} < 0, \tag{5.27}
$$

where

$$\bar{\Sigma}_0 \triangleq \begin{bmatrix} \bar{\Sigma}_{0,1} & -\bar{P}_{2,K_j(s,g)} & 0 & 0 \\ * & -\bar{P}_{3,K_j(s,g)} & 0 & 0 \\ * & * & -Q_{K_j^i(s,g+1)} & 0 \\ * & * & * & -\gamma^2\beta_j^i(s,g+1)I \end{bmatrix},$$

$$\bar{\Sigma}_1 \triangleq \begin{bmatrix} \Sigma_{1,1} & -\beta_j^i(s,g+1)\bar{C}_F & \Sigma_{1,2} & \Sigma_{1,3} \end{bmatrix},$$

$$\bar{\Sigma}_2 \triangleq \begin{bmatrix} \Sigma_{2,1} & \beta_j^i(s,g+1)\bar{A}_F & \Sigma_{2,2} & \Sigma_{2,3} \\ \Sigma_{2,4} & \beta_j^i(s,g+1)\bar{A}_F & \Sigma_{2,5} & \Sigma_{2,6} \end{bmatrix},$$

$$\bar{\Sigma}_{0,1} \triangleq -\bar{P}_{1,K_j^i(s,g+1)} + \tau Q_{K_j^i(s,g+1)}$$

and other notations are defined in (5.26). *Moreover, a suitable filter realization is given by* (5.17).

From Corollary 5.1, we give Algorithm 19, a delay-independent but polynomially parameter-dependent approach to designing suboptimal robust H_∞ filters for uncertain discrete time-delay systems with polytopic uncertainty.

Algorithm 19 Delay-Independent Robust H_∞ Filter Design for Discrete Time-Delay Systems

- Prescribe an integer $g \in \mathbb{N}_0^+$;
- Solve the minimization problem:

$$\min_{\substack{\bar{P}_{K_j(s,g)} = \begin{bmatrix} \bar{P}_{1,K_j(s,g)} & \bar{P}_{2,K_j(s,g)} \\ * & \bar{P}_{3,K_j(s,g)} \end{bmatrix} = \bar{P}^{\mathrm{T}}_{K_j(s,g)} > 0, \\ Q_{K_j(s,g)} = Q^{\mathrm{T}}_{K_j(s,g)} > 0,\ R_{K_j(s,g)},\ S_{K_j(s,g)},\ U_{K_j(s,g)}, \\ K_j(s,g) \in K(s,g),\ j=1,2,\ldots,J(s,g), \\ T,\ \bar{A}_F,\ \bar{B}_F,\ \bar{C}_F,\ \bar{D}_F,\ \mu}} \mu = \gamma^2 \quad \text{s.t. (5.27)}$$

- Compute the filter matrices from (5.17).

5.4 Illustrative Examples

In this section, we present several examples to illustrate the effectiveness and advantages of the presented filter design methods.

Example 5.1 Consider system (5.1) with

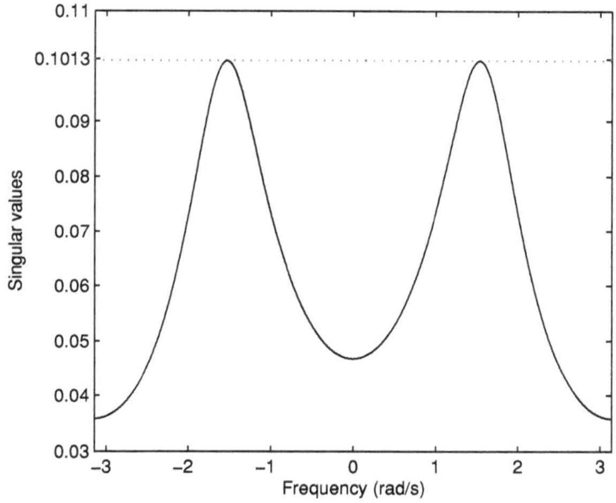

Fig. 5.1 Singular value curve of the filtering error transfer function with the filter in Eq. (5.29) in Example 5.1

$$\begin{aligned} A(\lambda) &= \begin{bmatrix} 0 & 0.3 \\ -0.2 & \rho \end{bmatrix}, A_d(\lambda) = \begin{bmatrix} 0 & 0 \\ 0.1 & \phi \end{bmatrix}, B(\lambda) = \begin{bmatrix} 0 \\ 1 \end{bmatrix}, \\ C(\lambda) &= \begin{bmatrix} 1 & 0 \end{bmatrix}, C_d(\lambda) = \begin{bmatrix} 0.2 & 0 \end{bmatrix}, D(\lambda) = 1, \\ L(\lambda) &= \begin{bmatrix} 1 & 2 \end{bmatrix}, L_d(\lambda) = \begin{bmatrix} 0 & 0 \end{bmatrix}, G(\lambda) = 0. \end{aligned} \tag{5.28}$$

The uncertain parameters satisfy $-0.5 \le \rho \le 0.4$ and $|\phi| \le 0.1$.

First, we assume that the given system is perfectly known, i.e., $\rho = \phi = 0$. In this situation, the minimum noise attenuation level obtained from Algorithm 18 with $d = \tau = 1$ is $\gamma^* = 0.1013$, with the associated filter matrices

$$\begin{bmatrix} A_F & B_F \\ C_F & D_F \end{bmatrix} = \left[\begin{array}{cc|c} 0.0000 & 0.3000 & -0.0000 \\ -1.2000 & -0.000 & -0.9990 \\ \hline -0.9998 & -2.0000 & -8.8722 \times 10^{-5} \end{array}\right]. \tag{5.29}$$

The singular value plots of the filtering error system by connecting the obtained filter (5.29) to the nominal system are shown in Fig. 5.1. The effectiveness of the guaranteed H_∞ noise attenuation level is apparent.

Now, we consider the uncertain case for system (5.28). When $d(k)$ is time varying, the results in [1] are inapplicable to this case. Using Algorithm 18, the achieved H_∞ performances γ^* of the filtering error system are listed in Table 5.1 for different lower bounds $\underline{d}$ when $\bar{d} = 5$. It is clearly shown from this table that γ^* decreases as $\underline{d}$ increases.

We now suppose that the delay $d \triangleq d(k)$ is time invariant. We calculate the H_∞ performance γ^* of the filtering error system corresponding to different delay upper bounds $\bar{d}$ by using Algorithm 18 in this chapter and the method in [1]. The obtained results are listed in Table 5.2. From this table one can clearly see that Algorithm 18 obtains much less conservative results than that in [1]. When $\bar{d} = 2$, Algorithm 18

Table 5.1 Achieved minimum H_∞ level γ^* when $\bar{d} = 5$ in Example 5.1

Index	$\underline{d} = 1$	$\underline{d} = 2$	$\underline{d} = 3$	$\underline{d} = 4$	$\underline{d} = 5$
γ^* $(g = 1)$	2.1274	1.9647	1.8053	1.6412	1.4653
γ^* $(g = 2)$	2.0913	1.9458	1.7963	1.6388	1.4645

Table 5.2 Achieved minimum H_∞ level γ^* by different methods in Example 5.1

Method	γ^*	
	$\underline{d} = 1$	$\underline{d} = 2$
[1]	2.0165	2.0188
Algorithm 18 $(g = 1)$	1.4645	1.4647
Algorithm 18 $(g = 2)$	1.4304	1.4646

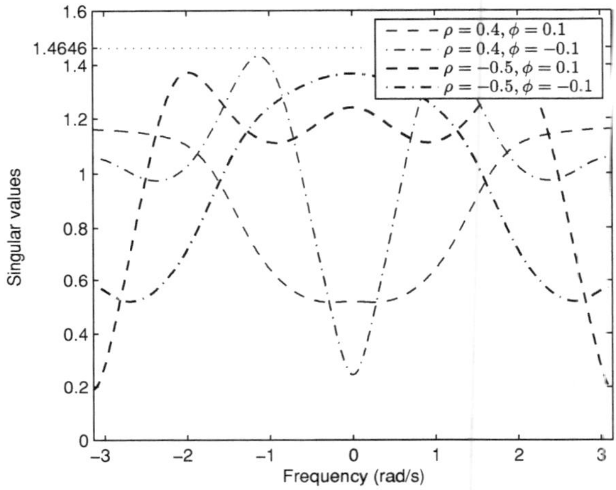

Fig. 5.2 Singular value curves of the filtering error transfer function with the filter in Eq. (5.30) in Example 5.1

yields the corresponding filter parameters for $g = 2$ as

$$\begin{bmatrix} A_F & B_F \\ C_F & D_F \end{bmatrix} = \left[\begin{array}{cc|c} 0.0495 & 0.2893 & -0.0160 \\ -0.1060 & -0.0241 & -1.0454 \\ \hline -0.9658 & -2.0052 & -0.0122 \end{array}\right]. \tag{5.30}$$

Connecting filter parameters (5.30) to the system in (5.28), we depict the singular value curves of the transfer functions at the four vertices, which are shown in Fig. 5.2. Clearly, all of the singular values are less than 1.4646, which demonstrates the effectiveness of Algorithm 18.

Example 5.2 To illustrate the potentials of the proposed delay-independent H_∞ filter design method, Algorithm 19, consider the uncertain linear discrete time-delay system in (5.1) with

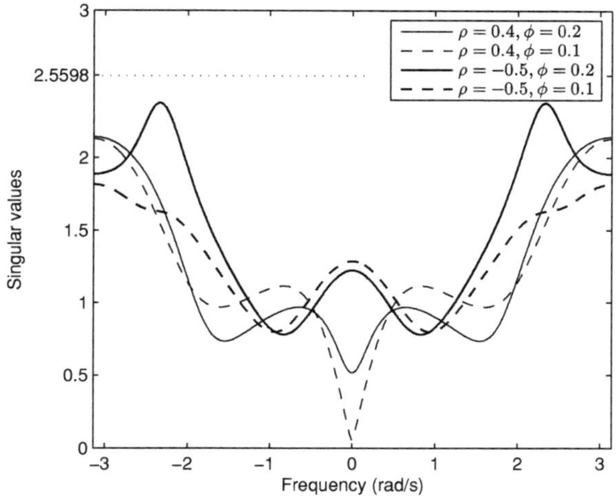

Fig. 5.3 Singular value curves of the filtering error transfer function with the filter in Eq. (5.31) ($d = 2$)

$$A(\lambda) = \begin{bmatrix} 0 & 0.3 \\ -0.2 & \rho \end{bmatrix}, A_d(\lambda) = \begin{bmatrix} 0 & 0.1 \\ 0 & \phi \end{bmatrix}, B(\lambda) = \begin{bmatrix} 0 \\ 1 \end{bmatrix},$$
$$C(\lambda) = \begin{bmatrix} 1 & 0 \end{bmatrix}, C_d(\lambda) = \begin{bmatrix} 0.3 & 0 \end{bmatrix}, D(\lambda) = 1,$$
$$L(\lambda) = \begin{bmatrix} 1 & 2 \end{bmatrix}, L_d(\lambda) = \begin{bmatrix} 0 & 0 \end{bmatrix}, G(\lambda) = 0,$$

where ρ and ψ are uncertain real parameters satisfying $-0.5 \leq \rho \leq 0.4$ and $0.1 \leq \phi \leq 0.2$.

The minimum disturbance attenuation level obtained by Algorithm 19 is $\gamma^* = 2.5598$ for $g = 1$, and the state-space realization of the corresponding filter is given by

$$\begin{bmatrix} A_F & B_F \\ C_F & D_F \end{bmatrix} = \left[\begin{array}{cc|c} -0.1286 & 0.1444 & 0.1001 \\ -0.3451 & -0.1350 & -1.2836 \\ \hline -1.0973 & -1.9591 & -0.0910 \end{array}\right]. \tag{5.31}$$

To illustrate the performance of the filter in (5.31), Fig. 5.3 shows the maximum singular value plots of the corresponding filtering error transfer function at the four vertices of the uncertainty domain $\mathcal{R}$ and for $d = 2$. Figure 5.4 displays the maximum singular value plots of the filtering error transfer functions for $d = 25$. The effectiveness of the guaranteed disturbance attenuation level γ^* is apparent. However, with the technique proposed in [4], one can obtain the minimum noise attenuation level bound $\gamma^* = 7.3731$, much higher than that obtained by Algorithm 19, showing the advantage of the presented approach.

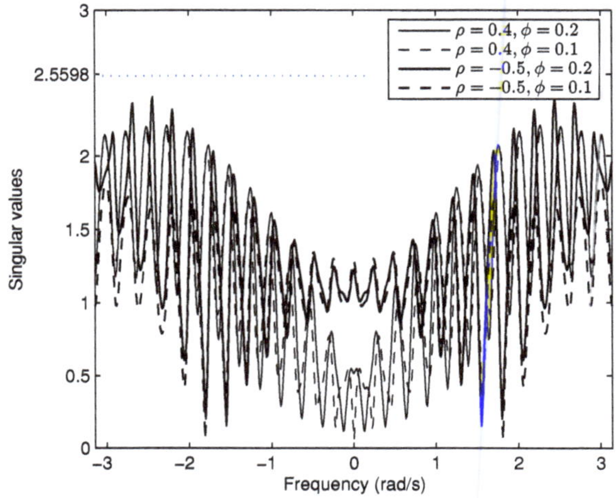

Fig. 5.4 Singular value curves of the filtering error transfer function with the filter in Eq. (5.31) in Example 5.1 ($d = 25$)

5.5 Summary and Notes

5.5.1 Summary

The problem of robust H_∞ filtering for uncertain discrete-time systems with time-varying delay and parameter uncertainties has been considered in this chapter. Using the polynomially parameter-dependent method, an innovative design for robust H_∞ filtering by means of LMI has been proposed, which guarantees the asymptotic stability of the filtering error system as well as a prescribed H_∞ performance requirement for all admissible parameter uncertainties. The results reported in this chapter are both parameter-dependent and delay-dependent, significantly reducing the design conservatism. The delay-independent design has been obtained as a special case of the delay-dependent one. Numerical examples have been presented to demonstrate the performance and advantages of the proposed approaches.

5.5.2 Notes

Robust H_∞ filtering for uncertain discrete-time systems with state delay has been considered in [1, 4–9] by virtue of the LMI technique. The results in [4] were based on a delay-independent and parameter-independent analysis condition, hence fell into the quadratic stability-type approaches. Other mentioned references all derived delay-dependent and parameter-dependent methods for robust H_∞ filtering. [1, 5, 7] utilized the technique in [3] to introduce slack matrices, and then adopted linearly parameter-dependent Lyapunov matrices to linearize the filter synthesis conditions.

Instead, [6, 8] applied the linearly parameter-dependent form to both Lyapunov matrices and slack matrices, where full slack matrices in light of Projection Lemma are introduced in [6]. A most recent development using polynomially parameter-dependent matrices can be found in [9].

References

1. Gao, H., Wang, C.: A delay-dependent approach to robust H_∞ filtering for uncertain discrete-time state-delayed systems. IEEE Trans. Signal Process. **52**(6), 1631–1640 (2004)
2. Xu, S., Lam, J.: On equivalence and efficiency of certain stability criteria for time-delay systems. IEEE Trans. Autom. Control **52**(1), 95–101 (2007)
3. de Oliveira, M.C., Bernussou, J., Geromel, J.C.: A new discrete-time robust stability condition. Syst. Control Lett. **37**, 261–265 (1999)
4. Palhares, R.M., de Souza, C.E., Peres, P.L.D.: Robust H_∞ filtering for uncertain discrete-time state-delayed systems. IEEE Trans. Signal Process. **49**(8), 1696–1703 (2001)
5. Zhang, X., Han, Q.: Delay-dependent robust H_∞ filtering for uncertain discrete-time systems with time-varying delay based on a finite sum inequality. IEEE Trans. Circuits Syst. II, Express. Briefs **53**(12), 1466–1470 (2006)
6. Qiu, J., Feng, G., Yang, J.: Improved delay-dependent H_∞ filtering design for discrete-time polytopic linear delay systems. IEEE Trans. Circuits Syst. II, Express. Briefs **55**(2), 178–182 (2008)
7. He, Y., Liu, G., Rees, D., Wu, M.: H_∞ filtering for discrete-time systems with time-varying delay. Signal Process. **89**(3), 275–282 (2009)
8. Li, X., Li, Z., Gao, H.: Further results on H_∞ filtering for discrete-time systems with state delay. Int. J. Robust Nonlinear Control **21**(3), 248–270 (2011)
9. Lacerda, M.J., Leite, V.J.S., Oliveira, R.C.L.F., Peres, P.L.D.: Delay-dependent robust H_∞ filter design for state-delayed discrete-time linear systems via homogeneous polynomial matrices. IET Control Theory Appl. **7**(1), 125–135 (2013)

Chapter 6
Robust Filtering for Uncertain 2-D Systems

6.1 Introduction

In recent years, the filtering problem for 2-D systems has drawn considerable attention. 2-D systems have important applications in the areas of multidimensional digital filtering, image data processing and transmission, and thermal process modeling, etc. A number of important results have been obtained so far. To mention a few, the problem of H_∞ filtering has been solved for 2-D systems with parameter uncertainties [1], time delays [2], and state-dependent noises [3].

Earlier results on the filtering problem of uncertain 2-D systems were mostly based on the notion of quadratic stability. Methods based on quadratic stability are inevitably conservative as the same parameter-independent Lyapunov function must be used for the entire uncertainty domain. As shown in the previous sections, one way to overcome this conservatism that has been well recognized is to consider a parameter-dependent Lyapunov function.

In the previous chapters, polynomially parameter-dependent approaches to robust filter design have been systematically developed for 1-D systems. In this chapter, to overcome the conservatism of quadratic approaches, we apply the polynomially parameter-dependent method to derive robust approaches to robust H_∞ filter design for uncertain 2-D systems described by the FM model and the Roesser model. The presented results are based on homogeneously polynomially parameter-dependent matrices of an arbitrary degree. It is still proven that as the degree grows, test of increasing precision is obtained, providing less conservative filter designs. The proposed conditions include results in the quadratic framework (that entail fixed matrices for the entire uncertainty domain), and the linearly parameter-dependent framework (that use linear convex combinations of matrices) as special cases of the zeroth degree and the first degree, respectively.

H. Gao and X. Li, *Robust Filtering for Uncertain Systems*,
Communications and Control Engineering, DOI: 10.1007/978-3-319-05903-7_6,

6.2 FM Model

In this section, we first consider the design of robust H_∞ filters for uncertain 2-D systems described by the FM model.

6.2.1 Problem Formulation

Consider discrete-time 2-D systems described by the FM model [4]:

$$\begin{aligned} x(i+1,j+1) &= A_1(\lambda)\,x(i,j+1) + A_2(\lambda)\,x(i+1,j) \\ &\quad + B_1(\lambda)\,v(i,j+1) + B_2(\lambda)\,v(i+1,j), \\ y(i,j) &= C(\lambda)\,x(i,j) + D(\lambda)\,v(i,j), \\ z(i,j) &= H(\lambda)\,x(i,j). \end{aligned} \tag{6.1}$$

Here $x(i,j) \in \mathbb{R}^n$ is the state vector; $y(i,j) \in \mathbb{R}^m$ is the measured output vector; $z(i,j) \in \mathbb{R}^p$ is the signal to be estimated; and $v(i,j) \in \mathbb{R}^q$ is the disturbance input vector. The system matrices are supposed to be unknown but belong to a given convex bounded polyhedral domain, namely,

$$\begin{aligned} \Omega(\lambda) &\triangleq (A_1(\lambda), A_2(\lambda), B_1(\lambda), B_2(\lambda), C(\lambda), D(\lambda), H(\lambda)) \in \mathcal{R}, \\ \mathcal{R} &\triangleq \left\{ \Omega(\lambda) \middle| \Omega(\lambda) = \sum_{i=1}^{s} \lambda_i \Omega_i;\ \lambda \in \Gamma \right\}, \end{aligned} \tag{6.2}$$

with $\Omega_i \triangleq (A_{1i}, A_{2i}, B_{1i}, B_{2i}, C_i, D_i, H_i)$ denoting the vertices of the polytope, and Γ denoting the unit simplex:

$$\Gamma \triangleq \left\{ (\lambda_1, \lambda_2, \ldots, \lambda_s) : \sum_{i=1}^{s} \lambda_i = 1, \lambda_i \geq 0 \right\}. \tag{6.3}$$

The boundary condition of the state vector is supposed to satisfy

$$\lim_{N\to\infty} \sum_{k=1}^{N} \left(x(0,k)^{\mathrm{T}} x(0,k) + x(k,0)^{\mathrm{T}} x(k,0) \right) < \infty.$$

The objective is to estimate the signal $z(i,j)$ by a robust filter as

$$\begin{aligned} \tilde{x}(i+1,j+1) &= A_{F1}\tilde{x}(i,j+1) + A_{F2}\tilde{x}(i+1,j) \\ &\quad + B_{F1}y(i,j+1) + B_{F2}y(i+1,j), \\ \tilde{z}(i,j) &= C_F\tilde{x}(i,j), \end{aligned} \tag{6.4}$$

where $\tilde{x}(i, j) \in \mathbb{R}^n$ is the filter state, and $\tilde{z}(i, j)$ is the estimate of $z(i, j)$.

Letting $\xi(i, j) = \left[x(i, j)^{\mathrm{T}}\ \tilde{x}(i, j)^{\mathrm{T}}\right]^{\mathrm{T}}$ and $e(i, j) \triangleq z(i, j) - \tilde{z}(i, j)$, the overall filtering error system can be expressed as:

$$\begin{aligned}\xi(i+1, j+1) &= \bar{A}_1(\lambda)\xi(i, j+1) + \bar{A}_2(\lambda)\xi(i+1, j) \\ &\quad + \bar{B}_1(\lambda)v(i, j+1) + \bar{B}_2(\lambda)v(i+1, j), \\ e(i, j) &= \bar{C}(\lambda)\xi(i, j),\end{aligned} \tag{6.5}$$

where

$$\begin{aligned}\bar{A}_1(\lambda) &= \begin{bmatrix} A_1(\lambda) & 0 \\ B_{F1}C(\lambda) & A_{F1} \end{bmatrix}, \ \bar{A}_2(\lambda) = \begin{bmatrix} A_2(\lambda) & 0 \\ B_{F2}C(\lambda) & A_{F2} \end{bmatrix}, \\ \bar{B}_1(\lambda) &= \begin{bmatrix} B_1(\lambda) \\ B_{F1}D(\lambda) \end{bmatrix}, \ \bar{B}_2(\lambda) = \begin{bmatrix} B_2(\lambda) \\ B_{F2}D(\lambda) \end{bmatrix}, \\ \bar{C}(\lambda) &= \begin{bmatrix} H(\lambda) & -C_F \end{bmatrix}.\end{aligned}$$

The transfer function of the filtering error system is

$$T_{ev}(z_1, z_2, \lambda) = \bar{C}(\lambda)\left[z_1z_2I_{2n} - z_2\bar{A}_1(\lambda) - z_1\bar{A}_2(\lambda)\right]^{-1}\left[z_2\bar{B}_1(\lambda) + z_1\bar{B}_2(\lambda)\right].$$

Then, the robust H_∞ filtering problem for uncertain 2-D FM model to be addressed in this section is described as follows.

Robust H_∞ Filtering for Uncertain 2-D FM Systems: Given the system in (6.1) subject to parameter uncertainty in (6.2), determine a filter of the form in (6.4), such that the filter error system (6.5) is robustly asymptotically stable for all $\lambda \in \Gamma$ and satisfies the following robust H_∞ performance

$$\|T_{ev}(z_1, z_2, \lambda)\|_\infty < \gamma, \ \forall\lambda \in \Gamma \tag{6.6}$$

where γ is a given positive scalar.

6.2.2 Filter Design

To solve the above problem, we need to introduce the following lemma [5], that is, a 2-D BRL for the FM model, which gives a sufficient H_∞ performance criterion for 2-D FM systems in terms of parameter-dependent LMI.

Lemma 6.1 *Given the 2-D FM system in* (6.1) *and the filter in* (6.4), *for any fixed* $\lambda \in \Gamma$, *the filtering error system in* (6.5) *is asymptotically stable and satisfies* (6.6) *if there exist real matrices* $P(\lambda) = P(\lambda)^{\mathrm{T}} > 0$ *and* $Q(\lambda) = Q(\lambda)^{\mathrm{T}} > 0$ *satisfying*

$$\begin{bmatrix} -P(\lambda) & 0 & 0 & P(\lambda)\bar{A}_1(\lambda) & P(\lambda)\bar{A}_2(\lambda) & P_\lambda \bar{B}_1(\lambda) & P(\lambda)\bar{B}_2(\lambda) \\ * & -I & 0 & \bar{C}(\lambda) & 0 & 0 & 0 \\ * & * & -I & 0 & \bar{C}(\lambda) & 0 & 0 \\ * & * & * & Q(\lambda)-P(\lambda) & 0 & 0 & 0 \\ * & * & * & * & -Q(\lambda) & 0 & 0 \\ * & * & * & * & * & -\gamma^2 I & 0 \\ * & * & * & * & * & * & -\gamma^2 I \end{bmatrix} < 0. \tag{6.7}$$

In the sequel, based on Lemma 6.1, we focus on the design of robust H_∞ filters. We first give the following lemma to guarantee the existence of a robust H_∞ filter through some parameter-dependent LMI conditions.

Lemma 6.2 *Given the 2-D FM system in* (6.1), *for any known and fixed* $\lambda \in \Gamma$, *there exist real matrices* $P(\lambda) = P(\lambda)^{\mathrm{T}} > 0$ *and* $Q(\lambda) = Q(\lambda)^{\mathrm{T}} > 0$ *and filter matrices* G_1, G_2, K_1, K_2 *and* L *satisfying* (6.7) *if and only if there exist matrices* $\bar{P}(\lambda) \triangleq \begin{bmatrix} \bar{P}_1(\lambda) & \bar{P}_2(\lambda) \\ * & \bar{P}_3(\lambda) \end{bmatrix} = \bar{P}(\lambda)^{\mathrm{T}} > 0$, $\bar{Q}(\lambda) \triangleq \begin{bmatrix} \bar{Q}_1(\lambda) & \bar{Q}_2(\lambda) \\ * & \bar{Q}_3(\lambda) \end{bmatrix} = \bar{Q}(\lambda)^{\mathrm{T}} > 0$, $R(\lambda)$, $S(\lambda)$, T *and matrices* $\bar{A}_{F1}$, $\bar{A}_{F2}$, $\bar{B}_{F1}$, $\bar{B}_{F2}$, $\bar{C}_F$ *satisfying*

$$\Delta(\lambda) \triangleq \begin{bmatrix} \Sigma_1(\lambda) & 0 & \Sigma_2(\lambda) & \Sigma_4(\lambda) & \Sigma_6(\lambda) \\ * & -I & \Sigma_3(\lambda) & \Sigma_5(\lambda) & 0 \\ * & * & \bar{Q}(\lambda)-\bar{P}(\lambda) & 0 & 0 \\ * & * & * & -\bar{Q}(\lambda) & 0 \\ * & * & * & * & -\gamma^2 I \end{bmatrix} < 0, \tag{6.8}$$

where

$$\Sigma_1(\lambda) \triangleq \begin{bmatrix} \bar{P}_1(\lambda) - R(\lambda) - R(\lambda)^{\mathrm{T}} & \bar{P}_{2\lambda} - S(\lambda) - T \\ * & \bar{P}_{3\lambda} - T^{\mathrm{T}} - T \end{bmatrix},$$

$$\Sigma_2(\lambda) \triangleq \begin{bmatrix} R(\lambda)^{\mathrm{T}} A_1(\lambda) + \bar{B}_{F1} C(\lambda) & \bar{A}_{F1} \\ S(\lambda)^{\mathrm{T}} A_1(\lambda) + \bar{B}_{F1} C(\lambda) & \bar{A}_{F1} \end{bmatrix},$$

$$\Sigma_3(\lambda) \triangleq \begin{bmatrix} H(\lambda) & -\bar{C}_F \\ 0 & 0 \end{bmatrix},$$

$$\Sigma_4(\lambda) \triangleq \begin{bmatrix} R(\lambda)^{\mathrm{T}} A_2(\lambda) + \bar{B}_{F2} C(\lambda) & \bar{A}_{F2} \\ S(\lambda)^{\mathrm{T}} A_2(\lambda) + \bar{B}_{F2} C(\lambda) & \bar{A}_{F2} \end{bmatrix},$$

$$\Sigma_5(\lambda) \triangleq \begin{bmatrix} 0 & 0 \\ H(\lambda) & -\bar{C}_F \end{bmatrix},$$

$$\Sigma_6(\lambda) \triangleq \begin{bmatrix} R(\lambda)^{\mathrm{T}} B_1(\lambda) + \bar{B}_{F1} D(\lambda) & R(\lambda)^{\mathrm{T}} B_2(\lambda) + \bar{B}_{F2} D(\lambda) \\ S(\lambda)^{\mathrm{T}} B_1(\lambda) + \bar{B}_{F1} D(\lambda) & S(\lambda)^{\mathrm{T}} B_2(\lambda) + \bar{B}_{F2} D(\lambda) \end{bmatrix}.$$

Moreover, if the above condition is satisfied, the matrices for an admissible realization of the filter are given by

$$\begin{bmatrix} A_{F1} & B_{F1} \\ A_{F2} & B_{F2} \\ C_F & 0 \end{bmatrix} = \begin{bmatrix} T^{-1} & 0 & 0 \\ 0 & T^{-1} & 0 \\ 0 & 0 & I \end{bmatrix} \begin{bmatrix} \bar{A}_{F1} & \bar{B}_{F1} \\ \bar{A}_{F2} & \bar{B}_{F2} \\ \bar{C}_F & 0 \end{bmatrix}. \tag{6.9}$$

Proof First, by the method for introducing slack matrix in [6], it is not difficult to show that there exist matrices $P(\lambda) > 0$ and $Q(\lambda) > 0$ satisfying (6.7) if and only if there exist matrices $P(\lambda) > 0$, $Q(\lambda) > 0$ and $V(\lambda)$ satisfying

$$\begin{bmatrix} \varXi & 0 & 0 & V(\lambda)^{\mathrm{T}} \bar{A}_1(\lambda) & V(\lambda)^{\mathrm{T}} \bar{A}_2(\lambda) & V(\lambda)^{\mathrm{T}} \bar{B}_1(\lambda) & V(\lambda)^{\mathrm{T}} \bar{B}_2(\lambda) \\ * & -I & 0 & \bar{C}(\lambda) & 0 & 0 & 0 \\ * & * & -I & 0 & \bar{C}(\lambda) & 0 & 0 \\ * & * & * & Q(\lambda) - P(\lambda) & 0 & 0 & 0 \\ * & * & * & * & -Q(\lambda) & 0 & 0 \\ * & * & * & * & * & -\gamma^2 I & 0 \\ * & * & * & * & * & * & -\gamma^2 I \end{bmatrix} < 0, \tag{6.10}$$

where $\varXi = P(\lambda) - V(\lambda)^{\mathrm{T}} - V(\lambda)$. In the following, we prove that (6.10) is equivalent to (6.8).

(6.10) $\Rightarrow$ (6.8): Suppose there exist matrices $P(\lambda) > 0$, $Q(\lambda) > 0$ and $V(\lambda)$ and filter parameter matrices A_{F1}, A_{F2}, B_{F1}, B_{F2} and C_F satisfying (6.7). Let matrices $P(\lambda)$, $Q(\lambda)$ and $V(\lambda)$ be partitioned as:

$$P(\lambda) = \begin{bmatrix} P_1(\lambda) & P_2(\lambda) \\ P_2(\lambda)^{\mathrm{T}} & P_3(\lambda) \end{bmatrix},$$

$$Q(\lambda) = \begin{bmatrix} Q_1(\lambda) & Q_2(\lambda) \\ Q_2(\lambda)^{\mathrm{T}} & Q_3(\lambda) \end{bmatrix},$$

$$V(\lambda) = \begin{bmatrix} V_1(\lambda) & V_2(\lambda) \\ V_4 & V_3 \end{bmatrix}.$$

Without loss of generality, we assume that V_4 and V_3 are invertible. Define

$$\phi \triangleq \begin{bmatrix} I & 0 \\ 0 & V_3^{-1} V_4 \end{bmatrix},$$

$$\bar{P}(\lambda) \triangleq \begin{bmatrix} \bar{P}_1(\lambda) & \bar{P}_2(\lambda) \\ * & \bar{P}_3(\lambda) \end{bmatrix} = \phi^{\mathrm{T}} P(\lambda) \phi,$$

$$\bar{Q}(\lambda) \triangleq \begin{bmatrix} \bar{Q}_1(\lambda) & \bar{Q}_2(\lambda) \\ * & \bar{Q}_3(\lambda) \end{bmatrix} = \phi^{\mathrm{T}} Q(\lambda) \phi. \tag{6.11}$$

Performing a congruence transformation to (6.10) by $\mathrm{diag}\{\phi, I, I, \phi, \phi, I, I\}$ and taking into account (6.5), we obtain

$$\begin{bmatrix} \Psi_1(\lambda) & 0 & 0 & \Psi_2(\lambda) & \Psi_4(\lambda) & \Psi_5(\lambda) & \Psi_6(\lambda) \\ * & -I & 0 & \Psi_3(\lambda) & 0 & 0 & 0 \\ * & * & -I & 0 & \Psi_3(\lambda) & 0 & 0 \\ * & * & * & \bar{Q}(\lambda) - \bar{P}(\lambda) & 0 & 0 & 0 \\ * & * & * & * & -\bar{Q}(\lambda) & 0 & 0 \\ * & * & * & * & * & -\gamma^2 I & 0 \\ * & * & * & * & * & * & -\gamma^2 I \end{bmatrix} < 0, \tag{6.12}$$

where

$$\begin{aligned} \Psi_1(\lambda) &= \bar{P}(\lambda) - \mathrm{sym}\left\{\begin{bmatrix} V_1(\lambda) & V_2(\lambda) V_3^{-1} V_4 \\ V_4^{\mathrm{T}} V_3^{-\mathrm{T}} V_4 & V_4^{\mathrm{T}} V_3^{-\mathrm{T}} V_4 \end{bmatrix}\right\}, \\ \Psi_2(\lambda) &= \begin{bmatrix} V_1(\lambda)^{\mathrm{T}} A_1(\lambda) + V_4^{\mathrm{T}} B_{F1} C(\lambda) & V_4^{\mathrm{T}} A_{F1} V_3^{-1} V_4 \\ V_4^{\mathrm{T}} V_3^{-\mathrm{T}} V_2(\lambda)^{\mathrm{T}} A_1(\lambda) + V_4^{\mathrm{T}} B_{F1} C(\lambda) & V_4^{\mathrm{T}} A_{F1} V_3^{-1} V_4 \end{bmatrix}, \\ \Psi_3(\lambda) &= \begin{bmatrix} H(\lambda) & -C_F V_3^{-1} V_4 \end{bmatrix}, \\ \Psi_4(\lambda) &= \begin{bmatrix} V_1(\lambda)^{\mathrm{T}} A_2(\lambda) + V_4^{\mathrm{T}} B_{F2} C(\lambda) & V_4^{\mathrm{T}} A_{F2} V_3^{-1} V_4 \\ V_4^{\mathrm{T}} V_3^{-\mathrm{T}} V_2(\lambda)^{\mathrm{T}} A_2(\lambda) + V_4^{\mathrm{T}} B_{F2} C(\lambda) & V_4^{\mathrm{T}} A_{F2} V_3^{-1} V_4 \end{bmatrix}, \\ \Psi_5(\lambda) &= \begin{bmatrix} V_1(\lambda)^{\mathrm{T}} B_1(\lambda) + V_{4\lambda}^{\mathrm{T}} B_{F1} D(\lambda) \\ V_4^{\mathrm{T}} V_3^{-\mathrm{T}} V_2(\lambda)^{\mathrm{T}} B_1(\lambda) + V_4^{\mathrm{T}} B_{F1} D(\lambda) \end{bmatrix}, \\ \Psi_6(\lambda) &= \begin{bmatrix} V_1(\lambda)^{\mathrm{T}} B_2(\lambda) + V_4^{\mathrm{T}} B_{F2} D(\lambda) \\ V_4^{\mathrm{T}} V_3^{-\mathrm{T}} V_2(\lambda)^{\mathrm{T}} B_2(\lambda) + V_4^{\mathrm{T}} B_{F2} D(\lambda) \end{bmatrix}. \end{aligned}$$

Let matrices $R(\lambda)$, $S(\lambda)$, $T(\lambda)$, $\bar{A}_{F1}$, $\bar{A}_{F2}$, $\bar{B}_{F1}$, $\bar{B}_{F2}$ and $\bar{C}_F$ be defined as

$$\begin{aligned} &R(\lambda) \triangleq V_1(\lambda), \ S(\lambda) \triangleq V_2(\lambda) V_3^{-1} V_4, \ T \triangleq V_4^{\mathrm{T}} V_3^{-1} V_4, \\ &\begin{bmatrix} \bar{A}_{F1} & \bar{B}_{F1} \\ \bar{A}_{F2} & \bar{B}_{F2} \\ \bar{C}_F & 0 \end{bmatrix} \triangleq \begin{bmatrix} V_4^{\mathrm{T}} & 0 & 0 \\ 0 & V_4^{\mathrm{T}} & 0 \\ 0 & 0 & I \end{bmatrix} \begin{bmatrix} A_{F1} & B_{F1} \\ A_{F2} & B_{F2} \\ C_F & 0 \end{bmatrix} \begin{bmatrix} V_3^{-1} V_4 & 0 \\ 0 & I \end{bmatrix}. \end{aligned} \tag{6.13}$$

Comparing (6.12) with (6.8), one knows that there exist matrices $\bar{P}(\lambda) > 0$ and $\bar{Q}(\lambda) > 0$ in (6.11) and $R(\lambda)$, $S(\lambda)$, T, $\bar{A}_{F1}$, $\bar{A}_{F2}$, $\bar{B}_{F1}$, $\bar{B}_{F2}$ and $\bar{C}_F$ in (6.13) satisfying (6.8). Hence, (6.10) implies (6.8).

(6.10) $\Leftarrow$ (6.8): (6.8) implies that $\bar{P}_{3\lambda} - T^{\mathrm{T}} - T < 0$. Hence T is invertible and we can always find square and nonsingular matrices V_3 and V_4 satisfying $T = V_4^{\mathrm{T}} V_3^{-1} V_4$. Let matrices $P(\lambda) > 0$, $Q(\lambda) > 0$, A_{F1}, A_{F2}, B_{F1}, B_{F2} and C_F be obtained from $\bar{P}(\lambda) > 0$, $\bar{Q}(\lambda) > 0$, $R(\lambda)$, $S(\lambda)$, T, $\bar{A}_{F1}$, $\bar{A}_{F2}$, $\bar{B}_{F1}$, $\bar{B}_{F2}$ and $\bar{C}_F$ by the converse definitions in (6.11) and (6.13), and denote $V(\lambda) = \begin{bmatrix} V_1(\lambda) & V_2(\lambda) \\ V_4 & V_3 \end{bmatrix}$. Using these assignments, one knows that (6.8) can be rewritten as (6.12), and (6.10) is a congruence transformation from (6.12) by $\mathrm{diag}\{\phi^{-1}, I, I, \phi^{-1}, \phi^{-1}, I, I\}$. Hence, (6.10) is implied by (6.8). Consequently, (6.7) is equivalent to (6.8).

Now, we prove (6.7) gives an acceptable state space realization of the H_∞ filter (6.4). Denote the transfer function of the filter (6.4) as

$$T_{\tilde{z}y}(z_1, z_2) = C_F\left(z_1 z_2 I_n - z_2 A_{F1} - z_1 A_{F2}\right)^{-1}\left(z_2 B_{F1} + z_1 B_{F2}\right).$$

From (6.13), we have

$$\begin{aligned}
T_{\tilde{z}y}(z_1, z_2) &= \bar{C}_F V_4^{-1} V_3 \left(z_1 z_2 I_n - z_2 V_4^{-\mathrm{T}} \bar{A}_{F1} V_4^{-1} V_3 - z_1 V_4^{-\mathrm{T}} \bar{A}_{F2} V_4^{-1} V_3\right)^{-1} \\
&\quad \times \left(z_2 V_4^{-\mathrm{T}} \bar{B}_{F1} + z_1 V_4^{-\mathrm{T}} \bar{B}_{F2}\right) \\
&= \bar{C}_F \left[V_4^{-1} V_3 \left(z_1 z_2 I_n - z_2 V_4^{-\mathrm{T}} \bar{A}_{F1} V_4^{-1} V_3 - z_1 V_4^{-\mathrm{T}} \bar{A}_{F2} V_4^{-1} V_3\right)\right. \\
&\quad \left. \times V_3^{-1} V_4\right]^{-1} V_4^{-1} V_3 \left(z_2 V_4^{-\mathrm{T}} \bar{B}_{F1} + z_1 V_4^{-\mathrm{T}} \bar{B}_{F2}\right) \\
&= \bar{C}_F \left[z_1 z_2 I_n - z_2 T^{-1} \bar{A}_{F1} - z_1 T^{-1} \bar{A}_{F2}\right]^{-1} \\
&\quad \times \left[z_2 T^{-1} B_{F1} + z_1 T^{-1} B_{F2}\right],
\end{aligned}$$

which completes the entire proof. □

Lemma (6.2) provides a method for designing H_∞ filters for known and fixed λ, which casts the NLMI in Lemma 6.1 into a set of LMIs. It is noted that the condition in Lemma (6.2) is dependent on the parameter λ, therefore it cannot be used to design robust H_∞ filters due to the infinite-dimensional nature of the parameter λ. In what follows, based on Lemma (6.2), we propose a new method for designing robust H_∞ filters via the structural polynomial parameter-dependence method.

Define the notations $K(s, g)$, $K_j(s, g)$, $J(s, g)$, $I(s, g)$, $K_j^i(s, g)$ and $\beta_j^i(s, g+1)$ as in Sect. 3.4.1 and let the parameter-dependent matrices $\bar{P}(\lambda)$, $\bar{Q}(\lambda)$, $R(\lambda)$ and $S(\lambda)$ in Lemma 6.1 take the homogeneous polynomial of arbitrary degree g and depend polynomially on the uncertain parameters λ_i, $i = 1, \ldots, s$:

$$\bar{P}_g(\lambda) = \sum_{j=1}^{J(s,g)} \lambda_1^{k_1} \lambda_2^{k_2} \cdots \lambda_s^{k_s} \bar{P}_{K_j(s,g)},$$

$$\bar{Q}_g(\lambda) = \sum_{j=1}^{J(s,g)} \lambda_1^{k_1} \lambda_2^{k_2} \cdots \lambda_s^{k_s} \bar{Q}_{K_j(s,g)},$$

$$R_g(\lambda) = \sum_{j=1}^{J(s,g)} \lambda_1^{k_1} \lambda_2^{k_2} \cdots \lambda_s^{k_s} R_{K_j(s,g)},$$

$$S_g(\lambda) = \sum_{j=1}^{J(s,g)} \lambda_1^{k_1} \lambda_2^{k_2} \cdots \lambda_s^{k_s} S_{K_j(s,g)},$$
$$K_j(s,g) = (k_1, k_2, \ldots, k_s) \in K(s,g). \tag{6.14}$$

We present the main result for the 2-D FM model of this section, namely, a polynomially parameter-dependent approach to robust 2-D H_∞ filter design.

Theorem 6.1 *Given a stable 2-D FM system* (6.1) *and a scalar* $\gamma > 0$, *a filter* (6.1) *exists such that the filtering error system* (6.5) *is robustly asymptotically stable and satisfies* (6.6), *if there exist real matrices* $R_{K_j(s,g)}$, $S_{K_j(s,g)}$, $\bar{P}_{K_j(s,g)} = \begin{bmatrix} \bar{P}_{1,K_j(s,g)} & \bar{P}_{2,K_j(s,g)} \\ * & \bar{P}_{3,K_j(s,g)} \end{bmatrix} = \bar{P}^{\mathrm{T}} > 0$, $\bar{Q}_{K_j(s,g)} = \begin{bmatrix} \bar{Q}_{1,K_j(s,g)} & \bar{Q}_{2,K_j(s,g)} \\ * & \bar{Q}_{3,K_j(s,g)} \end{bmatrix} = \bar{Q}^{\mathrm{T}}_{K_j(s,g)} > 0$, $K_j(s,g) \in K(s,g)$, $j = 1, 2, \ldots, J(s,g)$ *and* T, $\bar{A}_{F1}$, $\bar{A}_{F2}$, $\bar{B}_{F1}$, $\bar{B}_{F2}$, $\bar{C}_F$ *such that the following LMIs hold for all* $K_l(s, g+1) \in K(s, g+1)$, $l = 1, 2, \ldots, J(s, g+1)$:

$$\Delta_l \triangleq \sum_{i \in I_l(s,g+1)} \begin{bmatrix} \Theta_1 & 0 & \Theta_2 & \Theta_5 & \Theta_7 \\ * & -\beta_l^i(s,g+1)I & \Theta_3 & \Theta_6 & 0 \\ * & * & \Theta_4 & 0 & 0 \\ * & * & * & -\bar{Q}_{K_l^i(s,g+1)} & 0 \\ * & * & * & * & -\beta_l^i(s,g+1)\gamma^2 I \end{bmatrix} < 0, \tag{6.15}$$

where

$$\Theta_1 = \bar{P}_{K_l^i(s,g+1)} - \begin{bmatrix} \mathrm{sym}\left\{R_{K_l^i(s,g+1)}\right\} & S_{K_l^i(s,g+1)} + \beta_l^i(s,g+1)T \\ * & \beta_l^i(s,g+1)\mathrm{sym}\{T\} \end{bmatrix},$$

$$\Theta_2 = \begin{bmatrix} R^{\mathrm{T}}_{K_l^i(s,g+1)} A_{1i} + \beta_l^i(s,g+1)\bar{B}_{F1}C_i & \beta_l^i(s,g+1)\bar{A}_{F1} \\ S^{\mathrm{T}}_{K_l^i(s,g+1)} A_{1i} + \beta_l^i(s,g+1)\bar{B}_{F1}C_i & \beta_l^i(s,g+1)\bar{A}_{F1} \end{bmatrix},$$

$$\Theta_3 = \begin{bmatrix} \beta_l^i(s,g+1)H_i & -\beta_l^i(s,g+1)\bar{C}_F \\ 0 & 0 \end{bmatrix},$$

$$\Theta_4 = \bar{Q}_{K_l^i(s,g+1)} - \bar{P}_{K_l^i(s,g+1)},$$

$$\Theta_5 = \begin{bmatrix} R^{\mathrm{T}}_{K_l^i(s,g+1)} A_{2i} + \beta_l^i(s,g+1)\bar{B}_{F2}C_i & \beta_l^i(s,g+1)\bar{A}_{F2} \\ S^{\mathrm{T}}_{K_l^i(s,g+1)} A_{2i} + \beta_l^i(s,g+1)\bar{B}_{F2}C_i & \beta_l^i(s,g+1)\bar{A}_{F2} \end{bmatrix},$$

$$\Theta_6 = \begin{bmatrix} 0 & 0 \\ \beta_l^i(s,g+1)H_i & -\beta_l^i(s,g+1)\bar{C}_F \end{bmatrix},$$

$$\Theta_7 = \begin{bmatrix} R^{\mathrm{T}}_{K_l^i(s,g+1)} B_{1i} + \beta_l^i(s,g+1)\bar{K}_1 D_i & R^{\mathrm{T}}_{K_l^i(s,g+1)} B_{2i} + \beta_l^i(s,g-1)\bar{K}_2 D_i \\ S^{\mathrm{T}}_{K_l^i(s,g+1)} B_{1i} + \beta_l^i(s,g+1)\bar{K}_1 D_i & S^{\mathrm{T}}_{K_l^i(s,g+1)} B_{2i} + \beta_l^i(s,g-1)\bar{K}_2 D_i \end{bmatrix}.$$

If the LMIs in (6.15) *are satisfied, the matrices of an admissible realization of the filter are given by* (6.8). *Moreover, if the LMIs in* (6.15) *are satisfied for a given degree $\hat{g}$, then the LMIs corresponding to any degree $g > \hat{g}$ are also satisfied.*

Proof Since $\bar{P}_{K_j(s,g)} > 0$ and $\bar{Q}_{K_j(s,g)} > 0$, $K_j(s,g) \in K(s,g)$, $j = 1, 2, \ldots, J(s,g)$, then $\bar{P}_g(\lambda) > 0$ and $\bar{Q}_g(\lambda) > 0$ for all $\lambda \in \Gamma$. Now, note that $\Delta(\lambda)$ in (6.8) for $(A_1(\lambda), A_2(\lambda), B_1(\lambda), B_2(\lambda), C(\lambda), D(\lambda), H(\lambda)) \in \mathcal{R}$ and T, $\bar{P}(\lambda)$, $\bar{Q}(\lambda)$, $R(\lambda)$ and $S(\lambda)$ given by (6.14) are homogeneously polynomially matrix functions of degree $g + 1$ that can be written as

$$\Delta(\lambda) = \sum_{l=1}^{J(s,g+1)} \lambda_1^{k_1} \lambda_2^{k_2} \cdots \lambda_s^{k_s} \Delta_l, \ (k_1, k_2, \ldots, k_s) = K_l(s, g+1).$$

Note that $\sum_{i=1}^{s} = 1$ and $\lambda_i > 0$, $i = 1, 2, \ldots, s$. Hence, conditions in (6.15) imposed for all l, $l = 1, 2, \cdots, J(s, g+1)$ assure that $\Delta(\lambda) < 0$ for all $\lambda \in \Gamma$. Moreover, according to Lemmas 6.1 and 6.2, a filter (6.1) given by (6.8) guarantees that the filtering error system (6.5) is stable and satisfies (6.6)

Suppose that LMIs in (6.15) are fulfilled for a certain degree $\hat{g}$, that is, there exist $J(\hat{g})$ symmetric positive definite matrices $\bar{P}_{K_j(s,\hat{g})}$ and $\bar{Q}_{K_j(s,\hat{g})}$ and matrices $R_{K_j(s,\hat{g})}$ and $S_{K_j(s,\hat{g})}$, $j = 1, 2, \cdots, J(s, \hat{g})$ such that $\bar{P}_{\hat{g}}(\lambda)$, $\bar{Q}_{\hat{g}}(\lambda)$, $R_{\hat{g}}(\lambda)$, $S_{\hat{g}}(\lambda)$ are homogeneously polynomially parameter-dependent matrices assuring $\Delta(\lambda) < 0$. Then, the terms of the polynomial matrices

$$\begin{aligned}
\bar{P}_{\hat{g}+1}(\lambda) &= (\lambda_1 + \cdots + \lambda_s)\, \bar{P}_{\hat{g}}(\lambda), \\
\bar{Q}_{\hat{g}+1}(\lambda) &= (\lambda_1 + \cdots + \lambda_s)\, \bar{Q}_{\hat{g}}(\lambda), \\
R_{\hat{g}+1}(\lambda) &= (\lambda_1 + \cdots + \lambda_s)\, R_{\hat{g}}(\lambda), \\
S_{\hat{g}+1}(\lambda) &= (\lambda_1 + \cdots + \lambda_s)\, S_{\hat{g}}(\lambda),
\end{aligned}$$

satisfy the LMIs in Theorem 6.1 corresponding to the degree $\hat{g} + 1$, which can be obtained by linear combination of the LMIs of Theorem 6.1 for $\hat{g}$. □

Note that these conditions in (6.15) are LMIs not only in the matrix variables, but also in the scalar $\mu = \gamma^2$. This implies that the scalar $\mu = \gamma^2$ can be included as an optimization variable so as to directly optimize the guaranteed H_∞ filtering performance bound. Then based on Theorem 6.1, Algorithm 20 is readily obtained for designing suboptimal robust H_∞ filters for uncertain 2-D FM systems.

Algorithm 20 Robust 2-D H_∞ Filter Design for 2-D FM model

- Prescribe an integer $g \in \mathbb{N}_0^+$;
- Solve the minimization problem:

$$\min \quad \mu = \gamma^2 \quad \text{s.t. (6.15)}$$

$$\bar{P}_{K_j(s,g)} = \begin{bmatrix} \bar{P}_{1,K_j(s,g)} & \bar{P}_{2,K_j(s,g)} \\ * & \bar{P}_{3,K_j(s,g)} \end{bmatrix} = \bar{P}^{\mathrm{T}}_{K_j(s,g)},$$
$$\bar{Q}_{K_j(s,g)} = \begin{bmatrix} \bar{Q}_{1,K_j(s,g)} & \bar{Q}_{2,K_j(s,g)} \\ * & \bar{Q}_{3,K_j(s,g)} \end{bmatrix} = \bar{Q}^{\mathrm{T}}_{K_j(s,g)},$$
$$R_{K_j(s,g)},\ S_{K_j(s,g)},\ K_j(s,g) \in K(s,g),$$
$$j = 1, 2, \ldots, J(s,g),\ T,\ \bar{A}_{F1}, \bar{A}_{F2}, \bar{B}_{F1}, \bar{B}_{F2}, \bar{C}_F,\ \mu$$

- Compute the filter matrices from (6.9).

6.3 Roesser Model

In this section, we discuss the design of robust H_∞ filters for uncertain 2-D systems described by the Roesser model.

6.3.1 Problem Formulation

It is known that discrete-time 2-D systems can also be represented by another commonly used state space form, i.e., the Roesser model [7]. Specifically, let us consider a 2-D discrete system given by model:

$$\begin{aligned} \begin{bmatrix} x^h(i+1,j) \\ x^v(i,j+1) \end{bmatrix} &= A(\lambda) \begin{bmatrix} x^h(i,j) \\ x^v(i,j) \end{bmatrix} + B(\lambda) w(i,j), \\ y(i,j) &= C(\lambda) \begin{bmatrix} x^h(i,j) \\ x^v(i,j) \end{bmatrix} + D(\lambda) w(i,j), \\ z(i,j) &= H(\lambda) \begin{bmatrix} x^h(i,j) \\ x^v(i,j) \end{bmatrix}, \end{aligned} \tag{6.16}$$

where $x^h(i,j) \in \mathbb{R}^{n_1}$ is the sate vector in the horizontal direction, $x^v(i,j) \in \mathbb{R}^{n_2}$ the state vector in the vertical direction, $y(i,j) \in \mathbb{R}^m$ the measured signal vector, $z(i,j) \in \mathbb{R}^p$ the to-be-estimated signal vector, and $w(i,j) \in \mathbb{R}^q$ the disturbance signal vector. It is assumed that $w(i,j)$ belongs to $l_2\{[0,\infty), [0,\infty)\}$, that is, it is energy-bounded. Matrices in (6.16) have the block form as follows:

$$A(\lambda) \triangleq \begin{bmatrix} A_{11}(\lambda) & A_{12}(\lambda) \\ A_{21}(\lambda) & A_{22}(\lambda) \end{bmatrix}, \quad B(\lambda) \triangleq \begin{bmatrix} B_1(\lambda) \\ B_2(\lambda) \end{bmatrix},$$
$$C(\lambda) \triangleq \begin{bmatrix} C_1(\lambda) & C_2(\lambda) \end{bmatrix}, \quad H(\lambda) \triangleq \begin{bmatrix} H_1(\lambda) & H_2(\lambda) \end{bmatrix},$$

where the dimensions of the blocks are compatible with the vectors. System matrices are assumed to be uncertain and bounded in a polyhedral domain:

$$\Omega(\lambda) \triangleq \left(A(\lambda), B(\lambda), C(\lambda), D(\lambda), H(\lambda)\right) \in \mathcal{R}, \tag{6.17}$$

where $\mathcal{R}$ denotes a polytope defined as $\mathcal{R} \triangleq \left\{\Omega(\lambda) \,\middle|\, \Omega(\lambda) = \sum_{i=1}^{s} \lambda_i \Omega_i;\ \lambda \in \Gamma\right\}$ with $\Omega_i \triangleq (A_i, B_i, C_i, D_i, H_i)$ denoting the vertices of $\mathcal{R}$ and

$$\Gamma \triangleq \left\{(\lambda_1, \lambda_2, \ldots, \lambda_s) : \sum_{i=1}^{s} \lambda_i = 1,\ \lambda_i \geq 0\right\} \tag{6.18}$$

denoting a unit simplex.

To reconstruct the signal $z(i, j)$ from the measured signal $y(i, j)$, consider a filter that is also represented in the Roesser model form:

$$\begin{aligned}
\begin{bmatrix} \bar{x}^h(i+1, j) \\ \bar{x}^v(i, j+1) \end{bmatrix} &= A_F \begin{bmatrix} \bar{x}^h(i, j) \\ \bar{x}^v(i, j) \end{bmatrix} + B_F y(i, j), \\
\bar{z}(i, j) &= C_F \begin{bmatrix} \bar{x}^h(i, j) \\ \bar{x}^v(i, j) \end{bmatrix}, \\
\bar{x}^h(0, k) &= 0, \bar{x}^v(k, 0) = 0,\ \ \forall k \in \mathbb{N}_0^+,
\end{aligned} \tag{6.19}$$

where $\bar{x}^h(i, j) \in \mathbb{R}^{n_1}$ is the filter state vector in the horizontal direction, $\bar{x}^v(i, j) \in \mathbb{R}^{n_2}$ the filter state vector in the vertical direction, $y(i, j)$ the measured signal of system (6.16), and $\bar{z}(i, j) \in \mathbb{R}^p$ the estimate of $z(i, j)$, respectively. Real-valued matrices in the block form

$$A_F \triangleq \begin{bmatrix} A_{F11} & A_{F12} \\ A_{F21} & A_{F22} \end{bmatrix}, B_F \triangleq \begin{bmatrix} B_{F1} \\ B_{F2} \end{bmatrix}, C_F \triangleq \begin{bmatrix} C_{F1} & C_{F2} \end{bmatrix}$$

have dimensions compatible with the vectors. Define the estimation error and the augmented state vectors, respectively, as

$$e(i, j) \triangleq z(i, j) - \bar{z}(i, j),\ \xi^h(i, j) = \begin{bmatrix} x^h(i, j) \\ \bar{x}^h(i, j) \end{bmatrix},\ \xi^v(i, j) = \begin{bmatrix} x^v(i, j) \\ \bar{x}^v(i, j) \end{bmatrix}.$$

The filtering error system is given by

$$\begin{aligned}
\begin{bmatrix} \xi^h(i+1, j) \\ \xi^v(i, j+1) \end{bmatrix} &= \bar{A}(\lambda) \begin{bmatrix} \xi^h(i, j) \\ \xi^v(i, j) \end{bmatrix} + \bar{B}(\lambda) w(i, j), \\
e(i, j) &= \bar{C}(\lambda) \begin{bmatrix} \xi^h(i, j) \\ \xi^v(i, j) \end{bmatrix},
\end{aligned} \tag{6.20}$$

where

$$\bar{A}(\lambda) \triangleq \Upsilon^{\mathrm{T}} \begin{bmatrix} A(\lambda) & 0 \\ B_F C(\lambda) & A_F \end{bmatrix} \Upsilon,$$
$$\bar{B}(\lambda) \triangleq \Upsilon^{\mathrm{T}} \begin{bmatrix} B(\lambda) \\ B_F D(\lambda) \end{bmatrix},$$
$$\bar{C}(\lambda) \triangleq \begin{bmatrix} H(\lambda) & -C_F \end{bmatrix} \Upsilon,$$

with Υ being a permutation matrix defined by

$$\Upsilon \triangleq \left[\begin{array}{cc|cc} I_{n_1} & 0 & 0 & 0 \\ 0 & 0 & I_{n_2} & 0 \\ \hline 0 & I_{n_1} & 0 & 0 \\ 0 & 0 & 0 & I_{n_2} \end{array}\right].$$

Accordingly, we can obtain the transfer function of the filtering error system as

$$T_{\mathrm{ew}}(z_1, z_2, \lambda) = \bar{C}(\lambda) \left[\mathrm{diag}\left\{z_1 I_{2\times n_1}, z_2 I_{2\times n_2}\right\} - \bar{A}(\lambda)\right]^{-1} \bar{B}(\lambda).$$

With the above preparations, the robust H_∞ filtering problem for uncertain 2-D Roesser model to be addressed in this section is formulated as follows.

Robust H_∞ Filtering for Uncertain 2-D Roesser Systems: Given the Roesser system in (6.16) with parameter uncertainty in (6.17), find a filter in (6.19) such that the filter error system in (6.20) is robustly asymptotically stable for all $\lambda \in \Gamma$ and satisfies the following robust H_∞ performance:

$$\|T_{\mathrm{ew}}(z_1, z_2, \lambda)\|_\infty < \gamma, \ \forall \lambda \in \Gamma \tag{6.21}$$

where γ is a given positive scalar.

6.3.2 Filter Design

Note that the filter error system in (6.20) is also a 2-D system of the Roesser model. First we introduce a 2-D BRL for analyzing the H_∞ performance of the filtering error system in (6.20) [8].

Lemma 6.3 *Given a positive scalar γ, the filtering error system in* (6.20) *is asymptotically stable and satisfies* (6.21) *for any fixed $\lambda \in \Gamma$, if there exists matrices $P_h(\lambda) = P_h(\lambda)^{\mathrm{T}} \in \mathbb{R}^{2n_1\times 2n_1}$ and $P_v(\lambda) = P_v(\lambda)^{\mathrm{T}} \in \mathbb{R}^{2n_2\times 2n_2}$ such that*

$$\begin{bmatrix} -P(\lambda) & P(\lambda)\bar{A}(\lambda) & P(\lambda)\bar{B}(\lambda) & 0 \\ \bar{A}(\lambda)^{\mathrm{T}} P(\lambda) & -P(\lambda) & 0 & \bar{C}(\lambda)^{\mathrm{T}} \\ \bar{B}^T P(\lambda) & 0 & -\gamma^2 I & 0 \\ 0 & \bar{C}(\lambda) & 0 & -I \end{bmatrix} < 0 \tag{6.22}$$

where $P(\lambda) = \operatorname{diag}\{P_h(\lambda), P_v(\lambda)\}$

The condition in (6.22) is a parameter-dependent LMI in block diagonal matrix function $P(\lambda)$. To reduce the conservatism of the above condition in robustness analysis, by the slack matrix technique in [6], we introduce a slack matrix to eliminate the product terms between $P(\lambda)$ and $\bar{A}(\lambda)$, $\bar{B}(\lambda)$ in (6.22). This manipulation results in the following condition: (6.22) holds for some $P(\lambda) > 0$ if and only if there exist matrices $P(\lambda) > 0$ and $V(\lambda)$ satisfying

$$\Xi(\lambda) \triangleq \begin{bmatrix} \Xi_1(\lambda) & V(\lambda)\bar{A}(\lambda) & V(\lambda)\bar{B}(\lambda) & 0 \\ \bar{A}(\lambda)^{\mathrm{T}} V(\lambda)^{\mathrm{T}} & -P(\lambda) & 0 & \bar{C}(\lambda)^{\mathrm{T}} \\ \bar{B}^{\mathrm{T}} V(\lambda)^{\mathrm{T}} & 0 & -\gamma^2 I & 0 \\ 0 & \bar{C}(\lambda) & 0 & -I \end{bmatrix} < 0. \tag{6.23}$$

where $\Xi_1(\lambda) = P(\lambda) - V(\lambda) - V(\lambda)^{\mathrm{T}}$ with $V(\lambda)$ being a parameter-dependent slack matrix. Partition $P(\lambda)$ and define $\bar{P}(\lambda)$ as

$$P(\lambda) = \operatorname{diag}\left\{\begin{bmatrix} P_{1h}(\lambda) & P_{2h}(\lambda) \\ P_{2h}(\lambda)^{\mathrm{T}} & P_{3h}(\lambda) \end{bmatrix}, \begin{bmatrix} P_{1v}(\lambda) & P_{2v}(\lambda) \\ P_{2v}(\lambda)^{\mathrm{T}} & P_{3v}(\lambda) \end{bmatrix}\right\},$$
$$\bar{P}(\lambda) \triangleq \Upsilon P(\lambda) \Upsilon^{\mathrm{T}} > 0. \tag{6.24}$$

The dimensions of the blocks can be easily determined by consistency. Moreover, let $V(\lambda)$ be with special structure such that the following notation is well-defined:

$$\bar{V}(\lambda) \triangleq \begin{bmatrix} \bar{V}_1(\lambda) & T \\ \bar{V}_2(\lambda) & T \end{bmatrix} \triangleq \Upsilon V(\lambda) \Upsilon^{\mathrm{T}}. \tag{6.25}$$

With the definitions in (6.24) and (6.25) and noting that $\Upsilon^{\mathrm{T}}\Upsilon = \Upsilon\Upsilon^{\mathrm{T}} = I$, one knows that (6.23) can be guaranteed by

$$\bar{\Xi}(\lambda) \triangleq \begin{bmatrix} \bar{\Xi}_1(\lambda) & \bar{\Xi}_2(\lambda) & \bar{\Xi}_3(\lambda) & 0 \\ * & -\bar{P}(\lambda) & 0 & \bar{\Xi}_4(\lambda)^{\mathrm{T}} \\ * & * & -\gamma^2 I & 0 \\ * & * & * & -I \end{bmatrix} < 0, \tag{6.26}$$

where

$$\begin{aligned} \bar{\Xi}_1(\lambda) &\triangleq \Upsilon \Xi_1(\lambda) \Upsilon^{\mathrm{T}} \\ &= \bar{P}(\lambda) - \bar{V}(\lambda) - \bar{V}(\lambda)^{\mathrm{T}}, \\ \bar{\Xi}_2(\lambda) &\triangleq \Upsilon V(\lambda) \bar{A}(\lambda) \Upsilon^{\mathrm{T}} \\ &= \begin{bmatrix} \bar{V}_1(\lambda) A(\lambda) + T B_F C(\lambda) & T A_F \\ \bar{V}_2(\lambda) A(\lambda) + T B_F C(\lambda) & T A_F \end{bmatrix}, \\ \bar{\Xi}_3(\lambda) &\triangleq \Upsilon V(\lambda) \bar{B}(\lambda) \end{aligned}$$

$$= \begin{bmatrix} \bar{V}_1(\lambda) B(\lambda) + T B_F D(\lambda) \\ \bar{V}_2(\lambda) B(\lambda) + T B_F D(\lambda) \end{bmatrix},$$
$$\bar{\Xi}_4(\lambda) \triangleq \bar{C}(\lambda) \Upsilon^{\mathrm{T}}$$
$$= \begin{bmatrix} H(\lambda) & -C_F \end{bmatrix}.$$

Here, a congruence transformation $\bar{\Xi}(\lambda) = J \Xi(\lambda) J^{\mathrm{T}}$ with $J = \mathrm{diag}\{\Upsilon, \Upsilon, I, I\}$ has been applied. Matrix $\bar{P}(\lambda)$ in (6.24) can be denoted as

$$\bar{P}_1(\lambda) = \begin{bmatrix} \bar{P}_1(\lambda) & \bar{P}_2(\lambda) \\ \bar{P}_2(\lambda)^{\mathrm{T}} & \bar{P}_3(\lambda) \end{bmatrix}; \; \bar{P}_i(\lambda) = \mathrm{diag}\{P_{ih}(\lambda), P_{iv}(\lambda)\}, \; i = 1, 2, 3.$$

By the change of variables

$$\bar{A}_F = T A_F, \; \bar{B}_F = T B_F, \; \bar{C}_F = C_F,$$

(6.26) can be expressed linearly with respect to variables

$$\bar{P}_i(\lambda) = \mathrm{diag}\{P_{ih}(\lambda), P_{iv}(\lambda)\} \; i = 1, 2, 3,$$
$$\bar{V}_1(\lambda), \; \bar{V}_2(\lambda), \; T, \; \bar{A}_F, \; \bar{B}_F, \; \bar{C}_F.$$

Moreover, (6.26) implies that $\bar{P}_3(\lambda) - T - T^{\mathrm{T}} < 0$. If (6.26) is satisfied, T is invertible, that is, the above change of variables is also invertible and the filter matrices A_F, B_F and C_F can be obtained from $\bar{A}_F$, $\bar{B}_F$, $\bar{C}_F$ and T. Thus, we have the following lemma on the filter realization in terms of parameter-dependent LMI.

Lemma 6.4 *Given a positive scalar* γ*, a filter in* (6.19) *exists such that the filtering error system in* (6.20) *is asymptotically stable and satisfies* (6.21) *for any fixed* $\lambda \in \Gamma$*, if there exist real matrices* $\bar{P}_i(\lambda) = \mathrm{diag}\{\bar{P}_{ih}(\lambda), \bar{P}_{iv}(\lambda)\} = \bar{P}_i(\lambda)^{\mathrm{T}}$, $i = 1, 3$, $\bar{P}_2(\lambda) = \mathrm{diag}\{\bar{P}_{2h}(\lambda), \bar{P}_{2v}(\lambda)\}$, $\bar{V}_i(\lambda)$, $i = 1, 2$, T *and* $\bar{A}_F$, $\bar{B}_F$, $\bar{C}_F$ *satisfying*

$$\Delta(\lambda) \triangleq \begin{bmatrix} \Phi_1(\lambda) & \Phi_2(\lambda) & \Phi_3(\lambda) & 0 \\ * & -\bar{P}(\lambda) & 0 & \Phi_4(\lambda)^{\mathrm{T}} \\ * & * & -\gamma^2 I & 0 \\ * & * & * & -I \end{bmatrix} < 0, \tag{6.27}$$

where

$$\Phi_1(\lambda) \triangleq \bar{P}(\lambda) - \mathrm{sym}\left\{\begin{bmatrix} \bar{V}_1(\lambda) & T \\ \bar{V}_2(\lambda) & T \end{bmatrix}\right\},$$
$$\Phi_2(\lambda) \triangleq \begin{bmatrix} \bar{V}_1(\lambda) A(\lambda) + \bar{B}_F C(\lambda) & \bar{A}_F \\ \bar{V}_2(\lambda) A(\lambda) + \bar{B}_F C(\lambda) & \bar{A}_F \end{bmatrix},$$

$$\Phi_3(\lambda) \triangleq \begin{bmatrix} \bar{V}_1(\lambda) B(\lambda) + \bar{B}_F D(\lambda) \\ \bar{V}_2(\lambda) B(\lambda) + \bar{B}_F D(\lambda) \end{bmatrix},$$

$$\Phi_4(\lambda) \triangleq \begin{bmatrix} H(\lambda) & -\bar{C}_F \end{bmatrix}.$$

Moreover, if the above conditions are satisfied, the matrices for an admissible realization of the filter are given by

$$\begin{bmatrix} A_F & B_F \\ C_F & 0 \end{bmatrix} \triangleq \begin{bmatrix} T^{-1} & 0 \\ 0 & I \end{bmatrix} \begin{bmatrix} \bar{A}_F & \bar{B}_F \\ \bar{C}_F & 0 \end{bmatrix}. \tag{6.28}$$

Proof The proof can be completed according to the above discussions. □

By applying the structurally polynomially parameter-dependent method to the condition in (6.27), we obtain the following result for the 2-D Roesser model with unknown λ, that is, a polynomially parameter-dependent approach to robust 2-D H_∞ filter design. The proof is omitted.

Theorem 6.2 *Given a stable 2-D Roesser system* (6.16) *and a scalar* $\gamma > 0$, *a filter in* (6.19) *exists such that the filtering error system in* (6.20) *is stable and satisfying* (6.21), *if there exist real matrices*

$$\bar{P}_{i,K_j(s,g)} = \operatorname{diag}\left\{\bar{P}_{ih,K_j(s,g)}, \bar{P}_{iv,K_j(s,g)}\right\} = \bar{P}^{\mathrm{T}}_{i,K_j(s,g)},\ i = 1, 3,$$
$$\bar{P}_{2,K_j(s,g)} = \operatorname{diag}\left\{\bar{P}_{2h,K_j(s,g)}, \bar{P}_{2v,K_j(s,g)}\right\},$$

$\bar{V}_{i,K_j(s,g)}$, $i = 1, 2$, $K_j(s, g) \in K(s, g)$, $j = 1, 2, \ldots, J(s, g)$ *and* T, $\bar{A}_F$, $\bar{B}_F$, $\bar{C}_F$ *such that the following LMIs hold for all* $K_l(s, g+1) \in K(s, g+1)$, $l = 1, 2, \ldots, J(s, g+1)$*:*

$$\Delta_l \triangleq \sum_{i \in I_l(s,g+1)} \begin{bmatrix} \Psi_1 & \Psi_2 & \Psi_3 & 0 \\ * & -\bar{P}_{K_l^i(s,g+1)} & 0 & \Psi_4^{\mathrm{T}} \\ * & * & -\gamma^2 \beta_l^i(s, g+1) I & 0 \\ * & * & * & -\beta_l^i(s, g+1) I \end{bmatrix} < 0, \tag{6.29}$$

where

$$\Psi_1 \triangleq \begin{bmatrix} \bar{P}_{1,K_l^i(s,g+1)} & \bar{P}_{2,K_l^i(s,g+1)} \\ * & \bar{P}_{3,K_l^i(s,g+1)} \end{bmatrix} - \operatorname{sym}\left\{\begin{bmatrix} \bar{V}_{1,K_l^i(s,g+1)} & \beta_l^i(s, g+1) T \\ \bar{V}_{2,K_l^i(s,g+1)} & \beta_l^i(s, g+1) T \end{bmatrix}\right\},$$

$$\Psi_2 \triangleq \begin{bmatrix} \bar{V}_{1,K_l^i(s,g+1)} A_i + \beta_l^i(s, g+1) \bar{B}_F C_i & \beta_l^i(s, g+1) \bar{A}_F \\ \bar{V}_{2,K_l^i(s,g+1)} A_i + \beta_l^i(s, g+1) \bar{B}_F C_i & \beta_l^i(s, g+1) \bar{A}_F \end{bmatrix},$$

Algorithm 21 Robust 2-D H_∞ Filter Design for 2-D Roesser model

- Prescribe an integer $g \in \mathbb{N}_0^+$;
- Solve the minimization problem:

$$\min_{\substack{\bar{P}_{i,K_j(s,g)} = \text{diag}\left\{\bar{P}_{ih,K_j(s,g)},\ \bar{P}_{iv,K_j(s,g)}\right\} \\ = \bar{P}^{\mathrm{T}}_{i,K_j(s,g)}\ (i=1,3), \\ \bar{P}_{2,K_j(s,g)} = \text{diag}\left\{\bar{P}_{2h,K_j(s,g)},\ \bar{P}_{2v,K_j(s,g)}\right\}, \\ \bar{V}_{i,K_j(s,g)}\ (i=1,2),\ K_j(s,g) \in K(s,g), \\ j=1,2,\ldots,J(s,g),\ T,\ \bar{A}_F,\ \bar{B}_F,\ \bar{C}_F,\ \mu}} \mu = \gamma^2 \quad \text{s.t.}\ (6.29)$$

- Compute the filter matrices from (6.28).

$$\Psi_3 \triangleq \begin{bmatrix} \bar{V}_{1,K_l^i(s,g+1)} B_i + \beta_l^i(s,g+1)\bar{B}_F D_i \\ \bar{V}_{2,K_l^i(s,g+1)} B_i + \beta_l^i(s,g+1)\bar{B}_F D_i \end{bmatrix},$$
$$\Psi_4 \triangleq \left[\beta_l^i(s,g+1)H_i \ \ -\beta_l^i(s,g+1)\bar{C}_F\right].$$

If the LMIs in (6.29) *are satisfied, the matrices for an admissible realization of the filter are given by* (6.28). *Moreover, if the LMIs in* (6.29) *are satisfied for a given degree* $\hat{g}$*, then the LMIs corresponding to any degree* $g > \hat{g}$ *are also satisfied.*

Accordingly, a polynomially parameter-dependent approach to designing suboptimal robust 2-D H_∞ filters for uncertain 2-D Roesser systems is readily obtained, which is given in Algorithm 21.

6.4 Illustrative Examples

This section includes two numerical examples to demonstrate the effectiveness of the presented robust H_∞ filter design methods for 2-D systems.

Example 6.1 Consider the following 2-D static random field model described by difference equation [5]:

$$\eta(i+1,j+1) = a_1\eta(i,j+1) + a_2\eta(i+1,j) - a_1a_2\eta(i,j) + \omega_1(i,j+1), \tag{6.30}$$

where $\eta(i,j)$ is the state of coordinates (i,j), and a_1 and a_2 are the vertical and horizontal correlative coefficients, respectively, satisfying $a_1^2 < 1$ and $a_2^2 < 1$.

Defining the augmented state vector

$$x(i,j) = \left[\eta(i,j+1)^{\mathrm{T}} - a_2\eta(i,j)^{\mathrm{T}}\ \eta(i,j)^{\mathrm{T}}\right]^{\mathrm{T}},$$

and supposing that the measured signal and the signal to be estimated are

$$y(i,j) = a_1\eta(i,j+1) + (1 - a_1a_2)\eta(i,j) + \omega_2(i,j),$$
$$z(i,j) = \eta(i,j),$$

it is not difficult to transform the above difference equation into a 2-D FM model in the form of (6.1), with the corresponding system matrices given by

$$A_1(\lambda) = \begin{bmatrix} a_1 & 0 \\ 0 & 0 \end{bmatrix}, \; A_2(\lambda) = \begin{bmatrix} 0 & 0 \\ 1 & a_2 \end{bmatrix}, \; B_1(\lambda) = \begin{bmatrix} 1 & 0 \\ 0 & 0 \end{bmatrix},$$
$$B_2(\lambda) = \begin{bmatrix} 0 & 0 \\ 0 & 0 \end{bmatrix}, \; C(\lambda) = \begin{bmatrix} a_1 \\ 1 \end{bmatrix}^{\mathrm{T}}, \; D(\lambda) = H(\lambda) = \begin{bmatrix} 0 \\ 1 \end{bmatrix}^{\mathrm{T}}.$$

Suppose that a_1 and a_2 are uncertain but bounded by $0.15 \le a_1 \le 0.45$, and $0.35 \le a_2 \le 0.85$, respectively, and thus the above system can be represented by a four-vertex polytopic system. We apply the suboptimal H_∞ filter design method, Algorithm 20, in the quadratic framework (that is, $g = 0$), and the achieved H_∞ disturbance attenuation level is $\gamma^* = 3.8709$. Moreover, the obtained guaranteed performances for $g = 1$ (linearly parameter-dependent approach) and $g = 2$ are 2.5450 and 2.5028, respectively. The obtained results show that as g increases, less conservative filter designs are achieved. For $g = 2$, the state space matrices of the designed filter are given by

$$\begin{bmatrix} A_{F1} & B_{F1} \\ A_{F2} & B_{F2} \\ C_F & 0 \end{bmatrix} = \left[\begin{array}{cc|c} 0.4465 & -0.1486 & -0.1681 \\ 0.0321 & -0.0105 & -0.0151 \\ \hline -0.1094 & 0.0142 & 0.0248 \\ 0.2995 & 0.2501 & -0.5925 \\ \hline 0.0062 & -1.0022 & 0 \end{array}\right]. \tag{6.31}$$

By connecting the above filter to the original FM system, Figs. 6.1, 6.2, 6.3 and 6.4 display the frequency responses of the filtering error system at the four vertices of the uncertainty domain. It is shown that all the singular values of the filtering error system at the four vertices are below the guaranteed H_∞ disturbance attenuation level bound $\gamma^* = 2.5028$.

Example 6.2 Still consider the static random field model in (6.30). In this example, remodel it by the 2-D Roesser description as in (6.16).

Similar to the manipulation in [9], denote the state vector

$$x^h(i,j) = \eta(i,j+1) - a_2\eta(i,j), \; x^v(i,j) = \eta(i,j),$$

and suppose that the measured signal and the signal to be estimated are

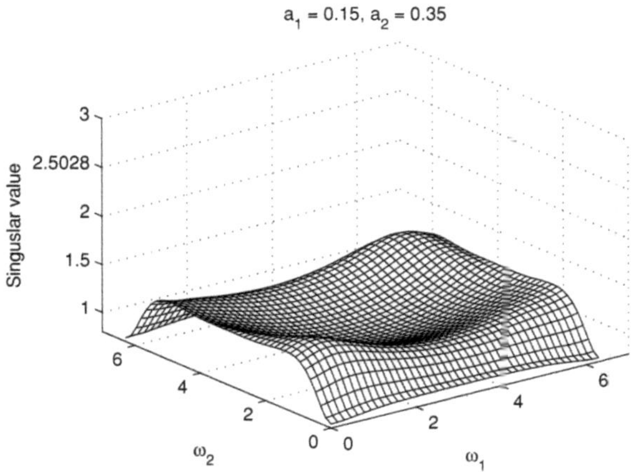

Fig. 6.1 Singular value curve of the filtering error transfer function with the filter in Eq. (6.31) in Example 6.1 for $(a_1, a_2) = (0.15, 0.35)$

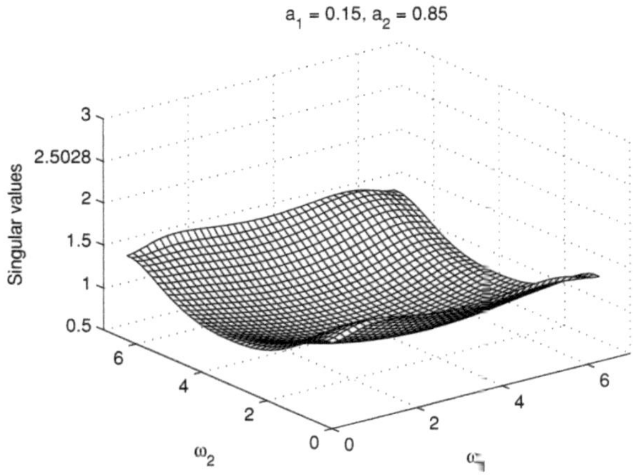

Fig. 6.2 Singular value curve of the filtering error transfer function with the filter in Eq. (6.31) in Example 6.1 for $(a_1, a_2) = (0.15, 0.85)$

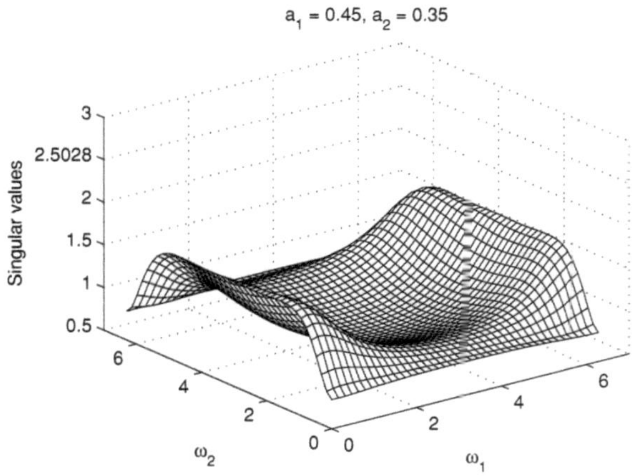

Fig. 6.3 Singular value curve of the filtering error transfer function with the filter in Eq. (6.31) in Example 6.1 for $(a_1, a_2) = (0.45, 0.35)$

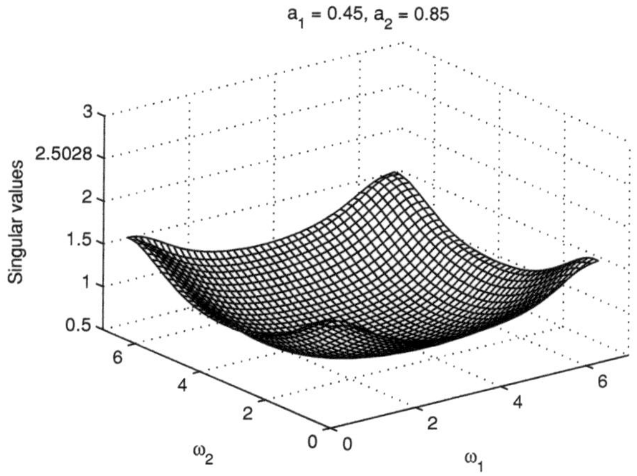

Fig. 6.4 Singular value curve of the filtering error transfer function with the filter in Eq. (6.31) in Example 6.1 for $(a_1, a_2) = (0.45, 0.85)$

$$
\begin{aligned}
y(i, j) &= a_1 \eta(i, j+1) + (1 - a_1 a_2)\eta(i, j) + \omega_2(i, j), \\
z(i, j) &= \eta(i, j).
\end{aligned}
$$

It is easy to express the original system in the form of the 2-D Roesser model as:

$$
\begin{aligned}
\begin{bmatrix} x^h(i+1, j) \\ x^v(i, j+1) \end{bmatrix} &= \begin{bmatrix} a_1 & 0 \\ 1 & a_2 \end{bmatrix} \begin{bmatrix} x^h(i, j) \\ x^v(i, j) \end{bmatrix} + \begin{bmatrix} 1 & 0 \\ 0 & 0 \end{bmatrix} \begin{bmatrix} w_1(i, j) \\ w_2(i, j) \end{bmatrix}, \\
y(i, j) &= \begin{bmatrix} a_1 & 1 \end{bmatrix} \begin{bmatrix} x^h(i, j) \\ x^v(i, j) \end{bmatrix} + \begin{bmatrix} 0 & 1 \end{bmatrix} \begin{bmatrix} w_1(i, j) \\ w_2(i, j) \end{bmatrix}, \\
z(i, j) &= \begin{bmatrix} 0 & 1 \end{bmatrix} \begin{bmatrix} x^h(i, j) \\ x^v(i, j) \end{bmatrix}.
\end{aligned}
$$

Also suppose that a_1 and a_2 are bounded by $0.15 \leq a_1 \leq 0.45$ and $0.35 \leq a_2 \leq 0.85$, respectively, such that the above system is subject to polytopic uncertainty with four vertices. Apply the suboptimal H_∞ filter design method, Algorithm 21, in the quadratic framework ($g = 0$) and the parameter-dependent framework ($g = 1$, $g = 2$). The achieved H_∞ disturbance attenuation levels are $\gamma^* = 2.4356$, 1.8586, and 1.8295, respectively, demonstrating that less conservative filter designs are obtained by increasing g. For $g = 2$, the state space matrices of the designed filter of the Roesser model form are given by

$$
\begin{bmatrix} A_F & B_F \\ C_F & 0 \end{bmatrix} = \left[\begin{array}{cc|c} 0.5305 & -0.1201 & -0.1045 \\ 0.1958 & 0.1839 & -0.6599 \\ \hline -0.0814 & -1.0059 & 0 \end{array}\right]. \tag{6.32}
$$

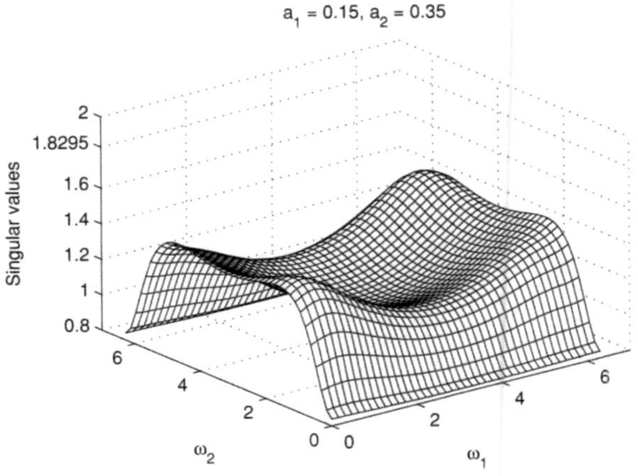

Fig. 6.5 Singular value curve of the filtering error transfer function with the filter in Eq. (6.32) in Example 6.2 for $(a_1, a_2) = (0.15, 0.35)$

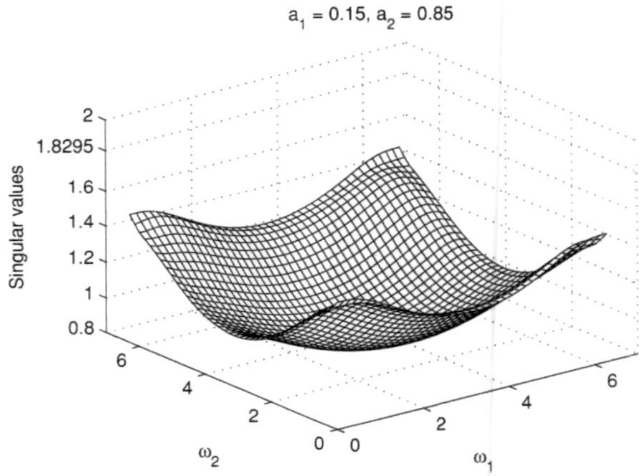

Fig. 6.6 Singular value curve of the filtering error transfer function with the filter in Eq. (6.32) in Example 6.2 for $(a_1, a_2) = (0.15, 0.85)$

By connecting the above filter to the original Roesser system, we depict the frequency responses of the filtering error system at the four vertices, which are displayed in Figs. 6.5, 6.6, 6.7, and 6.8. It is clearly seen that all the singular values of the filtering error system at the four vertices are smaller than the guaranteed H_∞ disturbance attenuation level bound $\gamma^* = 1.8295$, illustrating the effectiveness of the designed filter (6.32).

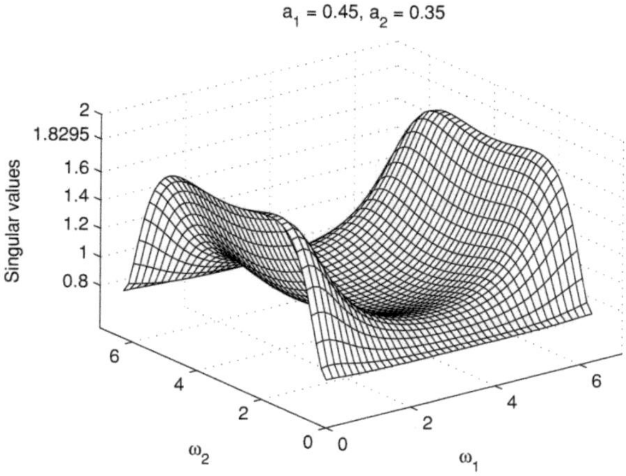

Fig. 6.7 Singular value curve of the filtering error transfer function with the filter in Eq. (6.32) in Example 6.2 for $(a_1, a_2) = (0.45, 0.35)$

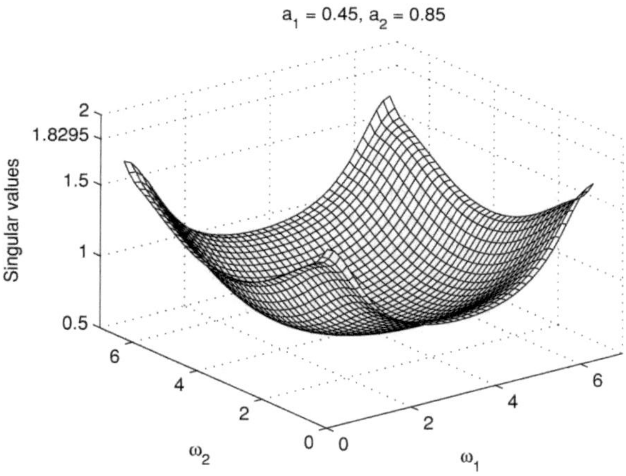

Fig. 6.8 Singular value curve of the filtering error transfer function with the filter in Eq. (6.32) in Example 6.2 for $(a_1, a_2) = (0.45, 0.85)$

6.5 Summary and Notes

6.5.1 Summary

In this chapter, we have considered the problem of robust H_∞ filter design for uncertain 2-D systems described by the FM model and the Roesser model, and by further extending the polynomially parameter-dependent idea arising in 1-D systems to 2-D systems, proposed new robust approaches to H_∞ filter design for the two models, respectively. It has been proven that as the degree grows, test of increasing precision is obtained and less conservative filter designs are generated. The proposed conditions include the results in the quadratic framework and the linearly parameter-dependent framework as special cases. Numerical examples has been provided to illustrate the effectiveness of the presented filter design method.

6.5.2 Notes

Using the LMI technique, the problem of H_∞ filtering has been addressed in [1, 2, 9–19], of which [2, 10, 11, 13, 14, 16, 19] are concerned with the FM model, [1, 9, 12, 15, 18] with the Roesser model, and [17] with linear repetitive processes that can be formulated as a 2-D model.

References

1. Xu, S., Lam, J., Zou, Y., Lin, Z., Paszke, W.: Robust H_∞ filtering for uncertain 2-D continuous systems. IEEE Trans. Signal Process. **53**(5), 1731–1738 (2005)
2. Chen, S.F., Fong, I.K.: Robust filtering for 2-D state-delayed systems with NFT uncertainties. IEEE Trans. Signal Process. **54**(1), 274–285 (2006)
3. Gao, H., Lam, J., Wang, C.: Robust H_∞ filtering for 2D stochastic systems. Circuits Syst. Signal Process. **23**(6), 479–505 (2004)
4. Fornasini, E., Marchesini, G.: Doubly indexed dynamical systems: state-space models and structual properties. Math. Syst. Theory **12**, 59–72 (1978)
5. Du, C., Xie, L.: H_∞ Control and Filtering of Two-dimensional Systems. Springer, Berlin (2002)
6. de Oliveira, M.C., Bernussou, J., Geromel, J.C.: A new discrete-time robust stability condition. Syst. Control Lett. **37**, 261–265 (1999)
7. Roesser, R.P.: A discrete state-space model for linear image processing. IEEE Trans. Autom. Control **20**, 1–10 (1975)
8. Du, C., Xie, L., Zhang, C.: H_∞ control and robust stabilization of two-dimensional systems in Roesser models. Automatica **37**, 205–211 (2001)
9. Du, C., Xie, L., Zhang, C.: Solutions for H_∞ filtering of two-dimensional systems. Multidimens. Syst. Signal Process. **11**(4), 301–320 (2000)
10. Du, C., Xie, L., Soh, Y.C.: H_∞ filtering of 2-d discrete systems. IEEE Trans. Signal Process. **48**(6), 1760–1768 (2000)
11. Tuan, H.D., Apkarian, P., Nguyen, T.Q.: Robust mixed H_2/H_∞ filtering of 2-D systems. IEEE Trans. Signal Process. **50**(7), 1759–1771 (2002)
12. Gao, C.Y., Duan, G.R., Meng, X.Y.: Robust H_∞ filter design for 2D discrete systems in Roesser model. Int. J. Autom. Control **05**(4), 413–418 (2008)
13. Gao, H., Meng, X., Chen, T.: New design of robust H_∞ filters for 2-D systems. IEEE Signal Process. Lett. **15**, 217–220 (2008)
14. de Souza, C.E., Xie, L., Coutinho, D.F.: Robust filtering for 2-D discrete-time linear systems with convex-bounded parameter uncertainty. Automatica **46**(4), 673–681 (2010)
15. Xu, H., Lin, Z., Makur, A.: Non-fragile h_2 and H_∞ filter designs for polytopic two-dimensional systems in Roesser model. Multidimens. Syst. Signal Process. **21**(3), 255–275 (2010)
16. Hoang, N.T., Tuan, H.D., Nguyen, T.Q., Hosoe, S.: Robust mixed generalized H_2/H_∞ filtering of 2-D nonlinear fractional transformation systems. IEEE Trans. Signal Process. **53**(12), 4697–4706 (2005)
17. Wu, L., Lam, J., Paszke, W., Galkowski, K., Rogers, E.: Robust H_∞ filtering for uncertain differential linear repetitive processes. Int. J. Adapt. Control Signal Process. **22**, 243–265 (2008)
18. Li, X., Gao, H.: Robust finite frequency H_∞ filtering for uncertain 2-D Roesser systems. Automatica **48**(6), 1163–1170 (2012)
19. Li, X., Gao, H.: Robust finite frequency H_∞ filtering for uncertain 2-D systems: The FM model case. Automatica **49**(8), 2446–2452 (2013)

Part III
New Problems in Robust Filtering

Chapter 7
Robust Estimation with Limited Communication Capacity

7.1 Introduction

The standard Kalman filtering theory and H_∞ filtering theory are based on an implicit assumption that there is a perfect communication channel between the measurement $y(t)$ of the system and the input $\tilde{y}(t)$ of the filter, giving rise to $y(t) = \tilde{y}(t)$. In many practical situations, however, the physical plant and the filter are difficult to be located at the same place, and thus measured signals are required to be transmitted from one place to another through communication channels. Networked systems have several advantages over traditional systems, such as low cost, reduced weight and power requirements, simple installation and maintenance, and high reliability. For the estimation problem, an immediate concern arising from the introduction of these network media is that $y(t)$ no longer equals to $\tilde{y}(t)$, which makes the estimation problem much more involved and challenging [1–3].

What aspects should be considered for a typical communication channel? Probably the most significant one is *signal transmission delay*, which is usually caused by the limited bit rate of the communication channel, by a node waiting to send out a packet via a busy channel, or by signal processing and propagation. The second interesting problem in a communication channel is the *data packet dropout* phenomenon, which is usually caused by the unavoidable errors or losses in the transmission. Though many networked systems employ automatic repeat request mechanisms, packet dropout phenomenon is still unavoidable. Moreover, packet dropouts may happen if one packet sampled at the sensor node reaches the destination later than its successors. In this situation, it is natural to use the most updated packet by dropping out the old ones, giving rise to the packet dropout phenomenon. Another important issue in a communication channel is the *quantization effect*. In a network environment, signals are usually quantized before being communicated, and the number of quantization levels is closely related to the information flow between the components of the system and thus to the capacity required to transmit the information. The classical estimation theory, which is based on the standard assumption that data transmission required by the system can be performed with infinite precision, may

H. Gao and X. Li, *Robust Filtering for Uncertain Systems*,
Communications and Control Engineering, DOI: 10.1007/978-3-319-05903-7_7,

not be valid in the presence of signal quantization, and therefore there is a need for developing tools for analysis and design of signal quantized systems. In summary, signal transmission delay, data packet dropout, and signal quantization are three important issues to be addressed for communication-based systems.

For NCSs with limited communication capacity, there have been a few important results reported in the literature. To mention a few, stability analysis is presented in [4]; stabilization problem is solved in [5–8]; H_∞ control is addressed in [9, 10]; and a moving horizon control is reported in [11]. It is worth noting that the results obtained for NCSs are still very limited: (i) most of the aforementioned results only consider one or two aspects of the communication issues, while few literature addresses the analysis or synthesis problems with simultaneous consideration of the three important communication issues mentioned above; (ii) most of the obtained results are concerned with the control problem, while little attention has been paid to the estimation problem with limited communication capacity, which is equally important in applications.

In this chapter, we investigate the problem of robust H_∞ filtering for uncertain linear systems subject to limited communication capacity. The parameter uncertainty is assumed to be of the polytopic type, and the information limitations include all the three important aspects analyzed above (that is, measurement quantization, signal transmission delay, and data packet dropout), which appear typically in a network environment. The physical plant and the filter are assumed to be in continuous time, and the measurement is sampled and transmitted over communication networks periodically. The sampling behavior, together with the signal transmission delay and data packet dropout, is dealt with via a delay system approach, and the measurement quantization is treated using a sector bound method. By using a Lyapunov-Krasovskii approach, the problem of H_∞ filter design is first solved for a nominal system, which is then extended to the uncertain case based on the notion of quadratic stability. To further reduce the overdesign in the quadratic framework, we also propose a parameter-dependent filter design procedure, which is much less conservative than the quadratic approach. The quadratic and parameter-dependent approaches provide alternatives for designing robust H_∞ filters with different degrees of conservativeness and computational complexity. Two examples, including a mass-spring system, are utilized to illustrate the design procedures proposed in this chapter.

7.2 Problem Formulation

The filtering problem for NCSs with limited communication capacity is shown in Fig. 7.1. Suppose the physical plant is given by

$$\begin{aligned} \mathcal{S}: \dot{x}(t) &= Ax(t) + Bw(t), \\ y(t) &= Cx(t) + Dv(t), \\ z(t) &= Lx(t), \end{aligned} \tag{7.1}$$

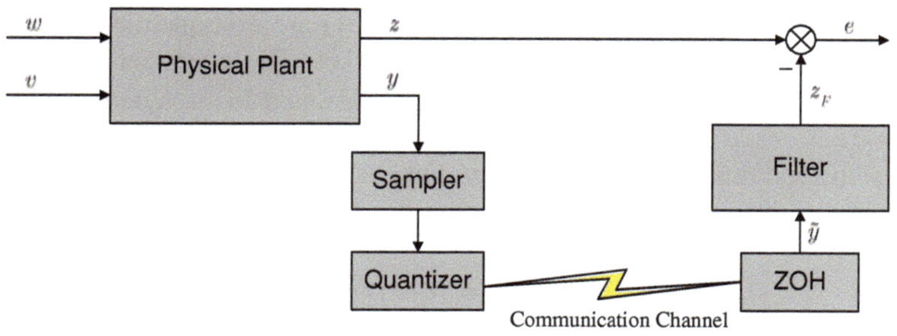

Fig. 7.1 Diagram for filtering problem with limited communication capacity

where $x(t) \in \mathbb{R}^n$ is the state vector; $y(t) \in \mathbb{R}^m$ is the measured output; $z(t) \in \mathbb{R}^p$ is the signal to be estimated; $w(t) \in \mathbb{R}^l$ and $v(t) \in \mathbb{R}^q$ are the input signal and measurement noise respectively, which belong to $L_2[0, \infty)$. The system matrices A, B, C, D, and L are appropriately dimensioned with partially unknown parameters. It is assumed that

$$\Omega \triangleq (A, B, C, D, L) \in \mathcal{R}, \tag{7.2}$$

where $\mathcal{R}$ is a given convex bounded polyhedral domain described by s vertices:

$$\mathcal{R} \triangleq \left\{ \Omega \,\middle|\, \Omega = \sum_{i=1}^{s} \lambda_i \Omega_i; \quad \sum_{i=1}^{s} \lambda_i = 1, \quad \lambda_i \geq 0 \right\},$$

with $\Omega_i \triangleq (A_i, B_i, C_i, D_i, L_i)$ denoting the vertices of the polytope. In addition, without loss of generality it is assumed that $w(t) = 0$, $v(t) = 0$, $x(t) = 0$ for $t < 0$.

The problem considered in this chapter is to estimate the signal $z(t)$ by a filter of a general structure described by

$$\begin{aligned} \mathcal{F} : \dot{x}_F(t) &= A_F x_F(t) + B_F \tilde{y}(t), \\ z_F(t) &= C_F x_F(t) + D_F \tilde{y}(t), \end{aligned} \tag{7.3}$$

where $x_F(t) \in \mathbb{R}^n$ is the filter state vector, $\tilde{y}(t)$ is the input of the filter, and A_F, B_F, C_F and D_F are appropriately dimensioned filter matrices to be determined.

From Fig. 7.1, we can see that there is a communication channel between $y(t)$ and $\tilde{y}(t)$. As discussed above, three effects are generally needed to be taken into consideration: measurement quantization, signal transmission delay, and data packet dropout. In what follows, we model them mathematically.

- *Measurement quantization*: It is assumed that the sampler and quantizer are clock-driven, while the zero-order hold (ZOH) is event-driven. The sampling period is assumed to be h where h is a positive real constant and we denote the *sampling instants* of the sampler as s_k, $k = 1, 2, \ldots, \infty$. In addition, it is assumed that the sampled measurements of $y(t)$ are first quantized via a quantizer, and then transmitted with a single packet. The quantizer is denoted as $f(\cdot) = \left[f_1(\cdot)\ f_2(\cdot) \cdots f_m(\cdot) \right]^{\mathrm{T}}$, which is assumed to be symmetric, that is, $f_j(-v) = -f_j(v)$, $j = 1, 2, \ldots, m$. Here, let us only consider the logarithmic static and time-invariant quantizer. For each $f_j(\cdot)$, the set of quantized levels is described by

$$\mathcal{U}_j = \left\{ \pm u_i^{(j)},\quad i = 0, \pm 1, \pm 2, \cdots \right\} \cup \{0\}. \tag{7.4}$$

According to [12, 13], a quantizer is called *logarithmic* if the set of quantized levels is characterized by

$$\begin{aligned} \mathcal{U}_j = & \left\{ \pm u_i^{(j)},\ u_i^{(j)} = \rho_j^i u_0^{(j)},\ i = \pm 1, \pm 2, \cdots \right\} \\ & \cup \left\{ \pm u_0^{(j)} \right\} \cup \{0\},\ 0 < \rho_j < 1,\ u_0^{(j)} > 0. \end{aligned} \tag{7.5}$$

Each of the quantization level $u_i^{(j)}$ corresponds to a segment such that the quantizer maps the whole segment to this quantization level. In addition, these segments form a partition of $\mathbb{R}$, that is, they are disjoint and their union equals to $\mathbb{R}$. For the logarithmic quantizer, the associated quantizer $f_j(\cdot)$ is defined as follows:

$$f_j(v) = \begin{cases} u_i^{(j)}, & \text{if } \frac{1}{1+\sigma_j} u_i^{(j)} < v \le \frac{1}{1-\sigma_j} u_i^{(j)},\ v > 0, \\ 0, & \text{if } v = 0, \\ -f_j(-v), & \text{if } v < 0, \end{cases} \tag{7.6}$$

where

$$\sigma_j = \frac{1 - \rho_j}{1 + \rho_j}. \tag{7.7}$$

Then, at the sampling instant s_k, we have

$$\bar{y}(s_k) = f(y(s_k)) = \begin{bmatrix} f_1(y_1(s_k)) \\ \vdots \\ f_m(y_m(s_k)) \end{bmatrix},$$

where $\bar{y}(s_k)$ denotes the measurements after quantization.

- *Signal transmission delay*: Denote the *updating instants* of the ZOH as t_k, $k = 1, \ldots, \infty$, and suppose that the updating signal (successfully transmitted signal from the sampler to the ZOH) at the instant t_k has experienced signal transmission delay η_k. Therefore, at the updating instant t_k, we have the following relationship:

$$\tilde{y}(t_k) = f(y(t_k - \eta_k)). \tag{7.8}$$

Thus, considering the behavior of the ZOH, we have

$$\tilde{y}(t) = f\left(y\left(t_k - \eta_k\right)\right), \quad t_k \leq t < t_{k+1}, \tag{7.9}$$

with t_{k+1} being the next updating instant of the ZOH after t_k. A natural assumption on the signal transmission delay η_k can be made as follows:

$$\eta_m \leq \eta_k \leq \eta_M, \tag{7.10}$$

where η_m and η_M denote the minimum and maximum delays, respectively.

- *Data packet dropout*: At the updating instant t_k, the number of accumulated data packet dropouts since the last updating instant t_{k-1} is denoted as δ_k. We assume that the maximum number of data packet dropouts is $\bar{\delta}$, that is,

$$0 \leq \delta_k \leq \bar{\delta}. \tag{7.11}$$

Then, it can be seen from (7.10) and (7.11) that

$$t_{k+1} - t_k = \left(\delta_{k+1} + 1\right) h + \eta_{k+1} - \eta_k. \tag{7.12}$$

Then, the robust H_∞ filtering problem for NCSs with limited communication capacity to be addressed in this chapter can be expressed as follows.

Robust H_∞ Filtering with Limited Communication Capacity: Consider the filtering problem in Fig. 7.1. Given system $\mathcal{S}$ in 7.1, determine the matrices A_F, B_F, C_F, and D_F of filter $\mathcal{F}$ in (7.3), such that the filtering error system from $w(t)$ and $v(t)$ to $z(t) - z_F(t)$ is asymptotically stable, and the estimation error $z(t) - z_F(t)$ satisfies a prescribed H_∞ performance γ (that is, $\|z - z_F\|_{L_2}^2 \leq \gamma^2 \left(\|w\|_{L_2}^2 + \|v\|_{L_2}^2\right)$ under zero initial conditions) for all admissible uncertain parameters, measurement quantizations, signal transmission delays, and data packet dropouts.

Remark 7.1 Few result has been reported for the problem of H_∞ filtering with simultaneous consideration of signal transmission delay, data packet dropout, and signal quantization. As a partial problem, the problem of state estimation for state-delayed systems has been addressed in [14–17] as well as in Chaps. 4 and 5 of this book. It is worth noting that, however, the problems addressed therein are much simpler than the one we formulated above.

Remark 7.2 The parameter uncertainties considered in this chapter are assumed to be of the polytopic type, which has been widely used in the problems of robust control and filtering for uncertain systems (see, [18–20] and the references therein), and many practical systems possess parameter uncertainties which can be either exactly modeled or overbounded by a polytope of the form in $\mathcal{R}$.

7.3 H_∞ Filtering

This section is devoted to the problem of H_∞ filtering for system $\mathcal{S}$ in (7.1) with exactly known matrices, that is, there is no uncertain parameters in the system matrices (A, B, C, D, L).

7.3.1 Filter Analysis

We are first concerned with the filtering analysis problem. More specifically, assuming that the filter matrices (A_F, B_F, C_F, D_F) are known, we shall study the conditions under which the filtering error system from $w(t)$ and $v(t)$ to $z(t) - z_F(t)$ is asymptotically stable, and the estimation error $z(t) - z_F(t)$ satisfies a prescribed H_∞ performance γ for all admissible measurement quantizations, signal transmission delays, and data packet dropouts. The following H_∞ performance criterion shows that the H_∞ performance of the filtering error system can be guaranteed if there exist some matrix variables satisfying certain LMIs. This theorem will play an instrumental role in the filter design problems.

Lemma 7.1 *Consider the filtering problem in Fig. 7.1, and suppose that the matrices $(A, B, C, D, L) \in \mathcal{R}$ represent a fixed system. Given the filter matrices A_F, B_F, C_F, and D_F and a positive constant γ, the filtering error system is asymptotically stable with an H_∞ disturbance attenuation level γ if there exist real matrices $P = P^{\mathrm{T}} > 0$ $Q = Q^{\mathrm{T}} \geq 0$, $M_i = M_i^{\mathrm{T}} > 0$, $i = 1, 2$, U and V, and a scalar $\epsilon > 0$ satisfying*

$$\begin{bmatrix} \Xi & \sqrt{\eta_m}U & \sqrt{\kappa}V & \Xi_4^{\mathrm{T}} & \sqrt{\eta_m}\Xi_5^{\mathrm{T}}K^{\mathrm{T}}M_1 & \sqrt{\kappa}\Xi_5^{\mathrm{T}}K^{\mathrm{T}}M_2 & \Xi_6 \\ * & -M_1 & 0 & 0 & 0 & 0 & 0 \\ * & * & -M_2 & 0 & 0 & 0 & 0 \\ * & * & * & -I & 0 & 0 & -D_F \\ * & * & * & * & -M_1 & 0 & \sqrt{\eta_m}M_1KH_1 \\ * & * & * & * & * & -M_2 & \sqrt{\kappa}M_2KH_1 \\ * & * & * & * & * & * & -\epsilon I \end{bmatrix} < 0 \tag{7.13}$$

where

$$\begin{aligned} \Xi &= \Xi_1 + \Xi_2 + \Xi_2^{\mathrm{T}} + \epsilon\Xi_3^{\mathrm{T}}\Xi_3, \\ \Xi_1 &= \begin{bmatrix} P\bar{A} + \bar{A}^{\mathrm{T}}P + K^{\mathrm{T}}QK & 0 & P\bar{E} & P\bar{B} \\ * & -Q & 0 & 0 \\ * & * & 0 & 0 \\ * & * & * & -\gamma^2 I \end{bmatrix}, \\ \Xi_2 &= \begin{bmatrix} UK & -U + V & -V & 0 \end{bmatrix}, \\ \Xi_3 &= \begin{bmatrix} 0 & 0 & \Lambda C & \Lambda H_2 \end{bmatrix}, \\ \Xi_4 &= \begin{bmatrix} \bar{C} & 0 & \bar{F} & \bar{D} \end{bmatrix}, \end{aligned}$$

$$
\begin{aligned}
\Xi_5 &= \left[\bar{A}\ 0\ \bar{E}\ \bar{B}\right],\\
\Xi_6 &= \left[H_1^{\mathrm{T}} P\ 0\ 0\ 0\right]^{\mathrm{T}},\\
\kappa &= \eta_M - \eta_m + \left(\bar{\delta} + 1\right) h,\\
\Lambda &= \mathrm{diag}\left\{\sigma_1, \ldots, \sigma_m\right\},\ K = \left[I\ 0\right],\\
\bar{A} &= \begin{bmatrix} A & 0 \\ 0 & A_F \end{bmatrix},\ \bar{E} = \begin{bmatrix} 0 \\ B_F C \end{bmatrix},\\
\bar{B} &= \begin{bmatrix} B & 0 \\ 0 & B_F D \end{bmatrix},\ H_1 = \begin{bmatrix} 0 \\ B_F \end{bmatrix},\\
\bar{C} &= \left[L\ -C_F\right],\ \bar{F} = -D_F C,\\
\bar{D} &= \left[0\ -D_F D\right],\ H_2 = \left[0\ D\right].
\end{aligned}
\tag{7.14}
$$

Proof The whole filtering problem involves not only continuous-time systems but also discrete signals, which makes the filtering error system hybrid. Our key step in the proof is to formulate the filtering error system in the continuous-time framework, with two successive delay components appearing in the state and norm-bounded uncertainties in the system matrices. First, by substituting (7.9) into (7.3), we obtain

$$
\begin{aligned}
\dot{x}_F(t) &= A_F x_F(t) + B_F f\left(y\left(t_k - \eta_k\right)\right),\\
z_F(t) &= C_F x_F(t) + D_F f\left(y\left(t_k - \eta_k\right)\right),\ t_k \le t < t_{k+1}.
\end{aligned}
\tag{7.15}
$$

Considering the quantization behavior shown in (7.4)–(7.7) and according to [12, 13], (7.15) can be expressed as

$$
\begin{aligned}
\dot{x}_F\left(t\right) &= A_F x_F\left(t\right) + B_F\left(I + \Lambda\left(t\right)\right) y\left(t_k - \eta_k\right),\\
z_F\left(t\right) &= C_F x_F\left(t\right) + D_F\left(I + \Lambda\left(t\right)\right) y\left(t_k - \eta_k\right),\ t_k \le t < t_{k+1},
\end{aligned}
\tag{7.16}
$$

where

$$
\begin{aligned}
\Lambda\left(t\right) &= \mathrm{diag}\left\{\Lambda_1\left(t\right), \Lambda_2\left(t\right), \ldots, \Lambda_n\left(t\right)\right\},\\
\Lambda_j\left(t\right) &\in \left[-\sigma_j, \sigma_j\right],\quad j = 1, 2, \ldots, m.
\end{aligned}
\tag{7.17}
$$

The quantization behavior has been transformed into a kind of norm-bounded uncertainty. The second step is to absorb the discrete signals into the continuous-time framework. To this end, let us represent the time instant $t_k - \eta_k$ in (7.16) as

$$
t_k - \eta_k = t - \eta_m - \eta\left(t\right),
\tag{7.18}
$$

where

$$
\eta(t) = t - t_k + (\eta_k - \eta_m),\quad t_k \le t < t_{k+1}.
$$

Then, from (7.12), we have

$$0 \leq \eta(t) \leq \kappa, \tag{7.19}$$

where κ is given in (7.14). By substituting (7.18) into (7.16), we obtain

$$\begin{aligned} \dot{x}_F(t) &= A_F x_F(t) + B_F(I + \Lambda(t))y(t - \eta_m - \eta(t)), \\ z_F(t) &= C_F x_F(t) + D_F(I + \Lambda(t))y(t - \eta_m - \eta(t)). \end{aligned} \tag{7.20}$$

By connecting (7.1) and (7.20), the filtering error system can be written as

$$\begin{aligned} \dot{\xi}(t) &= \bar{A}\xi(t) + \left(\bar{E} + \Delta E\right) K\xi(t - \eta_m - \eta(t)) + \left(\bar{B} + \Delta B\right) \omega(t), \\ e(t) &= \bar{C}\xi(t) + \left(\bar{F} + \Delta F\right) K\xi(t - \eta_m - \eta(t)) + \left(\bar{D} + \Delta D\right) \omega(t), \end{aligned} \tag{7.21}$$

where $e(t) = z(t) - z_F(t)$,

$$\xi(t) = \begin{bmatrix} x(t) \\ x_F(t) \end{bmatrix}, \ \omega(t) = \begin{bmatrix} w(t) \\ v(t - \eta_m - \eta(t)) \end{bmatrix},$$
$$\begin{bmatrix} \Delta E & \Delta B \\ \Delta F & \Delta D \end{bmatrix} = \begin{bmatrix} H_1 \\ -D_F \end{bmatrix} \Lambda(t) \begin{bmatrix} C & H_2 \end{bmatrix},$$

and $\bar{A}$, $\bar{E}$, $\bar{B}$, $\bar{C}$, $\bar{F}$, $\bar{D}$, H_1 and H_2 are given in (7.14). Now, the filtering error system has been formulated as a class of continuous time-delay systems with two successive delay components and norm-bounded uncertainties. In particular, the quantization behavior is represented with $\Lambda(t)$, and the signal transmission delay and data packet dropout are absorbed into the delays appearing in the state. Therefore, if system (7.21) is asymptotically stable with a prescribed H_∞ performance for $\eta(t)$ satisfying (7.19) and $\Lambda(t)$ satisfying (7.17), the H_∞ filtering performance is guaranteed for all admissible measurement quantization, signal transmission delay and data packet dropout. In the following, we will prove that the condition in (7.13) guarantees the asymptotic stability and H_∞ performance of the system in (7.21) for $\eta(t)$ satisfying (7.19) and $\Lambda(t)$ satisfying (7.17).

Choose the following LKF:

$$\begin{aligned} V(t) &= \xi(t)^{\mathrm{T}} P\xi(t) + \int_{t-\eta_m}^{t} \xi(\alpha)^{\mathrm{T}} K^{\mathrm{T}} QK\xi(\alpha)\,\mathrm{d}\alpha \\ &\quad + \int_{-\eta_m}^{0}\int_{t+\beta}^{t} \dot{\xi}(\alpha)^{\mathrm{T}} K^{\mathrm{T}} M_1 K\dot{\xi}(\alpha)\mathrm{d}\alpha\mathrm{d}\beta \\ &\quad + \int_{-\eta_m-\kappa}^{-\eta_m}\int_{t+\beta}^{t} \dot{\xi}(\alpha)^{\mathrm{T}} K^{\mathrm{T}} M_2 K\dot{\xi}(\alpha)\mathrm{d}\alpha\mathrm{d}\beta, \end{aligned} \tag{7.22}$$

where $P > 0$, $Q \geq 0$ and $M_i > 0$ are matrices to be determined. Then, along the solution of the system in (7.21), the time derivative of $V(t)$ is given by

$$\begin{aligned}\dot{V}(t) \le\ & 2\xi(t)^{\mathrm{T}} P\dot{\xi}(t) - \xi(t-\eta_m)^{\mathrm{T}} K^{\mathrm{T}} Q K \xi(t-\eta_m) + \xi(t)^{\mathrm{T}} K^{\mathrm{T}} Q K \xi(t) \\ & + \dot{\xi}(t)^{\mathrm{T}} K^{\mathrm{T}} \Psi K \dot{\xi}(t) - \int_{t-\eta_m}^{t} \dot{\xi}(\alpha)^{\mathrm{T}} K^{\mathrm{T}} M_1 K \dot{\xi}(\alpha) \mathrm{d}\alpha \\ & - \int_{t-\eta_m-\eta(t)}^{t-\eta_m} \dot{\xi}(\alpha)^{\mathrm{T}} K^{\mathrm{T}} M_2 K \dot{\xi}(\alpha) \mathrm{d}\alpha, \end{aligned} \tag{7.23}$$

where $\Psi = \eta_m M_1 + \kappa M_2$. By Newton-Leibniz formula, we have

$$\begin{aligned} \int_{t-\eta_m}^{t} \dot{\xi}(\alpha)\,\mathrm{d}\alpha &= \xi(t) - \xi(t-\eta_m), \\ \int_{t-\eta_m-\eta(t)}^{t-\eta_m} \dot{\xi}(\alpha)\mathrm{d}\alpha &= \xi(t-\eta_m) - \xi(t-\eta_m-\eta(t)). \end{aligned}$$

Then, for any appropriately dimensioned matrices,

$$U = \begin{bmatrix} U_1 \\ U_2 \\ U_3 \\ U_4 \end{bmatrix}, \quad V = \begin{bmatrix} V_1 \\ V_2 \\ V_3 \\ V_4 \end{bmatrix},$$

we have

$$\begin{aligned} \Delta_1 &\triangleq 2\phi(t)^{\mathrm{T}} U K \left[\xi(t) - \xi(t-\eta_m) - \int_{t-\eta_m}^{t} \dot{\xi}(\alpha)\,\mathrm{d}\alpha \right] = 0, \\ \Delta_2 &\triangleq 2\phi(t)^{\mathrm{T}} V K \left[\xi(t-\eta_m) - \xi(t-\eta_m-\eta(t)) \right. \\ & \qquad\qquad \left. - \int_{t-\eta_m-\eta(t)}^{t-\eta_m} \dot{\xi}(\alpha)\mathrm{d}\alpha \right] = 0, \end{aligned} \tag{7.24}$$

where

$$\phi(t) = \left[\xi(t)^{\mathrm{T}} \ \xi(t-\eta_m)^{\mathrm{T}} K^{\mathrm{T}} \ \xi(t-\eta_m-\eta(t))^{\mathrm{T}} K^{\mathrm{T}} \ \omega(t)^{\mathrm{T}} \right]^{\mathrm{T}}.$$

Then, from (7.21), (7.23) and (7.24), we obtain

$$\begin{aligned} & \dot{V}(t) + e(t)^{\mathrm{T}} e(t) - \gamma^2 \omega(t)^{\mathrm{T}} \omega(t) \\ \le\ & 2\xi(t)^{\mathrm{T}} P\dot{\xi}(t) + \xi(t)^{\mathrm{T}} K^{\mathrm{T}} Q K \xi(t) - \xi^{\mathrm{T}}(t-\eta_m) K^{\mathrm{T}} Q K \xi(t-\eta_m) \\ & + \dot{\xi}(t)^{\mathrm{T}} K^{\mathrm{T}} \Psi K \dot{\xi}(t) - \int_{t-\eta_m}^{t} \dot{\xi}(\alpha)^{\mathrm{T}} K^{\mathrm{T}} M_1 K \dot{\xi}(\alpha)\mathrm{d}\alpha \\ & - \int_{t-\eta_m-\eta(t)}^{t-\eta_m} \dot{\xi}(\alpha)^{\mathrm{T}} K^{\mathrm{T}} M_2 K \dot{\xi}(\alpha)\mathrm{d}\alpha + e^{\mathrm{T}}(t) e(t) - \gamma^2 \omega^{\mathrm{T}}(t)\omega(t) \\ & + \Delta_1 + \Delta_2 \end{aligned}$$

$$\leq \phi(t)^{\mathrm{T}} \left(\Gamma + \eta_m U M_1^{-1} U^{\mathrm{T}} + \kappa V M_2^{-1} V^{\mathrm{T}} \right) \phi(t)$$
$$- \int_{t-\eta_m}^{t} \left\{ \left[\phi(t)^{\mathrm{T}} U + \dot{\xi}(\alpha)^{\mathrm{T}} K^{\mathrm{T}} M_1 \right] M_1^{-1} \left[U^{\mathrm{T}} \phi(t) + M_1 K \dot{\xi}(\alpha) \right] \right\} \mathrm{d}\alpha$$
$$- \int_{t-\eta_m-\eta(t)}^{t-\eta_m} \left\{ \begin{array}{c} \left[\phi(t)^{\mathrm{T}} V + \dot{\xi}(\alpha)^{\mathrm{T}} K^{\mathrm{T}} M_2 \right] M_2^{-1} \\ \times \left[V^{\mathrm{T}} \phi(t) + M_2 K \dot{\xi}(\alpha) \right] \end{array} \right\} \mathrm{d}\alpha, \tag{7.25}$$

where

$$\Gamma = \Gamma_1 + \Xi_2 + \Xi_2^{\mathrm{T}} + \Gamma_2^{\mathrm{T}} \Gamma_2 + \Gamma_3^{\mathrm{T}} K^{\mathrm{T}} \Psi K \Gamma_3,$$
$$\Gamma_1 = \begin{bmatrix} P\bar{A} + \bar{A}^{\mathrm{T}} P + K^{\mathrm{T}} Q K & 0 & P\left(\bar{E} + \Delta E\right) & P\left(\bar{B} + \Delta B\right) \\ * & -Q & 0 & 0 \\ * & * & 0 & 0 \\ * & * & * & -\gamma^2 I \end{bmatrix},$$
$$\Gamma_2 = \left[\bar{C} \; 0 \; \bar{F} + \Delta F \; \bar{D} + \Delta D \right],$$
$$\Gamma_3 = \left[\bar{A} \; 0 \; \bar{E} + \Delta E \; \bar{B} + \Delta B \right].$$

By noticing that $M_1 > 0$ and $M_2 > 0$, we have

$$\left[\phi(t)^{\mathrm{T}} U + \dot{\xi}(\alpha)^{\mathrm{T}} K^{\mathrm{T}} M_1 \right] M_1^{-1} \left[U^{\mathrm{T}} \phi(t) + M_1 K \dot{\xi}(\alpha) \right] \geq 0,$$
$$\left[\phi(t)^{\mathrm{T}} V + \dot{\xi}(\alpha)^{\mathrm{T}} K^{\mathrm{T}} M_2 \right] M_2^{-1} \left[V^{\mathrm{T}} \phi(t) + M_2 K \dot{\xi}(\alpha) \right] \geq 0.$$

Therefore, from (7.25) we know that $\dot{V}(t) + e(t)^{\mathrm{T}} e(t) - \gamma^2 \omega(t)^{\mathrm{T}} \omega(t) \leq 0$ if

$$\Gamma + \eta_m U M_1^{-1} U^{\mathrm{T}} + \kappa V M_2^{-1} V^{\mathrm{T}} < 0, \tag{7.26}$$

which, by the Schur complement, is equivalent to

$$\begin{bmatrix} \Delta_3 & \sqrt{\eta_m} U & \sqrt{\kappa} V & \Gamma_2^{\mathrm{T}} & \sqrt{\eta_m} \Gamma_3^{\mathrm{T}} K^{\mathrm{T}} M_1 & \sqrt{\kappa} \Gamma_3^{\mathrm{T}} K^{\mathrm{T}} M_2 \\ * & -M_1 & 0 & 0 & 0 & 0 \\ * & * & -M_2 & 0 & 0 & 0 \\ * & * & * & -I & 0 & 0 \\ * & * & * & * & -M_1 & 0 \\ * & * & * & * & * & -M_2 \end{bmatrix} < 0, \tag{7.27}$$
$$\Delta_3 = \Gamma_1 + \Xi_2 + \Xi_2^{\mathrm{T}}.$$

Now, rewrite (7.27) in the form of (1.11) with

$$\Sigma_1 = \begin{bmatrix} \Xi_1 + \Xi_2 + \Xi_2^{\mathrm{T}} & \sqrt{\eta_m}U & \sqrt{\kappa}V & \Xi_4^{\mathrm{T}} & \sqrt{\eta_m}\Xi_5^{\mathrm{T}}K^{\mathrm{T}}M_1 & \sqrt{\kappa}\Xi_5^{\mathrm{T}}K^{\mathrm{T}}M_2 \\ * & -M_1 & 0 & 0 & 0 & 0 \\ * & * & -M_2 & 0 & 0 & 0 \\ * & * & * & -I & 0 & 0 \\ * & * & * & * & -M_1 & 0 \\ * & * & * & * & * & -M_2 \end{bmatrix},$$

$$\Sigma_3 = \begin{bmatrix} \Xi_6 \\ 0 \\ 0 \\ -D_F \\ \sqrt{\eta_m}M_1KH_1 \\ \sqrt{\kappa}M_2KH_1 \end{bmatrix}, \ \Sigma_2 = \begin{bmatrix} \Xi_3 \ 0 \ 0 \ 0 \ 0 \ 0 \end{bmatrix}, \ \Delta(t) = \Lambda(t)\Lambda^{-1}.$$

By Lemma 1.4 together with a Schur complement operation, (7.27) holds if, for some $\epsilon > 0$, (7.13) holds. Thus, (7.13) guarantees $\dot{V}(t) + e(t)^{\mathrm{T}}e(t) - \gamma^2\omega(t)^{\mathrm{T}}\omega(t) \le 0$. Integrating both sides with respect to t over the time period $[0, \infty)$, we have

$$V(\infty) - V(0) + \int_0^\infty \left[e(t)^{\mathrm{T}}e(t) - \gamma^2\omega(t)^{\mathrm{T}}\omega(t)\right]\mathrm{d}t \le 0. \qquad (7.28)$$

Under zero initial conditions, we have $V(0) = 0$ and $V(\infty) \ge 0$, thus (7.28) guarantees $\|z - z_F\|_{L_2}^2 \le \gamma^2\left(\|w\|_{L_2}^2 + \|v\|_{L_2}^2\right)$.

The second task is to show the asymptotic stability of the filtering error system when $\omega(t) = 0$. By following the similar procedures as above, we obtain

$$\begin{aligned} \dot{V}(t) \le{}& \bar{\phi}(t)^{\mathrm{T}}\left(\tilde{\Gamma} + \eta_m\tilde{U}M_1^{-1}\tilde{U}^{\mathrm{T}} + \kappa\tilde{V}M_2^{-1}\tilde{V}^{\mathrm{T}}\right)\bar{\phi}(t) \\ & - \int_{t-\eta_m}^{t}\left\{\begin{array}{c}\left[\bar{\phi}(t)^{\mathrm{T}}\tilde{U} + \dot{\xi}(\alpha)^{\mathrm{T}}K^{\mathrm{T}}M_1\right]M_1^{-1} \\ \times\left[\tilde{U}^{\mathrm{T}}\bar{\phi}(t) + M_1K\dot{\xi}(\alpha)\right]\end{array}\right\}\mathrm{d}\alpha \\ & - \int_{t-\eta_m-\eta(t)}^{t-\eta_m}\left\{\begin{array}{c}\left[\bar{\phi}(t)^{\mathrm{T}}\tilde{V} + \dot{\xi}(\alpha)^{\mathrm{T}}K^{\mathrm{T}}M_2\right]M_2^{-1} \\ \times\left[\tilde{V}^{\mathrm{T}}\bar{\phi}(t) + M_2K\dot{\xi}(\alpha)\right]\end{array}\right\}\mathrm{d}\alpha, \end{aligned}$$

where

$$\bar{\phi}(t) = \begin{bmatrix} \xi(t) \\ K\xi(t-\eta_m) \\ K\xi(t-\eta_m-\eta(t)) \end{bmatrix}, \ \tilde{U} = \begin{bmatrix} U_1 \\ U_2 \\ U_3 \end{bmatrix}, \ \tilde{V} = \begin{bmatrix} V_1 \\ V_2 \\ V_3 \end{bmatrix},$$

$$\tilde{\Gamma} = \tilde{\Gamma}_1 + \tilde{\Xi}_2 + \tilde{\Xi}_2^{\mathrm{T}} + \tilde{\Gamma}_3^{\mathrm{T}}K^{\mathrm{T}}\Psi K\tilde{\Gamma}_3,$$

$$\tilde{\Gamma}_1 = \begin{bmatrix} P\bar{A} + \bar{A}^{\mathrm{T}}P + K^{\mathrm{T}}QK & 0 & P\left(\bar{E} + \Delta E\right) \\ * & -Q & 0 \\ * & * & 0 \end{bmatrix},$$

$$\tilde{\Xi}_2 = \begin{bmatrix} \tilde{U}K & -\tilde{U}+\tilde{V} & -\tilde{V} \end{bmatrix}, \ \tilde{\Gamma}_3 = \begin{bmatrix} \bar{A} & 0 & \bar{E}+\Delta E \end{bmatrix}.$$

By following similar steps as (7.26) and (7.27), and by utilizing Lemma 1.4 and Schur complement, we can show that (7.13) guarantees

$$\tilde{\Gamma} + \eta_m \tilde{U} M_1^{-1} \tilde{U}^{\mathrm{T}} + \kappa \tilde{V} M_2^{-1} \tilde{V}^{\mathrm{T}} < 0.$$

Thus, we have $\dot{V}(t) < -\nu\xi(t)^{\mathrm{T}}\xi(t)$ for a sufficiently small positive constant ν and $\xi(t) \neq 0$, and the asymptotic stability of the system in (7.21) is established [21]. The proof is completed. □

Remark 7.3 The basic idea behind Lemma 7.1 is to formulate the filtering error system as a linear system with two successive delay components (η_m and $\eta(t)$ respectively) in the state and with norm-bounded uncertainty, as seen in (7.21). From (7.16) and (7.18), we can see that the sample behavior is dealt with via a delay system approach, and the quantization is treated using a sector bound approach. It is worth noting that η_m and $\eta(t)$ are constant and non-differentiable respectively, and thus it is not reasonable to combine them together as a single delay (otherwise we have to regard η_m as a non-differentiable delay as well instead of a constant one). The H_∞ performance of the filtering error system is then analyzed by defining a new LKF, which treats the two successive delay components separately [22]. Another significant feature is that no model transformation has been performed to deal with the delays in the formulated filtering error system in (7.21). This helps us avoid using bounding techniques for seeking upper bounds of the inner product between two vectors, and has the potential to give rise to less conservative results than those using model transformations.

7.3.2 Filter Design

It is noted that if the filter matrices (A_F, B_F, C_F, D_F) are given, the conditions in Lemma 7.1 are LMIs over the decision variables $P > 0$, $Q \geq 0$, $M_i > 0$, $i = 1, 2$, U, and V, and scalars $\epsilon > 0$ and $\gamma > 0$. However, since our eventual purpose is to determine the filter matrices (A_F, B_F, C_F, D_F), the above conditions are actually NLMIs. Our main objective hereafter is to transform them into LMI conditions. Based on Lemma 7.1, the following result gives a solution to H_∞ filter design in terms of LMI for NCSs with limited communication capacity.

Theorem 7.1 *Consider the filtering problem in Fig. 7.1, and suppose the matrices $(A, B, C, D, L) \in \mathcal{R}$ represent a fixed system. For a given positive constant γ, there exist real matrices $P = P^{\mathrm{T}} > 0$, $Q = Q^{\mathrm{T}} \geq 0$, $M_i = M_i^{\mathrm{T}} > 0$, $i = 1, 2$, U, V, A_F, B_F, C_F and D_F, and a scalar $\epsilon > 0$ satisfying (7.13) if and only if there exist real matrices $R = R^{\mathrm{T}} > 0$, $S = S^{\mathrm{T}} > 0$, $Q = Q^{\mathrm{T}} \geq 0$, $M_i = M_i^{\mathrm{T}} > 0$, $i = 1, 2$, $\bar{U}$, $\bar{V}$, $\bar{A}_F$, $\bar{B}_F$, $\bar{C}_F$, and $\bar{D}_F$, and a scalar $\epsilon > 0$ satisfying*

$$\Theta \triangleq \begin{bmatrix} \Omega & \sqrt{\eta_m}\bar{U} & \sqrt{\kappa}\bar{V} & \Omega_4 & \sqrt{\eta_m}\Omega_5 M_1 & \sqrt{\kappa}\Omega_5 M_2 & \Omega_6 \\ * & -M_1 & 0 & 0 & 0 & 0 & 0 \\ * & * & -M_2 & 0 & 0 & 0 & 0 \\ * & * & * & -I & 0 & 0 & -\bar{D}_F \\ * & * & * & * & -M_1 & 0 & 0 \\ * & * & * & * & * & -M_2 & 0 \\ * & * & * & * & * & * & -\epsilon I \end{bmatrix} < 0 \tag{7.29}$$

and

$$R - S > 0, \tag{7.30}$$

where

$$\Omega = \Omega_1 + \Omega_2 + \Omega_2^{\mathrm{T}} + \epsilon \Omega_3^{\mathrm{T}} \Omega_3,$$

$$\Omega_1 = \begin{bmatrix} RA + A^{\mathrm{T}}R + Q & \bar{A}_F + A^{\mathrm{T}}S & 0 & \bar{B}_F C & RB & \bar{B}_F D \\ * & \bar{A}_F + \bar{A}_F^{\mathrm{T}} & 0 & \bar{B}_F C & SB & \bar{B}_F D \\ * & * & -Q & 0 & 0 & 0 \\ * & * & * & 0 & 0 & 0 \\ * & * & * & * & -\gamma^2 I & 0 \\ * & * & * & * & * & -\gamma^2 I \end{bmatrix},$$

$$\Omega_2 = \begin{bmatrix} \bar{U} \; 0 \; -\bar{U} + \bar{V} \; -\bar{V} \; 0 \; 0 \end{bmatrix},$$

$$\Omega_3 = \begin{bmatrix} 0 \; 0 \; 0 \; \Lambda C \; 0 \; \Lambda D \end{bmatrix},$$

$$\Omega_4 = \begin{bmatrix} L^{\mathrm{T}} \\ -\bar{C}_F^{\mathrm{T}} \\ 0 \\ -C^{\mathrm{T}}\bar{D}_F^{\mathrm{T}} \\ 0 \\ -D^{\mathrm{T}}\bar{D}_F^{\mathrm{T}} \end{bmatrix}, \quad \Omega_5 = \begin{bmatrix} A^{\mathrm{T}} \\ 0 \\ 0 \\ 0 \\ B^{\mathrm{T}} \\ 0 \end{bmatrix}, \quad \Omega_6 = \begin{bmatrix} \bar{B}_F \\ \bar{B}_F \\ 0 \\ 0 \\ 0 \\ 0 \end{bmatrix},$$

and κ *and* Λ *are given in* (7.14). *Moreover, if the above conditions are satisfied, the matrices for an admissible* H_∞ *filter are given by*

$$\begin{bmatrix} A_F & B_F \\ C_F & D_F \end{bmatrix} = \begin{bmatrix} S^{-1} & 0 \\ 0 & I \end{bmatrix} \begin{bmatrix} \bar{A}_F & \bar{B}_F \\ \bar{C}_F & \bar{D}_F \end{bmatrix}. \tag{7.31}$$

Proof (Proof of the first part) We prove the first part by showing the equivalence between (7.13) and (7.29) and (7.30).

(7.13) $\Rightarrow$ (7.29) *and* (7.30): Suppose there exist matrices $P > 0$, $Q \geq 0$, $M_i > 0$, $i = 1, 2$, U, V, A_F, B_F, C_F and D_F, and a scalar $\epsilon > 0$ satisfying (7.13). Let the matrix P be partitioned as

$$P \triangleq \begin{bmatrix} P_1 & P_2 \\ P_2^{\mathrm{T}} & P_3 \end{bmatrix}. \tag{7.32}$$

By invoking a small perturbation if necessary, we can assume that P_2 and P_3 are nonsingular. Define the following invertible matrix:

$$J \triangleq \begin{bmatrix} I & 0 \\ 0 & P_3^{-1} P_2^{\mathrm{T}} \end{bmatrix}. \tag{7.33}$$

Then, performing a congruence transformation to (7.13) by

$$J_2 \triangleq \mathrm{diag}\{J_1, I, I, I, I, I, I\}$$

with $J_1 \triangleq \mathrm{diag}\{J, I, I, I\}$ together with (7.14) yields

$$\begin{bmatrix} \bar{\Xi} & \sqrt{\eta_m} J_1^{\mathrm{T}} U & \sqrt{\kappa} J_1^{\mathrm{T}} V & \bar{\Xi}_4 & \sqrt{\eta_m}\Omega_5 M_1 & \sqrt{\kappa}\Omega_5 M_2 & \bar{\Xi}_6 \\ * & -M_1 & 0 & 0 & 0 & 0 & 0 \\ * & * & -M_2 & 0 & 0 & 0 & 0 \\ * & * & * & -I & 0 & 0 & -D_F \\ * & * & * & * & -M_1 & 0 & 0 \\ * & * & * & * & * & -M_2 & 0 \\ * & * & * & * & * & * & -\epsilon I \end{bmatrix} < 0, \tag{7.34}$$

where

$$\bar{\Xi} = \bar{\Xi}_1 + \bar{\Xi}_2 + \bar{\Xi}_2^{\mathrm{T}} + \epsilon \Omega_3^{\mathrm{T}} \Omega_3,$$

$$\bar{\Xi}_1 = \begin{bmatrix} P_1 A + A^{\mathrm{T}} P_1 + Q & P_2 A_F P_3^{-1} P_2^{\mathrm{T}} + A^{\mathrm{T}} P_2 P_3^{-1} P_2^{\mathrm{T}} & 0 & P_2 B_F C & P_1 B & P_2 B_F D \\ * & P_2 A_F P_3^{-1} P_2^{\mathrm{T}} + P_2 P_3^{-1} A_F^{\mathrm{T}} P_2^{\mathrm{T}} & 0 & P_2 B_F C & P_2 P_3^{-1} P_2^{\mathrm{T}} B & P_2 B_F D \\ * & * & -Q & 0 & 0 & 0 \\ * & * & * & 0 & 0 & 0 \\ * & * & * & * & -\gamma^2 I & 0 \\ * & * & * & * & * & -\gamma^2 I \end{bmatrix},$$

$$\bar{\Xi}_2 = J_1^{\mathrm{T}} \begin{bmatrix} U & 0 & -U + V & -V & 0 & 0 \end{bmatrix} J_1,$$

$$\bar{\Xi}_4 = \begin{bmatrix} L^{\mathrm{T}} \\ -P_2 P_3^{-1} C_F^{\mathrm{T}} \\ 0 \\ -C^{\mathrm{T}} D_F^{\mathrm{T}} \\ 0 \\ -D^{\mathrm{T}} D_F^{\mathrm{T}} \end{bmatrix}, \quad \bar{\Xi}_6 = \begin{bmatrix} P_2 B_F \\ P_2 B_F \\ 0 \\ 0 \\ 0 \\ 0 \end{bmatrix}.$$

By defining

$$\begin{aligned}
[\,\bar{U}\ \bar{V}\,] &\triangleq J_1^{\mathrm{T}} [\,U\ V\,], \\
R &\triangleq P_1, \\
S &\triangleq P_2 P_3^{-1} P_2^{\mathrm{T}}, \\
\begin{bmatrix} \bar{A}_F & \bar{B}_F \\ \bar{C}_F & \bar{D}_F \end{bmatrix} &\triangleq \begin{bmatrix} P_2 & 0 \\ 0 & I \end{bmatrix} \begin{bmatrix} A_F & B_F \\ C_F & D_F \end{bmatrix} \begin{bmatrix} P_3^{-1} P_2^{\mathrm{T}} & 0 \\ 0 & I \end{bmatrix}
\end{aligned} \tag{7.35}$$

from (7.34), we readily obtain (7.29).

In addition, $J^{\mathrm{T}} P J > 0$ and a Schur complement operation lead to (7.30).

(7.13) $\Leftarrow$ (7.29) *and* (7.30): Suppose that there exist matrices $R > 0$, $S > 0$, $Q \geq 0$, $M_i > 0$, $i = 1, 2$, $\bar{U}$, $\bar{V}$, $\bar{A}_F$, $\bar{B}_F$, $\bar{C}_F$ and $\bar{D}_F$, and a scalar $\epsilon > 0$ satisfying (7.29) and (7.30). Since $S > 0$, one can always find square and nonsingular matrices P_2 and P_3 satisfying $S = P_2 P_3^{-1} P_2^{\mathrm{T}}$. Now introduce the matrix P as in (7.32) with $P_1 \triangleq R$, J, J_1 and J_2 as above and

$$\begin{bmatrix} A_F & B_F \\ C_F & D_F \end{bmatrix} \triangleq \begin{bmatrix} P_2^{-1} & 0 \\ 0 & I \end{bmatrix} \begin{bmatrix} \bar{A}_F & \bar{B}_F \\ \bar{C}_F & \bar{D}_F \end{bmatrix} \begin{bmatrix} P_2^{-\mathrm{T}} P_3 & 0 \\ 0 & I \end{bmatrix}. \tag{7.36}$$

Then, the matrix P defined in (7.32) is positive definite due to the fact that $P_1 > 0$ and $P_1 - P_2 P_3^{-1} P_2^{\mathrm{T}} = R - S > 0$. In addition, the matrices A_F, B_F, C_F and D_F are defined in (7.36).

Now, by some matrix algebraic manipulations, it can be established that (7.29) and (7.30) are equivalent to

$$J_2^{\mathrm{T}} \Theta J_2 < 0, \tag{7.37}$$

$$J^{\mathrm{T}} P J > 0. \tag{7.38}$$

The equivalence between (7.37), (7.38) and (7.29), (7.30) can be verified in a reverse order. Then, performing congruence transformations to (7.37) by J_2^{-1} and to (7.38) by J^{-1} yields (7.13) and $P > 0$, respectively.

(Proof of the second part) From the above proof, we know that if conditions (7.29) and (7.30) have a set of feasible solution $R > 0$, $S > 0$, $Q \geq 0$, $M_i > 0$, $i = 1, 2$, $\bar{U}$, $\bar{V}$, $\bar{A}_F$, $\bar{B}_F$, $\bar{C}_F$ and $\bar{D}_F$, and a scalar $\epsilon > 0$, then the filter with a state space realization A_F, B_F, C_F, D_F defined in (7.36) guarantees the filtering error system to be asymptotically stable with an H_∞ performance γ. Now denote the operator from

$\tilde{y}(t)$ to $z_F(t)$ by $T_{z_F\tilde{y}}(s) = C_F(sI - A_F)^{-1}B_F + D_F$. By substituting the matrices with (7.36) and by considering the relationship $S = P_2 P_3^{-1} P_2^{\mathrm{T}}$, we have

$$\begin{aligned} T_{z_F\tilde{y}}(s) &= \bar{C}_F P_2^{-T} P_3 \left(sI - P_2^{-1}\bar{A}_F P_2^{-T} P_3\right)^{-1} P_2^{-1}\bar{B}_F + D_F \\ &= \bar{C}_F \left(sI - S^{-1}\bar{A}_F\right)^{-1} S^{-1}\bar{B}_F + D_F. \end{aligned}$$

Therefore, an admissible H_∞ filter can be given by (7.31), and the proof is completed. □

Theorem 7.1 tells us that admissible H_∞ filters, if exist, can be found by solving a convex optimization problem. Note that (7.29), (7.30) are LMIs not only over the matrix variables, but also over the square of the scalar γ. This implies that the scalar γ can be included as an optimization variable to obtain a reduction of the attenuation level bound. Then, based on Theorem 7.1, the suboptimal H_∞ filters can be readily calculated by the LMI technique, as stated in Algorithm 22.

Algorithm 22 H_∞ Filter Design with Limited Communication Capacity

- Solve the minimization problem:

$$\min_{\substack{R = R^{\mathrm{T}},\, S = S^{\mathrm{T}} > 0,\, Q = Q^{\mathrm{T}}, \\ M_i = M_i^{\mathrm{T}},\, i = 1,2,\, \bar{U},\, \bar{V}, \\ \bar{A}_F,\, \bar{B}_F,\, \bar{C}_F,\, \bar{D}_F,\, \epsilon,\, \mu}} \mu = \gamma^2 \quad \text{s.t. (7.29) and (7.30)}$$

- Compute the filter matrices from (7.31).

7.4 Robust H_∞ Filtering

In this section, we consider the problem of robust H_∞ filtering for system $\mathcal{S}$ in (7.1) with partially unknown data and limited communication, that is, uncertain parameters are present in the system matrices (A, B, C, D, L). For systems with polytopic uncertainties, one can usually solve the analysis and synthesis problems based on either the quadratic stability notion or the parameter-dependent stability notion. Consequently, we present two approaches for designing robust H_∞ filters which have different degrees of conservativeness and computational complexity.

7.4.1 Quadratic Approach

One can easily extend Theorem 7.1 to obtain the following theorem on H_∞ filter design for NCSs with limited communication capacity based on the notion of quadratic stability, thanks to the inherent properties of convex combination.

Theorem 7.2 *Consider the filtering problem in Fig. 7.1, and suppose that the system matrices $(A, B, C, D, L) \in \mathcal{R}$ represent an uncertain system. For a given positive constant γ, a robust H_∞ filter exists such that the filtering error system is asymptotically stable with an H_∞ performance γ if there exist real matrices $R = R^{\mathrm{T}} > 0$, $S = S^{\mathrm{T}} > 0$, $Q = Q^{\mathrm{T}} \geq 0$, $M_i = M_i^{\mathrm{T}} > 0$, $i = 1, 2$, $\bar{U}$, $\bar{V}$, $\bar{A}_F$, $\bar{B}_F$, $\bar{C}_F$ and $\bar{D}_F$, and a scalar $\epsilon > 0$ satisfying (7.30) and for $i = 1, 2, \ldots, s$,*

$$\Theta_i \triangleq \begin{bmatrix} \Upsilon_i & \sqrt{\eta_m}\bar{U} & \sqrt{\kappa}\bar{V} & \Omega_{4,i} & \sqrt{\eta_m}\Omega_{5,i}M_1 & \sqrt{\kappa}\Omega_{5,i}M_2 & \Omega_6 \\ * & -M_1 & 0 & 0 & 0 & 0 & 0 \\ * & * & -M_2 & 0 & 0 & 0 & 0 \\ * & * & * & -I & 0 & 0 & -\bar{D}_F \\ * & * & * & * & -M_1 & 0 & 0 \\ * & * & * & * & * & -M_2 & 0 \\ * & * & * & * & * & * & -\epsilon I \end{bmatrix} < 0 \quad (7.39)$$

where

$$\Upsilon_i = \Omega_{1,i} + \Omega_2 + \Omega_2^{\mathrm{T}} + \epsilon\Omega_{3,i}^{\mathrm{T}}\Omega_{3,i},$$

$$\Omega_{1,i} = \begin{bmatrix} RA_i + A_i^{\mathrm{T}}R + Q & \bar{A}_F + A_i^{\mathrm{T}}S & 0 & \bar{B}_F C_i & RB_i & \bar{B}_F D_i \\ * & \bar{A}_F + \bar{A}_F^{\mathrm{T}} & 0 & \bar{B}_F C_i & SB_i & \bar{B}_F D_i \\ * & * & -Q & 0 & 0 & 0 \\ * & * & * & 0 & 0 & 0 \\ * & * & * & * & -\gamma^2 I & 0 \\ * & * & * & * & * & -\gamma^2 I \end{bmatrix},$$

$$\Omega_{3,i} = \begin{bmatrix} 0 & 0 & 0 & \Lambda C_i & 0 & \Lambda D_i \end{bmatrix},$$

$$\Omega_{4,i} = \begin{bmatrix} L_i^{\mathrm{T}} \\ -\bar{C}_F^{\mathrm{T}} \\ 0 \\ -C_i^{\mathrm{T}}\bar{D}_F^{\mathrm{T}} \\ 0 \\ -D_i^{\mathrm{T}}\bar{D}_F^{\mathrm{T}} \end{bmatrix}, \quad \Omega_{5,i} = \begin{bmatrix} A_i^{\mathrm{T}} \\ 0 \\ 0 \\ 0 \\ B_i^{\mathrm{T}} \\ 0 \end{bmatrix},$$

and κ, Λ are given in (7.14), *Ω_2, Ω_6 in* (7.29). *Moreover, if the above conditions are satisfied, the matrices for an admissible robust H_∞ filter are given by* (7.31).

According to Theorem 7.2, we can immediately obtain Algorithm 23, a quadratic approach to robust H_∞ filter design for NCSs with limited communication capacity.

7.4.2 Parameter-Dependent Approach

The quadratic approach presented in the above subsection requires the matrix variables $R > 0$, $S > 0$, $Q \geq 0$, $M_i > 0$, $i = 1, 2$, $\bar{U}$, $\bar{V}$, $\bar{A}_F$, $\bar{B}_F$, $\bar{C}_F$, and $\bar{D}_F$, and a scalar $\epsilon > 0$ to be fixed for the entire uncertainty domain. This approach, though

Algorithm 23 Quadratic Approach to Robust H_∞ Filter Design with Limited Communication Capacity

- Solve the minimization problem:

$$\min_{\substack{R=R^{\mathrm{T}},\, S=S^{\mathrm{T}}>0,\, Q=Q^{\mathrm{T}},\\ M_i=M_j^{\mathrm{T}}\ (i=1,2),\, \bar{U},\, \bar{V},\\ \bar{A}_F,\, \bar{B}_F,\, \bar{C}_F,\, \bar{D}_F,\, \epsilon,\, \mu}} \mu=\gamma^2 \quad \text{s.t. (7.30) and (7.39)}$$

- Compute the filter matrices from (7.31).

simple in computation, may sometimes lead to conservative results. Parameter-dependent approach has been recognized to be an effective alternative for reducing conservativeness (see the previous chapters). In the following, a parameter-dependent approach is presented, which can be used for robust H_∞ filter design with reduced conservativeness for NCSs with limited communication capacity.

Theorem 7.3 *Consider the filtering problem in Fig. 7.1, and suppose that the system matrices $(A, B, C, D, L) \in \mathcal{R}$ represent an uncertain system. For a given positive constant γ, a robust H_∞ filter exists such that the filtering error system is asymptotically stable with an H_∞ performance γ if there exist real matrices $R_i = R_i^{\mathrm{T}} > 0$, $S = S^{\mathrm{T}} > 0$, $Q_i = Q_i^{\mathrm{T}} \geq 0$, $M_{1,i} = M_{1,i}^{\mathrm{T}} > 0$, $M_{2,i} = M_{2,i}^{\mathrm{T}} > 0$, $\bar{U}_i$, $\bar{V}_i$, $\bar{A}_F$, $\bar{B}_F$, $\bar{C}_F$ and $\bar{D}_F$, and a scalar $\epsilon > 0$ satisfying*

$$\Pi_{i,j} + \Pi_{j,i} < 0 \quad (1 \leq i \leq j \leq s), \tag{7.40}$$

$$R_i - S > 0 \quad (i = 1, 2, \ldots, s), \tag{7.41}$$

where

$$\Pi_{i,j} = \begin{bmatrix} \Psi_{i,j} & \sqrt{\eta_m}\bar{U}_i & \sqrt{\kappa}\bar{V}_i & \Phi_{4,j} & \sqrt{\eta_m}\Phi_{5,j}M_{1,i} & \sqrt{\kappa}\Phi_{5,j}M_{2,i} & \Phi_6 \\ * & -M_{1,i} & 0 & 0 & 0 & 0 & 0 \\ * & * & -M_{2,i} & 0 & 0 & 0 & 0 \\ * & * & * & -I & 0 & 0 & -\bar{D}_F \\ * & * & * & * & -M_{1,i} & 0 & 0 \\ * & * & * & * & * & -M_{2,i} & 0 \\ * & * & * & * & * & * & -\epsilon I \end{bmatrix},$$

$$\Psi_{i,j} = \Phi_{1,i,j} + \Phi_{2,i} + \Phi_{2,i}^{\mathrm{T}} + \epsilon\Phi_{3,j}^{\mathrm{T}}\Phi_{3,j},$$

$$\Phi_{1,i,j} = \begin{bmatrix} R_iA_j + A_j^{\mathrm{T}}R_i + Q_i & \bar{A}_F + A_j^{\mathrm{T}}S & 0 & \bar{B}_FC_j & R_iB_j & \bar{B}_FD_j \\ * & \bar{A}_F + \bar{A}_F^{\mathrm{T}} & 0 & \bar{B}_FC_j & SB_j & \bar{B}_FD_j \\ * & * & -Q_i & 0 & 0 & 0 \\ * & * & * & 0 & 0 & 0 \\ * & * & * & * & -\gamma^2 I & 0 \\ * & * & * & * & * & -\gamma^2 I \end{bmatrix},$$

$$\Phi_{2,i} = \begin{bmatrix} \bar{U}_i & 0 & -\bar{U}_i + \bar{V}_i & -\bar{V}_i & 0 & 0 \end{bmatrix},$$

$$\Phi_{3,j} = \begin{bmatrix} 0 & 0 & 0 & \Lambda C_j & 0 & \Lambda D_j \end{bmatrix},$$

$$\Phi_{4,j} = \begin{bmatrix} L_j^{\mathrm{T}} \\ -\bar{C}_F^{\mathrm{T}} \\ 0 \\ -C_j^{\mathrm{T}} \bar{D}_F^{\mathrm{T}} \\ 0 \\ -D_j^{\mathrm{T}} \bar{D}_F^{\mathrm{T}} \end{bmatrix}, \quad \Phi_{5,j} = \begin{bmatrix} A_j^{\mathrm{T}} \\ 0 \\ 0 \\ 0 \\ B_j^{\mathrm{T}} \\ 0 \end{bmatrix}, \quad \Phi_6 = \begin{bmatrix} \bar{B}_F \\ \bar{B}_F \\ 0 \\ 0 \\ 0 \\ 0 \end{bmatrix},$$

and κ, Λ *are given in* (7.14). *Moreover, if the above conditions are satisfied, the matrices for an admissible robust* H_∞ *filter are given by* (7.31).

Proof Suppose that there are matrices $R_i > 0$, $S > 0$, $Q_i \geq 0$, $M_{1,i} > 0$, $M_{2,i} > 0$, $\bar{U}_i$, $\bar{V}_i$, $\bar{A}_F$, $\bar{B}_F$, $\bar{C}_F$, $\bar{D}_F$, and a scalar $\epsilon > 0$ satisfying (7.40) and (7.41). To prove the theorem, first for any fixed $\lambda = (\lambda_1, \lambda_2, \ldots, \lambda_s)$ with $\sum_{i=1}^{s} \lambda_i = 1$ and $\lambda_i \geq 0$, we define the following matrices:

$$\begin{aligned} R &\triangleq \sum_{i=1}^{s} \lambda_i R_i, \quad Q \triangleq \sum_{i=1}^{s} \lambda_i Q_i, \quad M_1 \triangleq \sum_{i=1}^{s} \lambda_i M_{1,i}, \\ \bar{U} &\triangleq \sum_{i=1}^{s} \lambda_i \bar{U}_i, \quad \bar{V} \triangleq \sum_{i=1}^{s} \lambda_i \bar{V}_i, \quad M_2 \triangleq \sum_{i=1}^{s} \lambda_i M_{2,i}. \end{aligned} \tag{7.42}$$

Multiplying the ith inequality in (7.41) by λ_i, summing the s inequalities up, and considering (7.42) and $\sum_{i=1}^{s} \lambda_i = 1$, we get (7.30).

In addition, from (7.40), we have

$$\Pi_{i,i} < 0 \quad (i = 1, 2, \ldots, s), \tag{7.43}$$

$$\Pi_{i,j} + \Pi_{j,i} < 0 \quad (1 \leq i < j \leq s), \tag{7.44}$$

Therefore, we have

$$\sum_{j=1}^{s} \sum_{i=1}^{s} \lambda_i \lambda_j \Pi_{i,j} = \sum_{i=1}^{s} \lambda_i^2 \Pi_{i,i} + \sum_{i=1}^{s-1} \sum_{j=i+1}^{s} \lambda_i \lambda_j \left(\Pi_{i,j} + \Pi_{j,i} \right) < 0. \tag{7.45}$$

It is noted that (7.45) implies (7.29) by noticing (7.2) and (7.42), and the proof is completed. □

Remark 7.4 It is worth mentioning that Theorem 7.3 encompasses Theorem 7.2 as a special case. That is, by imposing additional constraints on the matrix variables, Theorem 7.3 will recover Theorem 7.2. To this end, let us set

$$R_i \equiv R, \quad Q_i \equiv Q, \quad M_{1,i} \equiv M_1, \quad M_{2,i} \equiv M_2, \quad \bar{U}_i \equiv \bar{U}, \quad \bar{V}_i \equiv \bar{V}.$$

Then, LMIs (7.40) and (7.41) are equivalent to the conditions in Theorem 7.2.

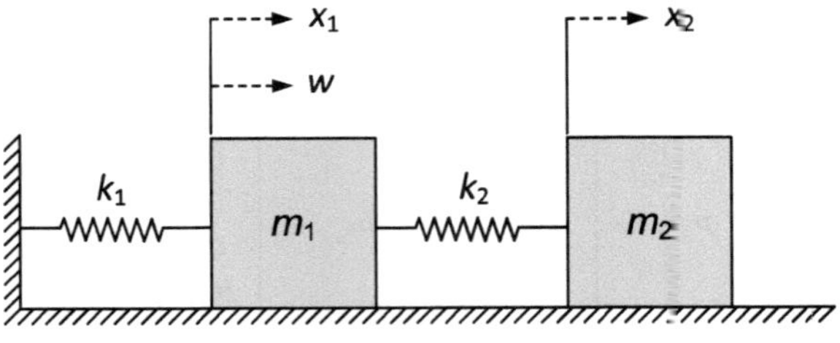

Fig. 7.2 Mass-spring system in Example 7.1

Accordingly, based on Theorem 7.3, a parameter-dependent approach Algorithm 24, can be obtained for robust suboptimal H_∞ filter design with limited communication capacity.

Algorithm 24 Parameter-Dependent Approach to Robust H_∞ Filter Design with Limited Communication Capacity

- Solve the minimization problem:

$$\min_{\substack{R_j = R_j^{\mathrm{T}},\ S = S^{\mathrm{T}} > 0,\ Q_j = Q_j^{\mathrm{T}},\\ M_{ij} = M_{ij}^{\mathrm{T}},\ i=1,2,\ \bar{U}_j,\ \bar{V}_j,\ j=1,2,\ldots,s,\\ \bar{A}_F,\ \bar{B}_F,\ \bar{C}_F,\ \bar{D}_F,\ \epsilon,\ \mu}} \mu = \gamma^2 \quad \text{s.t. (7.40) and (7.41)}$$

- Compute the filter matrices from (7.31).

7.5 Illustrative Examples

In this section, we provide two examples to illustrate the filter design methods proposed in this chapter.

Example 7.1 Consider a mechanical system with two masses and two springs, shown in Fig. 7.2. In this system, x_1 and x_2 are the positions of masses m_1 and m_2, respectively, and k_1 and k_2 are the spring constants. The viscous friction coefficient between the masses and the horizontal surface is denoted by c. A state space realization of this system is given by the first equation in (7.1) with

$$A = \begin{bmatrix} 0 & 0 & 1 & 0 \\ 0 & 0 & 0 & 1 \\ -\frac{k_1+k_2}{m_1} & \frac{k_2}{m_1} & -\frac{c}{m_1} & 0 \\ \frac{k_2}{m_2} & -\frac{k_2}{m_2} & 0 & -\frac{c}{m_2} \end{bmatrix}, \quad B = \begin{bmatrix} 0 \\ 0 \\ \frac{1}{m_1} \\ 0 \end{bmatrix}.$$

It is assumed that the position of mass m_1 is measured by a device with disturbance $v(t)$, given by the second equation in (7.1) with

$$C = \begin{bmatrix} 1\,0\,0\,0 \end{bmatrix}, \quad D = d,$$

where d is a constant. The purpose is to design an H_∞ filter to estimate the position of mass m_2.

(1) *Nominal Filter Design*

First, we consider the problem of H_∞ filter design for the nominal system. The parameters are chosen as: $m_1 = 1$ kg, $m_2 = 0.5$ kg, $k_1 = k_2 = 1$, $c = 0.5$, and $d = 0.1$. Thus, the matrices for the model in (7.1) are given by

$$A = \begin{bmatrix} 0 & 0 & 1 & 0 \\ 0 & 0 & 0 & 1 \\ -2 & 1 & -0.5 & 0 \\ 2 & -2 & 0 & -1 \end{bmatrix}, \quad B = \begin{bmatrix} 0 \\ 0 \\ 1 \\ 0 \end{bmatrix},$$
$$C = \begin{bmatrix} 1\,0\,0\,0 \end{bmatrix}, \quad D = 0.1, \quad L = \begin{bmatrix} 0\,1\,0\,0 \end{bmatrix}.$$

It is assumed that: the sampling period $h = 20$ ms; the network induced delay bounds in (7.10) are given by $\eta_m = 30$ ms and $\eta_M = 100$ ms; the maximum number of data packet dropouts $\bar{\delta} = 4$. In addition, the parameter for the quantizer $f(\cdot)$ is assumed to be $\rho = 0.9$. By solving the convex problem in Algorithm 22, we obtain the following matrices (we do not list all the obtained matrices here):

$$S = \begin{bmatrix} 1.0940 & -0.4325 & -0.3026 & -0.3338 \\ -0.4325 & 1.0786 & 0.0796 & 0.4059 \\ -0.3026 & 0.0796 & 0.1476 & 0.1040 \\ -0.3338 & 0.4059 & 0.1040 & 0.5769 \end{bmatrix},$$
$$\begin{bmatrix} \bar{A}_F & \bar{B}_F \\ \bar{C}_F & \bar{D}_F \end{bmatrix} = \left[\begin{array}{cccc|c} -1.8275 & 0.0372 & 0.7893 & -0.2237 & -1.9636 \\ 0.5883 & -0.6479 & -0.3537 & 0.7043 & 0 \\ -0.2072 & 0.0983 & -0.1549 & 0.0356 & 0 \\ 0.8626 & -0.9387 & -0.2306 & -0.1299 & 0 \\ \hline 0 & -1 & 0 & 0 & 0 \end{array}\right].$$

Thus, according to (7.31), the filter matrices are given by

$$\begin{bmatrix} A_F & B_F \\ C_F & D_F \end{bmatrix} = \left[\begin{array}{cccc|c} -5.0114 & -0.0067 & 0.9948 & -0.0052 & -4.8933 \\ -1.1743 & -0.0025 & -0.0015 & 0.9990 & -1.1459 \\ -12.1841 & 2.0636 & 0.9921 & 0.3988 & -9.1489 \\ 1.6185 & -2.0010 & -0.0019 & -1.0030 & -0.3755 \\ \hline 0 & -1 & 0 & 0 & 0 \end{array}\right],$$

and the obtained minimum guaranteed H_∞ performance is $\gamma^* = 0.3513$.

To illustrate the performance of the designed filter, let us select a set of input signals as follows:

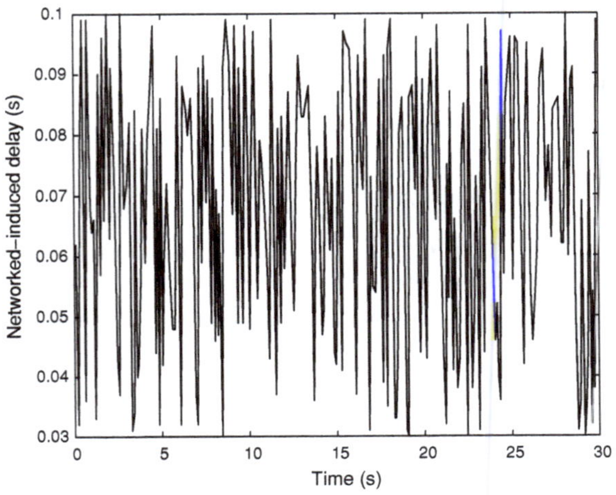

Fig. 7.3 Network-induced delay in Example 7.1

$$w(t) = \begin{cases} 1, & (5 \leq t \leq 10 \text{ s}) \\ -1, & (15 \leq t \leq 20 \text{ s}) \\ 0, & \text{otherwise} \end{cases} \tag{7.46}$$

The disturbance $v(t)$ is assumed to be uniformly distributed from -0.1 to 0.1 for the time interval $[0, 30]$ (zero elsewhere). In the simulation, the network induced delays and the data packet dropouts are generated randomly (uniformly distributed within their ranges) according to the above assumptions, and shown in Figs. 7.3 and 7.4. The measurement signal $y(t)$ and its quantized values arriving at the ZOH before the filter are shown in Fig. 7.5, where we can see the discontinuous holding behavior of the sampled measurements. Figure 7.6 depicts $z(t)$ and $z_F(t)$, and Fig. 7.7 depicts the estimation error $z(t) - z_F(t)$. By calculation, we have $\|w\|_{L_2}^2 = 9.9980$, $\|v\|_{L_2}^2 = 0.0999$ and $\|e\|_{L_2}^2 = 0.0376$, which yield

$$\sqrt{\frac{\|e\|_{L_2}^2}{\|w\|_{L_2}^2 + \|v\|_{L_2}^2}} = 0.0610 < \gamma^* = 0.3513,$$

showing the effectiveness of the H_∞ filter design.

(2) *Robust Filter Design*

The second task is to show the usefulness of the robust H_∞ filter design procedures. We still assume $m_1 = 1$ kg, $m_2 = 0.5$ kg, $k_1 = k_2 = 1$, $d = 0.1$, but the parameter c is assumed to be $c = 0.5 + \Delta c$, with $|\Delta c| \leq \bar{c}$. We first assume $\bar{c} = 0.1$ (that is, 20 % of the nominal value). In this case, the system can be represented as a two-vertex polytope. The network induced delay, data packet dropout, and measurement quantization are still assumed to be the same as above. By employing the quadratic

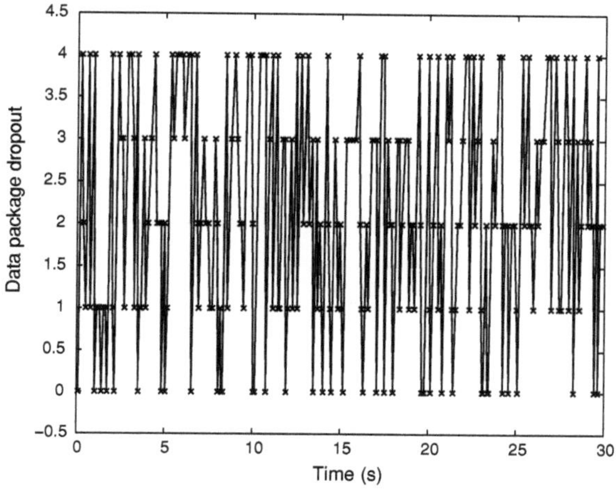

Fig. 7.4 Data packet dropout in Example 7.1

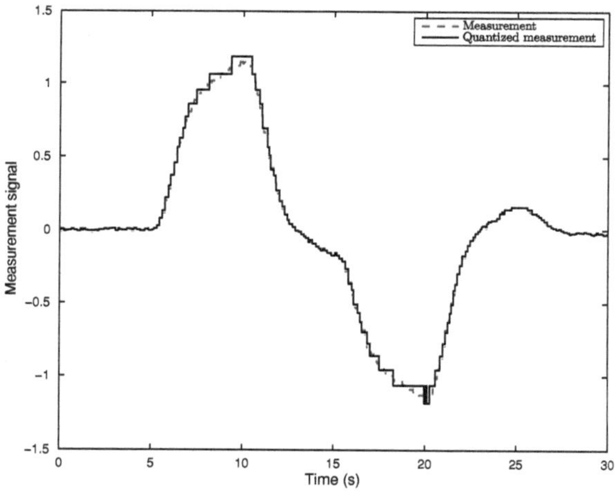

Fig. 7.5 Measurement and quantized measurement in Example 7.1

approach (Algorithm 23), the obtained minimum guaranteed H_∞ performance is $\gamma^* = 0.5069$, and the associated filter matrices are given by

$$\begin{bmatrix} A_F & B_F \\ C_F & D_F \end{bmatrix} = \left[\begin{array}{cccc|c} -5.7873 & 0 & 1 & 0 & -5.6671 \\ -3.3377 & 0 & 0 & 1 & -3.2684 \\ -13.4502 & 1.7676 & 0.8999 & 0.3196 & -10.6136 \\ 4.0084 & -1.9360 & 0.0008 & -0.8999 & 2.0671 \\ \hline -0.0026 & -1 & 0 & 0 & 0 \end{array}\right].$$

By connecting the above filter to the original system, we can use Lemma 7.1 to evaluate the H_∞ performance of the filtering error system in a point-to-point manner for the entire uncertainty domain, shown in Fig. 7.8. From Fig. 7.8, we can see

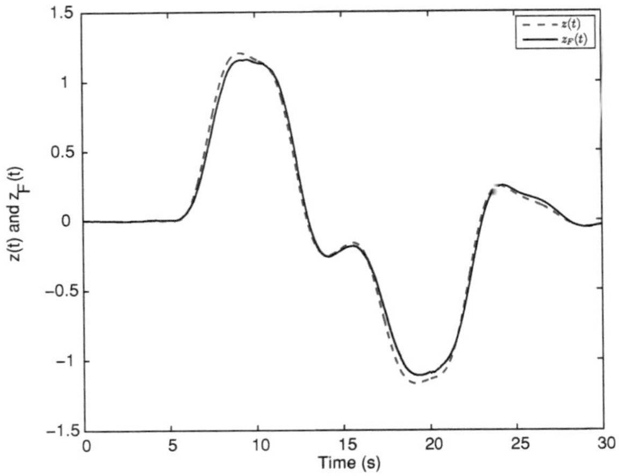

Fig. 7.6 Estimated signals in Example 7.1

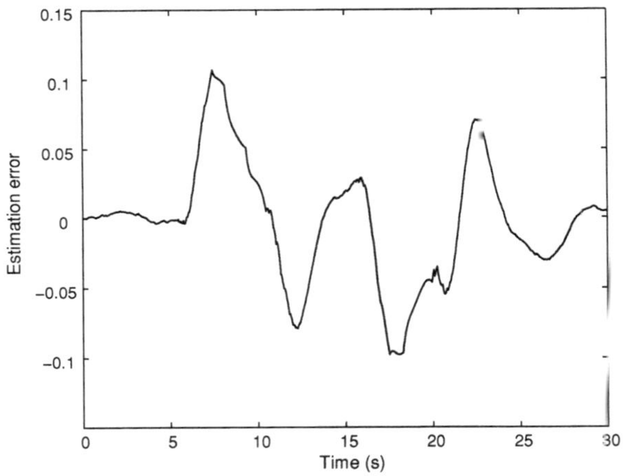

Fig. 7.7 Estimation error in Example 7.1

that the H_∞ performance for the entire uncertainty domain is below the guaranteed performance $\gamma^* = 0.5069$. Figure 7.9 depicts $z(t)$ and $z_F(t)$ for $c = 0.4$, $c = 0.5$ and $c = 0.6$, from which we can see that the designed robust filter estimates $z(t)$ pretty well for the whole uncertainty domain ($w(t)$ and $v(t)$ are the same as above).

By employing the parameter-dependent approach (Algorithm 24), the obtained minimum guaranteed H_∞ performance is $\gamma^* = 0.4694$, with the associated filter matrices given by

$$\begin{bmatrix} A_F & B_F \\ C_F & D_F \end{bmatrix} = \left[\begin{array}{cccc|c} -20.2811 & -0.0098 & 1.0856 & 0.3213 & -19.9303 \\ -13.4508 & -0.0065 & 0.0567 & 1.2131 & -13.2182 \\ -46.9236 & 1.7819 & 1.1114 & 1.0156 & -43.4862 \\ 13.1211 & -2.0040 & -0.0581 & -1.1008 & 10.9163 \\ \hline -0.0020 & -1 & 0 & -0.0020 & 0.0085 \end{array}\right]$$

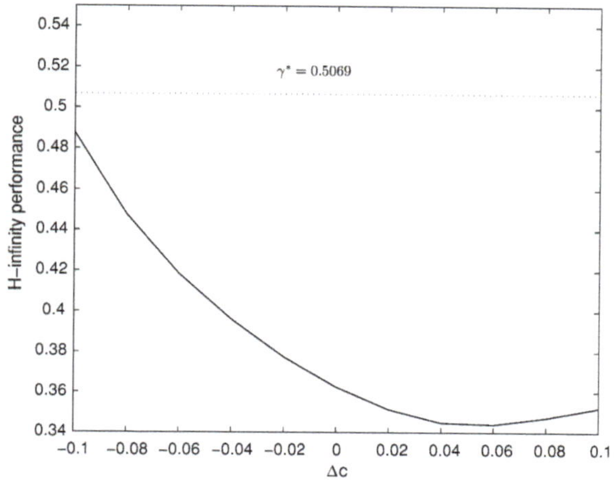

Fig. 7.8 Performance evaluation for the quadratic robust filter in Example 7.1

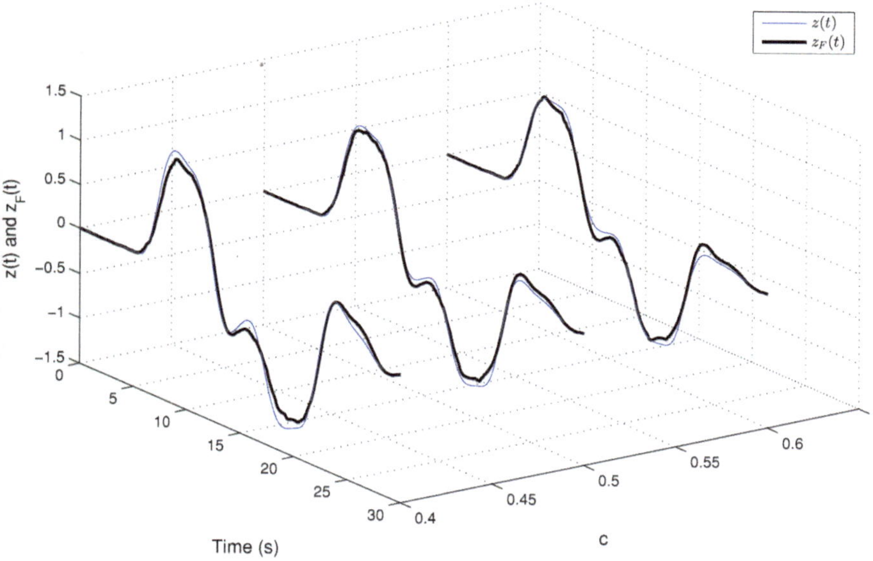

Fig. 7.9 Estimated signals for the quadratic robust filter in Example 7.1

The point-to-point evaluated H_∞ performance of the filtering error system for the entire uncertainty domain is shown in Fig. 7.10, and $z(t)$ and $z_F(t)$ for $c = 0.4$, $c = 0.5$ and $c = 0.6$ under the above filter designed by the parameter-dependent method are displayed in Fig. 7.11.

The minimum guaranteed H_∞ performances obtained above have shown that the parameter-dependent approach (Algorithm 24) is less conservative than the quadratic

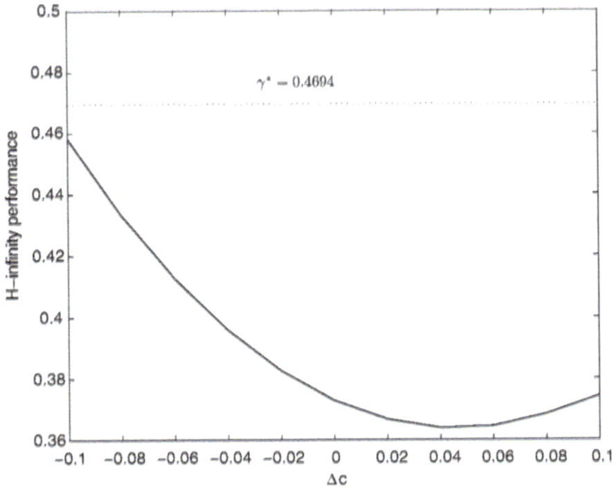

Fig. 7.10 Performance evaluation for the parameter-dependent robust filter in Example 7.1

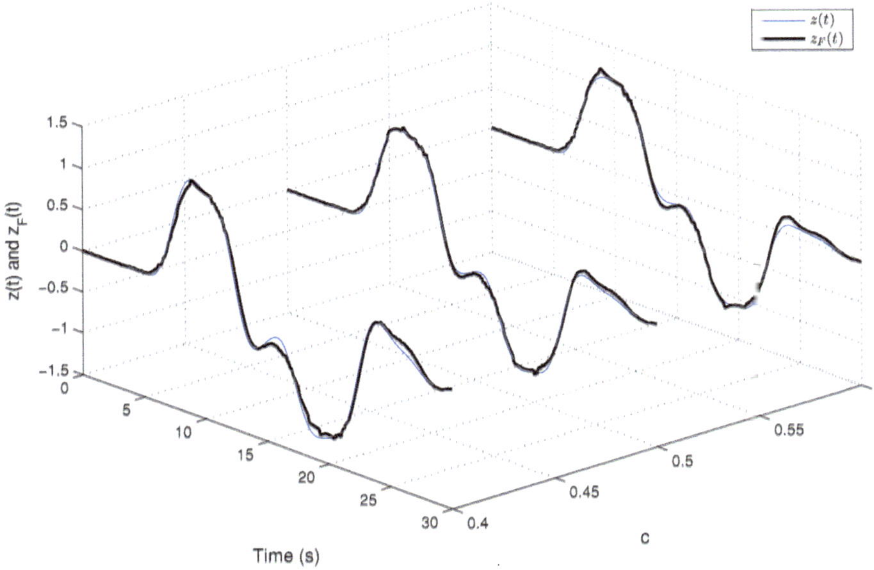

Fig. 7.11 Estimated signals for the parameter-dependent robust filter in Example 7.1

approach (Algorithm 23). To make this clearer, a more detailed comparison is provided in Table 7.1.

Example 7.2 Consider an unobservable but detectable system with the system matrices given by

Table 7.1 Minimum guaranteed H_∞ performance for different cases

	$\bar{c}=0.1$	$\bar{c}=0.15$	$\bar{c}=0.2$	$\bar{c}=0.25$	$\bar{c}=0.3$	$\bar{c}=0.35$	$\bar{c}=0.4$
Algorithm 23	0.5069	0.6030	0.7129	0.8443	1.0270	1.3116	1.8390
Algorithm 24	0.4694	0.5459	0.6343	0.7411	0.8849	1.1069	1.5432

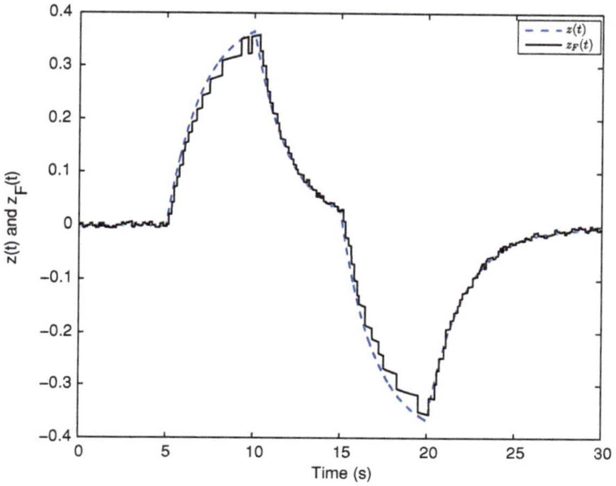

Fig. 7.12 Estimated signals in the unobservable case in Example 7.2

$$A=\begin{bmatrix}-0.6 & 0\\ 0 & -0.5\end{bmatrix},\ B=\begin{bmatrix}1\\ 0.2\end{bmatrix},$$
$$C=\begin{bmatrix}1 & 0\end{bmatrix},\ D=0.3,\ L=\begin{bmatrix}0 & 1\end{bmatrix}. \tag{7.47}$$

It is easy to verify that

$$\text{rank}\left(\begin{bmatrix}C\\ CA\end{bmatrix}\right)=\text{rank}\left(\begin{bmatrix}1 & 0\\ -0.6 & 0\end{bmatrix}\right)=1\neq 2.$$

Thus, system (7.47) is unobservable. In this example, we will show that the proposed filter design methods still work for this unobservable system. The network induced delay, data packet dropout, and measurement quantization are still assumed to be the same as in Example 7.1. By using Algorithm 22, the obtained minimum guaranteed H_∞ performance is $\gamma^*=0.0870$, and the associated filter matrices are given by

$$\begin{bmatrix}A_F & B_F\\ C_F & D_F\end{bmatrix}=\left[\begin{array}{cc|c}-390.7013 & -145.6723 & -403.2377\\ -78.1310 & -29.6479 & -80.7564\\ \hline 0 & 1 & 0\end{array}\right].$$

Simulation is carried out based on the above network assumptions and the disturbance inputs are generated in the same way as in Example 7.1. Figure 7.12 depicts signals $z(t)$ and $z_F(t)$, from which we can see that the designed filter produces a good estimate of $z(t)$. By calculation, we have $\|w\|_{L_2}^2 = 9.9980$, $\|v\|_{L_2}^2 = 0.1007$ and $\|e\|_{L_2}^2 = 0.0045$, which yield

$$\sqrt{\frac{\|e\|_{L_2}^2}{\|w\|_{L_2}^2 + \|v\|_{L_2}^2}} = 0.0210 < \gamma^* = 0.0870,$$

showing the effectiveness of the H_∞ filter design.

7.6 Summary and Notes

7.6.1 Summary

This chapter has solved the problem of robust H_∞ estimation for linear systems with polytopic uncertain parameters and subject to limited communication capacity. The information limitations include measurement quantization, signal transmission delay, and data packet dropout, which appear typically in a network environment. LMI-based conditions have been formulated for the existence of admissible filters, which ensure the filtering error systems to be asymptotically stable with a prescribed H_∞ disturbance attenuation level. For the robust filtering case, two filter design methods have been proposed based on quadratic and parameter-dependent stability ideas, which provide alternatives for designing robust H_∞ filters with different degrees of conservativeness and computational complexity. Two examples, including a mass-spring system, have been exploited to illustrate the applicability and effectiveness of the filter design methodologies proposed in this chapter.

7.6.2 Notes

Missing measurements The filtering problems with missing measurements or data packet dropout are investigated in [9, 23–32]. The phenomenon of data packet dropout in measurement is usually modeled as a random Bernoulli process [9, 24, 25, 30, 32], that is, either measurements are successfully received by the filter or just missing. In [23, 26, 31], this phenomenon was modeled as a random process with known probability density function on [0, 1], which covers the Bernoulli distribution as a special case. In [28, 29], the actual measurement was assumed to take a switching signal between the current measurement and the one-step delayed signal held in the filter, where the switching sequence satisfies a random Bernoulli distribution.

Quantized measurements The filtering problems with quantized measurements are addressed in [24, 31, 33–35]. The sector-bounded approach in [13] is often adopted to handle the quantization nonlinearity in the filtering problem [31, 33, 35].

Delayed measurements The filtering problems with delayed measurements are studied in [27, 34, 36–39].

References

1. Basin, M., Fridman, L.M., Skliar, M.: Optimal and robust sliding mode filter for systems with continuous and delayed measurements. In: Proceedings of the 41th Conference on Decision and Control, pp. 10–13. Las Vegas, NV (2002)
2. Zhang, H., Basin, M., Skliar, M.: Optimal state estimation for continuous stochastic state-space system with hybrid measurements. Int. J. Innov. Comput. Inf. Control **2**, 863–874 (2006)
3. Zhang, H., Basin, M., Skliar, M.: Itô-volterra optimal state estimation with continuous, multirate, randomly sampled, and delayed measurements. IEEE Trans. Autom. Control **52**(3), 401–416 (2007)
4. Walsh, G.C., Ye, H., Bushnell, L.: Stability analysis of networked control systems. IEEE Trans. Control Syst. Technol. **10**(3), 438–446 (2002)
5. Yue, D., Han, Q.L., Peng, C.: State feedback controller design of networked control systems. IEEE Trans. Circuits Syst. II Express Briefs **51**(11), 640–644 (2004)
6. Zhang, W., Branicky, M., Phillips, S.: Stability of networked control systems. IEEE Control Syst. Mag. **21**, 84–99 (2001)
7. Zhang, L., Shi, Y., Chen, T., Huang, B.: A new method for stabilization of networked control systems with random delays. IEEE Trans. Autom. Control **50**(8), 1177–1181 (2005)
8. Zhivoglyadov, P.V., Middleton, R.H.: Networked control design for linear systems. Automatica **39**(4), 743–750 (2003)
9. Wang, Z., Yang, F., Ho, D., Liu, X.: Robust finite-horizon filtering for stochastic systems with missing measurements. IEEE Signal Process. Lett. **12**(6), 437–440 (2005)
10. Yue, D., Han, Q.L., Lam, J.: Network-based robust H_∞ control of systems with uncertainty. Automatica **41**(6), 999–1007 (2005)
11. Goodwin, G.C., Haimovich, H., Quevedo, D.E., Welsh, J.S.: A moving horizon approach to networked control system design. IEEE Trans. Autom. Control **49**(9), 1427–1445 (2004)
12. Elia, N., Mitter, S.K.: Stabilization of linear systems with limited information. IEEE Trans. Autom. Control **46**(9), 1384–1400 (2001)
13. Fu, M., Xie, L.: The sector bound approach to quantized feedback control. IEEE Trans. Autom. Control **50**(11), 1698–1711 (2005)
14. Fridman, E., Shaked, U., Xie, L.: Robust H_∞ filtering of linear systems with time-varying delay. IEEE Trans. Autom. Control **48**(1), 159–165 (2003)
15. Gao, H., Wang, C.: A delay-dependent approach to robust H_∞ filtering for uncertain discrete-time state-delayed systems. IEEE Trans. Signal Process. **52**(6), 1631–1640 (2004)
16. de Souza, C.E., Shaked, U.: Robust H_2 filtering for uncertain systems with measurable inputs. IEEE Trans. Signal Process. **47**(8), 2286–2292 (1999)
17. Wang, Z., Lam, J., Liu, X.H.: Nonlinear filtering for state delayed systems with Markovian switching. IEEE Trans. Signal Process. **51**, 2321–2328 (2003)
18. He, Y., Wu, M., She, J.H., Liu, G.-P.: Parameter-dependent Lyapunov functional for stability of time-delay systems with polytopic-type uncertainties. IEEE Trans. Autom. Control **49**(5), 828–832 (2004)
19. Oliveira, R.C.L.F., Peres, P.L.D.: LMI conditions for robust stability analysis based on polynomially parameter-dependent Lyapunov functions. Syst. Control Lett. **55**(1), 52–61 (2006)

20. Palhares, R.M., Peres, P.L.D.: Robust filtering with guaranteed energy-to-peak performance-an LMI approach. Automatica **36**(6), 851–858 (2000)
21. Hale, J., Lunel, S.M.V.: Introduction to Functional Differential Equations. Springer, Berlin, Germany (1993)
22. Lam, J., Gao, H., Wang, C.: Stability analysis for continuous systems with two additive time-varying delay components. Syst. Control Lett. **56**(1), 16–24 (2007)
23. Dong, H., Wang, Z., Ho, D.W.C., Gao, H.: Variance-constrained H_∞ filtering for a class of nonlinear time-varying systems with multiple missing measurements: The finite-horizon case. IEEE Trans. Signal Process. **58**(5), 2534–2543 (2010)
24. Dong, H., Wang, Z., Gao, H.: Distributed filtering for a class of time-varying systems over sensor networks with quantization errors and successive packet dropouts. IEEE Trans. Signal Process. **60**(6), 3164–3173 (2012)
25. Gao, H., Zhao, Y., Lam, J., Chen, K.: H_∞ fuzzy filtering of nonlinear systems with intermittent measurements. IEEE Trans. Fuzzy Syst. **17**(2), 291–300 (2009)
26. Hu, J., Wang, Z., Gao, H., Stergioulas, L.K.: Extended Kalman filtering with stochastic non-linearities and multiple missing measurements. Automatica **48**(9), 2007–2015 (2012)
27. Moayedi, M., Foo, Y.K., Soh, Y.C.: Adaptive Kalman filtering in networked systems with random sensor delays, multiple packet dropouts and missing measurements. IEEE Trans. Signal Process. **58**(3), 1577–1588 (2010)
28. Sahebsara, M., Chen, T., Shah, S.L.: Optimal H_2 filtering in networked control systems with multiple packet dropout. IEEE Trans. Autom. Control **52**(8), 1508–1513 (2007)
29. Sahebsara, M., Chen, T., Shah, S.L.: Optimal H_∞ filtering in networked control systems with multiple packet dropouts. Syst. Control Lett. **57**(9), 696–702 (2008)
30. Wang, Z., Ho, D.W.C., Liu, X.: Variance-constrained filtering for uncertain stochastic systems with missing measurements. IEEE Trans. Autom. Control **48**(7), 1254–1258 (2003)
31. Wang, Z., Dong, H., Shen, B., Gao, H.: Finite-horizon H_∞ filtering with missing measurements and quantization effects. IEEE Trans. Autom. Control **58**(7), 1707–1018 (2013)
32. Yang, R., Shi, P., Liu, G.P.: Filtering for discrete-time networked nonlinear systems with mixed random delays and packet dropouts. IEEE Trans. Autom. Control **56**(56), 2655–2660 (2011)
33. Fu, M., de Souza, C.E.: State estimation for linear discrete-time systems using quantized measurements. Automatica **45**(12), 2937–2945 (2009)
34. Karimi, H.R.: Robust H_∞ filter design for uncertain linear systems over network with network-induced delays and output quantization. Modeling Identif. Control **30**(1), 27–37 (2009)
35. Liu, M., Ho, D.W.C., Niu, Y.: Robust filtering design for stochastic system with mode-dependent output quantization. IEEE Trans. Signal Process. **58**(12), 6410–6416 (2010)
36. Basin, M., Alcorta-Garcia, A., Rodrigues-Gonzalez, J.: Optimal filtering for linear systems with state and observation delays. Int. J. Robust Nonlinear Control **15**(17), 859–871 (2005)
37. Basin, M., Shi, P., Dario, C.A.: Central suboptimal H_∞ filter design for linear time-varying systems with state and measurement delays. Int. J. Syst. Sci. **41**(4), 411–421 (2010)
38. Zhang, H., Feng, G., Duan, G.R.: H_∞ filtering for multiple-time-delay measurements. IEEE Trans. Signal Process. **54**(5), 1681–1688 (2006)
39. Zhang, H., Lu, X., Cheng, D.: Optimal estimation for continuous-time systems with delayed measurements. IEEE Trans. Autom. Control **51**(5), 823–827 (2006)

Chapter 8
Finite Frequency H_∞ Filtering for Time-Delay Systems

8.1 Introduction

It is well known that actuators, sensors and transmission lines in feedback loops especially in networked controlled systems always introduce after affect more or less. Therefore, system models with delays will behave more like the real process [1]. In this context, Chaps. 4 and 5 have been devoted to robust H_∞ filtering for uncertain continuous and discrete time-delay systems, respectively, while Chap. 7 is concerned with H_∞ filtering with limited communication capacity, a more realistic and involved problem related to time-delay systems.

However, it deserves to be pointed out that the objectives of all the H_∞ filtering results presented in the previous chapters are to minimize the H_∞ norm (i.e., L_2- or l_2-induced gain) of the filtering error system in the entire frequency (EF) range. If the frequency of the external disturbance belongs to a known finite range, it is unnecessary to guarantee an H_∞ disturbance attenuation level over the EF range, and the H_∞ filtering results in the previous chapters would be too conservative due to overdesign. Naturally, to design filters according to the specified FF range will be more effective. A popular method for adjusting the discrepancy is the so-called frequency weighting functions, while the additional weighting functions tend to increase the system complexity and to search for good frequency weighting functions can be tedious and time-consuming [2]. Instead in recent years, with the presence of the Generalized Kalman–Yakubovich–Popov (GKYP) Lemma [3, 4], considerable attention has been paid to directly applying this fundamental tool, in the framework of LMI, to study analysis and synthesis problems of control systems with FF specifications [5–10]. In this chapter, we specifically call the H_∞ filtering results in the previous chapters as *EF* H_∞ *filtering* results so as to differentiate the problem and the results regarding *FF* H_∞ *filtering* in this chapter.

In this chapter, based on the GKYP Lemma, we study the problem of H_∞ filtering for continuous time-delay systems with FF specifications. Compared with most of the H_∞ filtering results in the literature as well as in the previous chapters of this paper, the remarkable difference of the results in this chapter is that the disturbance

H. Gao and X. Li, *Robust Filtering for Uncertain Systems*,
Communications and Control Engineering, DOI: 10.1007/978-3-319-05903-7_8,

is assumed to reside in a known low/middle/high frequency (LF/MF/HF) range, and the filter design methods can design H_∞ filters for time-delay systems according to the FF properties of the external disturbance. By virtue of the GKYP Lemma and the Projection Lemma, a delay-dependent FF BRL is first derived for analyzing the FF H_∞ performance of the filtering error system and then filter design methods are obtained in terms of solving a set of LMIs. To reduce the conservatism inherent to sufficient conditions for time-delay systems, the delay-partitioning idea [11] is incorporated into the derived results. Moreover, two approaches, respectively, starting from the transfer function and an LKF, will be presented to prove the new FF BRL. Finally, we will give two numerical examples to illustrate the effectiveness and advantage of the developed FF H_∞ filter design methods in this chapter

8.2 Problem Formulation

Consider a class of linear time-delay systems with a state delay:

$$\begin{aligned}\dot{x}(t) &= Ax(t) + A_d x(t-d) + Bw(t),\\ y(t) &= Cx(t) + C_d x(t-d) + Dw(t),\\ z(t) &= Hx(t) + H_d x(t-d) + Lw(t),\\ x(t) &= \phi(t),\ \forall t = [-d, 0],\end{aligned} \tag{8.1}$$

where $x(t) \in \mathbb{C}^{n_x}$ is the state vector; $y(t) \in \mathbb{C}^{n_y}$ is the measured output; $z(t) \in \mathbb{C}^{n_z}$ is the signal to be estimated; $w(t) \in \mathbb{C}^{n_w}$ is the noise input belonging to $L_2[0, \infty)$ and the frequency of $w(t)$ resides in a known but finite range $\boldsymbol{\Omega}$; d is a state delay, which is assumed to be time-invariant and known; $\phi(t)$ is the initial state vector function over $[-d, 0]$; and A, A_d, B, C, C_d, D, H, H_d and L are known complex system matrices with appropriate dimensions. The FF range $\boldsymbol{\Omega}$ is assumed to be the general LF/MF/HF range, that is,

$$\boldsymbol{\Omega} \triangleq \begin{cases} \{\omega \in \mathbb{R} |\ |\omega| \le \omega_l, \omega_l \ge 0\}, & \text{(LF)},\\ \{\omega \in \mathbb{R} |\ \omega_1 \le \omega \le \omega_2, 0 \le \omega_1 \le \omega_2\}, & \text{(MF)},\\ \{\omega \in \mathbb{R} |\ |\omega| \ge \omega_h, \omega_h \ge 0\}, & \text{(HF)}.\end{cases} \tag{8.2}$$

We desire to design a full-order filter with its output $z_F(t)$ being the estimate of the unknown signal $z(t)$. Consider a full-order filter of the following state space realization:

$$\begin{aligned}\dot{x}_F(t) &= A_F x_F(t) + B_F y(t),\ x_F(0) = 0,\\ z_F(t) &= C_F x_F(t) + D_F y(t),\end{aligned} \tag{8.3}$$

where $x_F(t) \in \mathbb{C}^n$ is the filter state vector; $y(t)$ is the measured output in original system in (8.1); and A_F, B_F, C_F and D_F with appropriate dimensions are complex matrices to be determined. Augmenting the system in (8.1) to include the filter states in (8.3), we have the following filtering error system:

$$\begin{aligned}\dot{\xi}(t) &= \bar{A}\xi(t) + \bar{A}_d K\xi(t-d) + \bar{B}w(t),\\ e(t) &= \bar{C}\xi(t) + \bar{C}_d K\xi(t-d) + \bar{D}w(t),\\ \xi(t) &= [\phi(t)^*, 0]^*, \ \forall t = [-d, 0],\end{aligned} \tag{8.4}$$

where $e(t) \triangleq z(t) - z_F(t)$ is the filtering error, $\bar{x}(t) \triangleq \begin{bmatrix} x(t)^* & x_F(t)^* \end{bmatrix}^*$ is the augmented state, and

$$\begin{aligned}\bar{A} &= \begin{bmatrix} A & 0 \\ B_F C & A_F \end{bmatrix}, \ \bar{A}_d = \begin{bmatrix} A_d \\ B_F C_d \end{bmatrix}, \ \bar{B} = \begin{bmatrix} B \\ B_F D \end{bmatrix},\\ \bar{C} &= \begin{bmatrix} H - D_F C & -C_F \end{bmatrix}, \ \bar{C}_d = H_d - D_F C_d,\\ \bar{D} &= L - D_F D, \ K = \begin{bmatrix} I_{n_x} & 0_{n_x \times n_x} \end{bmatrix}.\end{aligned}$$

Accordingly, the transfer function of the filtering error system in (8.4) is given by

$$G(s) \triangleq \left(\bar{C} + \mathrm{e}^{-ds}\bar{C}_d K\right)\left(sI - \bar{A} - \mathrm{e}^{-ds}\bar{A}_d K\right)^{-1}\bar{B} + \bar{D}. \tag{8.5}$$

We can always find a set $\mathbf{S}$ of complex number such that $s \in \mathbf{S}$ is the equivalent characterization of $\omega \in \boldsymbol{\Omega}$ in (8.2) with $s = \mathrm{j}\omega$. For brevity and convenience, both $s \in \mathbf{S}$ and $\omega \in \boldsymbol{\Omega}$ denote the finite frequency specifications in the context.

Since no control input is considered in the original system model, the asymptotic stability of the filtering error system must be based on the assumption that the original system in (8.1) is asymptotically stable.

The FF H_∞ filtering problem of continuous time-delay systems to be investigated in this chapter is formulated as follows.

FF H_∞ Filtering for Continuous Time-Delay Systems: Find a filter of the state space realization in (8.3) for the state-delayed system in (8.1) such that

(i) the filtering error system in (8.4) is asymptotically stable;
(ii) under the zero initial condition, the following FF H_∞ performance index holds:

$$\sigma_{\max}[G(s)] < \gamma, \ \forall s \in \mathbf{S} \tag{8.6}$$

for a given proper positive scalar γ, or equivalently, the following index holds:

$$\|e\|_2 < \gamma \|w\|_2 \tag{8.7}$$

for all nonzero $w(t) \in L_2[0, \infty)$ with the frequency of $w(t)$ satisfying $\omega \in \boldsymbol{\Omega}$ and a given proper positive scalar γ.

Remark 8.1 Some researchers considered the problem of standard H_∞ filtering for the case of multiple delays [12, 13]. Whereas addressing this case would make the notations much more complicated and cloud the main idea; hence the system under consideration in this chapter only contains a single delay in the state. However, it is straightforward to extend the results in the chapter to the multiple delay case. Additionally, we only consider a nominal system with exactly known parameters, while it is also not difficult to extend our results, either by the quadratic approach or the parameter-dependent approach, to time-delay systems subject to polytopic uncertainty as extensively discussed in the previous chapters.

Compared with the system model in (4.1) in Chap. 4, the frequency of the disturbance $w(t)$ in (8.1) is assumed to satisfy $\omega \in \boldsymbol{\Omega}$ with $\boldsymbol{\Omega}$ in (8.2), which is reasonable because the energy of many practical disturbance signals is mostly distributed in an FF range, not the EF domain. Thus, in the context of H_∞ filtering, it is unnecessary, like the results in Chaps. 4, 5 and 7, to still employ the standard H_∞ norm index or the induced L_2 gain as the measure of the noise attenuation level. Instead, the H_∞ index in (8.6) or (8.7) with FF specifications is a better metric of the filtering performance.

It should be noted that when $\boldsymbol{\Omega} = \mathbb{R} \cup \{\infty\}$, the indices in (8.6) and (8.7) reduce to the standard H_∞ index or the induced L_2 gain. Hence, the EF H_∞ filtering results in Chaps. 4, 5 and 7 are still applicable to the problem in this chapter. However, the "coarse" application of these results undoubtedly introduces much conservatism due to overdesign. In this chapter, by virtue of the GKYP Lemma (Lemma 1.5), we will take the FF specification into consideration and directly derive methods for designing filters satisfying (i) and (ii).

8.3 Filter Analysis

In this section, we will first propose a delay-dependent condition ensuring the specifications in (i) and (ii) for filtering analysis. We mainly focus on analyzing the FF H_∞ performance specification in (ii) by two equivalent approaches. Exactly speaking, these two approaches start from the frequency-domain description in (8.6) and the time-domain one in (8.7) of the FF H_∞ performance, respectively.

8.3.1 Delay-Dependent FF BRL

The following theorem gives a delay-dependent sufficient condition guaranteeing the asymptotic stability and an FF H_∞ performance level for the filtering error system in (8.4), namely, a delay-dependent FF BRL.

Theorem 8.1 *Consider the system in* (8.1) *and suppose that a scalar $\gamma > 0$, an integer $m > 0$ and a scalar $\tau > 0$ satisfying $d = m\tau$ are given. For a given filter, the filtering error system in* (8.4) *is asymptotically stable with an FF H_∞ performance bound γ in the frequency range* $\mathbf{S}$, *if there exist matrices $P_1 \in \mathbf{H}_{2n_x}$, $0 < P_2 \in \mathbf{H}_{2n_x}$,*

$0 < Q \in \mathbf{H}_{2n_x}$, $R_1 = \begin{bmatrix} R_{1,1} & R_{1,2} \\ R_{1,2}^* & R_{1,3} \end{bmatrix} \in \mathbf{H}_{mn_x}$, $0 < R_2 = \begin{bmatrix} R_{2,1} & R_{2,2} \\ R_{2,2}^* & R_{2,3} \end{bmatrix} \in \mathbf{H}_{mn_x}$, $0 < S_j \in \mathbf{H}_{n_x}$ *and* $Y_j \in \mathbb{C}^{4n_x \times 2n_x}$, $j = 1, 2$ *such that*

$$\begin{bmatrix} -I_{n_z} & M \\ M^* & \Sigma + \text{sym}\left\{X_1^* Y_1 Z_1\right\} \end{bmatrix} < 0, \tag{8.8}$$

$$\Xi_2 + \text{sym}\left\{X_2^* Y_2 Z_2\right\} < 0, \tag{8.9}$$

where

$$\begin{aligned}
M &\triangleq \begin{bmatrix} 0_{n_z \times 2n_x} & \bar{C} & \bar{C}_d E_d & \bar{D} \end{bmatrix}, \\
\Sigma &\triangleq \text{diag}\{\Xi_1 + \Xi_Q, -\gamma^2 I_{n_w}\}, \\
X_1 &\triangleq \begin{bmatrix} I_{4n_x} & 0_{4n_x \times (mn_x + n_w)} \end{bmatrix}, \\
X_2 &\triangleq \begin{bmatrix} I_{4n_x} & 0_{4n_x \times mn_x} \end{bmatrix}, \\
Z_1 &\triangleq \begin{bmatrix} -I_{2n_x} & \bar{A} & \bar{A}_d E_d & \bar{B} \end{bmatrix}, \\
Z_2 &\triangleq \begin{bmatrix} -I_{2n_x} & \bar{A} & \bar{A}_d E_d \end{bmatrix}, \\
\Xi_Q &\triangleq \begin{bmatrix} \Psi \otimes Q & 0_{4n_x \times mn_x} \\ 0_{mn_x \times 4n_x} & 0_{mn_x \times mn_x} \end{bmatrix}, \\
\Xi_j &\triangleq \begin{bmatrix} \tau^2 K^{\mathrm{T}} S_j K & P_j & 0_{2n_x \times mn_x} \\ P_j & \Xi_{j,1} & \Xi_{j,2} \\ 0_{mn_x \times 2n_x} & \Xi_{j,2}^* & \Xi_{j,3} \end{bmatrix}, \\
\Xi_{j,1} &\triangleq K^{\mathrm{T}}[-S_j + R_{j,1}]K, \\
\Xi_{j,2} &\triangleq K^{\mathrm{T}}[R_{j,2} E_R + S_j E_S], \\
\Xi_{j,3} &\triangleq E_R^{\mathrm{T}} R_{j,3} E_R - R_j - E_S^{\mathrm{T}} S_j E_S, \\
E_S &\triangleq [\, I_{n_x} \;\; 0_{n_x \times (m-1)n_x} \,], \\
E_R &\triangleq [\, I_{(m-1)n_x} \;\; 0_{(m-1)n_x \times n_x} \,], \\
E_d &\triangleq [\, 0_{n_x \times (m-1)n_x} \;\; I_{n_x} \,].
\end{aligned}$$

In the following two subsections, based on the FF H_∞ specification in (8.6) and (8.7), we give two approaches to prove the above result.

8.3.2 The First Proof

In order to introduce the delay partitioning idea, we fraction the delay d into m subintervals, i.e., $d = m\tau$ with m being a positive integer, and define the following complex matrices:

$$
\begin{aligned}
T(s) &\triangleq \begin{bmatrix} T_1(s) \\ T_2(s) \\ I_{n_w} \end{bmatrix}, \\
T_1(s) &\triangleq \left(sI - \bar{A} - \mathrm{e}^{-\tau s}\bar{A}_d K\right)^{-1}\bar{B}, \\
T_2(s) &\triangleq \begin{bmatrix} \mathrm{e}^{-\tau s} K T_1(s) \\ \vdots \\ \mathrm{e}^{-m\tau s} K T_{1s} \end{bmatrix}.
\end{aligned} \tag{8.10}
$$

It follows that

$$
G(s) = \underbrace{\begin{bmatrix} \bar{C} & \bar{C}_d E_d & \bar{D} \end{bmatrix}}_{M_1} T(s), \tag{8.11}
$$

$$
sT_1(s) = \underbrace{\begin{bmatrix} \bar{A} & \bar{A}_d E_d & \bar{B} \end{bmatrix}}_{M_2} T(s), \tag{8.12}
$$

where

$$
M_1 \triangleq \left[\bar{C}\bar{C}_d E_d \bar{D}\right], \ M_2 \triangleq \left[\bar{A}\bar{A}_d E_d \bar{B}\right].
$$

The frequency-domain description of the FF H_∞ performance specification in (8.6) can be rewritten into the following frequency-domain inequality (FDI):

$$
\upsilon\left(G(s), \Pi\right) < 0, \ \forall s \in \mathbf{S}, \tag{8.13}
$$

where

$$
\Pi \triangleq \begin{bmatrix} I_{n_z} & 0 \\ 0 & -\gamma^2 I_{n_w} \end{bmatrix}.
$$

By defining $E_1 \triangleq \begin{bmatrix} 0_{n_w \times (m+2)n_x} & I_{n_w} \end{bmatrix}$ and using Eq. (8.11), the FDI in (8.13) can be equivalently expressed as

$$
T(s)^* \left(M_1^* M_1 - \gamma^2 E_1^* E_1\right) T(s) < 0, \ \forall s \in \mathbf{S}. \tag{8.14}
$$

Define the following set of complex vectors:

$$
\mathbf{W}_s \triangleq \left\{\zeta_s \in \mathbb{C}^{(m+2)n_x + n_w} \,\middle|\, \zeta_s = T(s)\rho, \ \rho \in \mathbb{C}^{n_w}, \ \rho \neq 0, \omega \in \mathbb{R} \cup \{\infty\}\right\}, \tag{8.15}
$$

and then the FDI in (8.14) can be further equivalently expressed as

$$
\mathbf{W}_s^* \left(M_1^* M_1 - \gamma^2 E_1^* E_1\right) \mathbf{W}_s < 0, \ \forall s \in \mathbf{S}, \tag{8.16}
$$

which will be used for further derivation.

Now we present the first method to prove Theorem 8.1.

Proof (Proof of Theorem 8.1) *The first part*: We show that (8.16) follows from (8.8). Due to $s = \mathrm{j}\omega$, for any $P_1 \in \mathbf{H}_{2n_x}$, $R_1 \in \mathbf{H}_{mn_x}$, $0 < S_1 \in \mathbf{H}_{n_x}$ and nonzero $\rho \in \mathbb{C}^{n_w}$, we have the following relations:

$$\begin{aligned}
\Delta_1 &\triangleq (s^* + s)\rho^* T_{1s}^* P_1 T_1(s)\rho \\
&\equiv 0, && (8.17)\\
\Delta_2 &\triangleq \left[(\mathrm{e}^{\tau s})^* \mathrm{e}^{\tau s} - 1\right] \rho^* T_2(s)^* R_1 T_2(s)\rho \\
&\equiv 0, && (8.18)\\
\Delta_3 &\triangleq [\tau^2 s^* s - 1 + \left(\mathrm{e}^{-\tau s}\right)^* + \mathrm{e}^{-\tau s} - \left(\mathrm{e}^{-\tau s}\right)^* \mathrm{e}^{\mathrm{j}\tau s}] \\
&\quad \times \rho^* T_1(s)^* K^{\mathrm{T}} S_1 K T_1(s)\rho \\
&= \left[\tau^2\omega^2 - 2 + 2\cos(\tau\omega)\right] \rho^* T_1(s)^* K^{\mathrm{T}} S_1 K T_1(s)\rho \\
&= 4\left[\left(\frac{\tau\omega}{2}\right)^2 - \sin^2\left(\frac{\tau\omega}{2}\right)\right] \rho^* T_1(s)^* K^{\mathrm{T}} S_1 K T_1(s)\rho \\
&\geq 0. && (8.19)
\end{aligned}$$

Define some symbols as

$$\begin{aligned}
F_1 &\triangleq \begin{bmatrix} M_2 \\ E_2 \end{bmatrix}, \quad E_2 \triangleq \begin{bmatrix} I_{2n_x} & 0_{2n_x \times (mn_x + n_w)} \end{bmatrix}, \\
F_2 &\triangleq \begin{bmatrix} F_1 \\ E_3 \end{bmatrix}, \quad E_3 \triangleq \begin{bmatrix} 0_{mn_x \times 2n_x} & I_{mn_x} & 0_{mn_x \times n_w} \end{bmatrix}. && (8.20)
\end{aligned}$$

Combining Eqs. (8.17)–(8.19), we have, for all $\zeta_s \in \mathbf{W}_s$,

$$\zeta_s^* F_2^* \Xi_1 F_2 \zeta_s = \Delta_1 + \Delta_2 + \Delta_3 \geq 0. \tag{8.21}$$

Thus the FDI in (8.16) can be ensured by

$$\mathbf{W}_s^* \left(M_1^* M_1 - \gamma^2 E_1^* E_1 + F_2^* \Xi_1 F_2\right) \mathbf{W}_s < 0, \forall s \in \mathbf{S}. \tag{8.22}$$

The null space of $\Gamma_s F_1$ is given by

$$\mathbf{N}_s \triangleq \left\{\zeta_s \in \mathbb{C}^{(m+2)n_x + n_w} | \Gamma_s F_1 \zeta_s = 0, \zeta_s \neq 0\right\}, \tag{8.23}$$

where Γ_s is defined in (1.13) with $n_1 = 2n_x$. Due to the fact that $\forall \zeta_s \in \mathbf{W}_s$, $\Gamma_s F_1 \zeta_s = 0$, it follows that $\mathbf{W}_s \subset \mathbf{N}_s$, which means that the FDI in (8.22) holds if the following FDI is satisfied:

$$\mathbf{N}_s^* \left(M_1^* M_1 - \gamma^2 E_1^* E_1 + F_2^* \Xi_1 F_2 \right) \mathbf{N}_s < 0, \forall s \in \mathbf{S}. \tag{8.24}$$

By the GKYP Lemma, i.e., Lemma 1.5 with $\Theta = M_1^* M_1 - \gamma^2 E_1^* E_1 + F_2^* \Xi_1 F_2$, the condition in (8.24) is equivalent to that the LMI

$$\Theta + F_1^* \left(\Phi \otimes P + \Psi \otimes Q \right) F_1 < 0 \tag{8.25}$$

holds for some $P, Q \in \mathbf{H}_{2n_x}$ with $Q > 0$. Redefining matrix variables $P_1 = P_1 + P$ and performing some routine matrix manipulations to the LMI in (8.25), we can rewrite the condition in (8.25) exactly as

$$\begin{bmatrix} M_2 \\ I_{(2+m)n_x+n_w} \end{bmatrix}^* \left(\Sigma + M^* M \right) \begin{bmatrix} M_2 \\ I_{(2+m)n_x+n_w} \end{bmatrix} < 0, \tag{8.26}$$

with M in (8.8).

Note that the null space of X_1 and Z_1 in (8.8) can be chosen as

$$\mathcal{N}_{X_1} = \begin{bmatrix} 0_{4n_x \times (mn_x+n_w)} \\ I_{mn_x+n_w} \end{bmatrix}, \ \mathcal{N}_{Z_1} = \begin{bmatrix} M_2 \\ I_{(2+m)n_x+n_w} \end{bmatrix},$$

respectively. By virtue of Lemma 1.2, we know that the LMI in (8.26) and

$$\mathcal{N}_{X_1}^* \left(\Sigma + M^* M \right) \mathcal{N}_{X_1} < 0 \tag{8.27}$$

hold if and only if there exists $Y_1 \in \mathbb{C}^{4n_x \times 2n_x}$ such that

$$\Sigma + M^* M + \text{sym}\left\{ X_1^* Y_1 Z_1 \right\} < 0 \tag{8.28}$$

holds. By calculation, the inequality in (8.27) can be explicitly expressed as

$$\mathcal{N}_{X_1}^* \left(\Sigma + M^* M \right) \mathcal{N}_{X_1} = \begin{bmatrix} \Xi_{j,3} + \left(\bar{C}_d E_d \right)^* \bar{C}_d E_d & \left(\bar{C}_d E_d \right)^* \bar{D} \\ \bar{D}^* \bar{C}_d E_d & \bar{D}^* \bar{D} - \gamma^2 I_{n_w} \end{bmatrix} < 0. \tag{8.29}$$

Note that this inequality is implied by the LMI in (8.26), which means that the LMI in (8.26) is equivalent to the one in (8.28). Then by the Schur complement (Lemma 1.1), it can be verified that (8.28) is exactly (8.8). According to the above derivation, the LMI in (8.8) gives a sufficient condition ensuring the FF H_∞ performance in terms of the FDI in (8.16), which completes the first part of the proof.

The second part: We demonstrate that (8.9) is a sufficient condition for the asymptotic stability condition of the filtering error system in (8.4). Define vectors:

$$g(t) \triangleq \begin{bmatrix} \xi(t) \\ g_1(t) \\ w(t) \end{bmatrix}, \ g_1(k) \triangleq \begin{bmatrix} x(t-\tau) \\ \vdots \\ x(t-m\tau) \end{bmatrix}, \tag{8.30}$$

where m is the delay partitioning number with the same meaning as in the first part. Using M_1 and M_2 in (8.11) and (8.12), it is obvious that

$$\dot{\xi}(t) = \bar{A}\xi(t) + \bar{A}_d K\xi(t-d) + \bar{B}w(t) = M_2 g(t), \tag{8.31}$$

$$e(t) = \bar{C}\xi(t) + \bar{C}_d K\xi(t-d) + \bar{D}w(t) = M_1 g(t). \tag{8.32}$$

Motivated by the delay-partitioning idea in [11], construct an LKF candidate as

$$V_2(t) \triangleq V_{2,1}(t) + V_{2,2}(t) + V_{2,3}(t), \tag{8.33}$$

where

$$\begin{aligned} V_{2,1}(t) &\triangleq \xi(t)^* P_2 \xi(t), \\ V_{2,2}(t) &\triangleq \int_{t-\tau}^{t} g_1(\eta+\tau)^* R_2 g_1(\eta+\tau)\mathrm{d}\eta, \\ V_{2,3}(t) &\triangleq \tau \int_{-\tau}^{0}\int_{t+\theta}^{t} \dot{x}(\eta)^* S_2 \dot{x}(\eta)\mathrm{d}\eta\mathrm{d}\theta \end{aligned}$$

with P_2, R_2 and S_2 being positive-definite Hermitian matrices of appropriate dimensions. Taking the derivative of $V_2(t)$ with respect to t along the trajectory of the filtering error system in (8.4), we have

$$\dot{V}_{2,1}(t) = \dot{\xi}(t)^* P_2 \xi(t) + \xi(t)^* P_2 \dot{\xi}(t), \tag{8.34}$$

$$\dot{V}_{2,2}(t) = g_1(t+\tau)^* R_2 g_1(t+\tau) - g_1(t)^* R_2 g_1(t), \tag{8.35}$$

$$\dot{V}_{2,3}(t) = \tau^2 \dot{x}(t)^* S_2 \dot{x}(t) - \tau \int_{t-\tau}^{t} \dot{x}(\theta)^* S_2 \dot{x}(\theta)\mathrm{d}\theta. \tag{8.36}$$

Using Jensen inequality in (1.14) yields

$$\begin{aligned} -\tau \int_{t-\tau}^{t} \dot{x}(\theta)^* S_2 \dot{x}(\theta)\mathrm{d}\theta &\leq -\left(\int_{t-\tau}^{t} \dot{x}(\theta)\mathrm{d}\theta\right)^* S_2 \left(\int_{t-\tau}^{t} \dot{x}(\theta)\mathrm{d}\theta\right) \\ &= -\left[x(t) - x(t-\tau)\right]^* S_2 \left[x(t) - x(t-\tau)\right]. \end{aligned} \tag{8.37}$$

Combining Eq. (8.32), and Eqs. (8.34)–(8.37), we have

$$\dot{V}_2(t) = \dot{V}_{2,1}(t) + \dot{V}_{2,2}(t) + \dot{V}_{2,3}(t) \leq g(t)^* F_2^* \Xi_2 F_2 g(t) \tag{8.38}$$

with F_2 in (8.20) and Ξ_2 in (8.9). Assuming zero noise input, we have

$$g^*(t) F_2^* \Xi_2 F_2 g(t) = \begin{bmatrix} \xi(t) \\ g_1(t) \end{bmatrix}^* \begin{bmatrix} N \\ I_{(2+m)n_x} \end{bmatrix}^* \Xi_2 \begin{bmatrix} N \\ I_{(2+m)n_x} \end{bmatrix} \begin{bmatrix} \xi(t) \\ g_1(t) \end{bmatrix},$$

where $N \triangleq [\, \bar{A} \ \bar{A}_d E_d \,]$, which implies that $\dot{V}_2(t) < 0$ follows from

$$\begin{bmatrix} N \\ I_{(2+m)n_x} \end{bmatrix}^* \Xi_2 \begin{bmatrix} N \\ I_{(2+m)n_x} \end{bmatrix} < 0 \tag{8.39}$$

when $w(k) = 0$. Based on Lyapunov stability theory, (8.39) is an asymptotic stability condition for the filtering error system in (8.4).

Similar to the first part of the proof, with the definitions of X_2 and Z_2 in (8.9), we can choose the null spaces of X_2 and Z_2 as

$$\mathcal{N}_{X_2} = \begin{bmatrix} 0_{4n_x \times mn_x} \\ I_{mn_x} \end{bmatrix}, \; \mathcal{N}_{Z_2} = \begin{bmatrix} N \\ I_{(2+m)n_x} \end{bmatrix}.$$

Then by Lemma 1.2, we have that the LMI in (8.39) and

$$\mathcal{N}_{X_2}^* \Xi_2 \mathcal{N}_{X_2} < 0 \tag{8.40}$$

hold if and only if there exists $Y_2 \in \mathbb{C}^{4n_x \times 2n_x}$ such that (8.9) holds. We can also obtain the equivalence between (8.9) and (8.39) from the fact that (8.40) is implied by (8.39). Therefore, (8.9) is an asymptotic stability condition for the filtering error system, which completes the second part of the proof as well as the entire proof. □

Remark 8.2 The LMIs in (8.8) and (8.9) both depend on τ, which means that Theorem 8.1 is a delay-dependent sufficient condition for analyzing the FF H_∞ performance of the filtering error system in (8.4). Since τ is related to the partitioning number m, Theorem 8.1 incorporates the delay-partitioning idea as well. Two special cases are worth pointing out. First, if we take $Q = 0$, Theorem 8.1 reduces to a sufficient condition for analyzing the EF H_∞ performance of the filtering error system. Second, if we set $S_1 = 0$, scalar τ will disappear from Theorem 8.1, implying that Theorem 8.1 reduces to a delay-independent condition. □

8.3.3 The Second Proof

In this section, from a time-domain point of view, we apply an LKF-based method to prove the first part of Theorem 8.1. This method has been extensively used to solve the EF H_∞ filtering problems.

Construct another LKF candidate $V_1(t)$, similar to $V_2(t)$ in (8.33), as

$$V_1(t) \triangleq V_{1,1}(t) + V_{1,2}(t) + V_{1,3}(t), \tag{8.41}$$

where

$$\begin{aligned} V_{1,1}(t) &\triangleq \xi(t)^* P_1 \xi(t), \\ V_{1,2}(t) &\triangleq \int_{t-\tau}^{t} g_1(\eta+\tau)^* R_1 g_1(\eta+\tau)\mathrm{d}\eta, \end{aligned}$$

$$V_{1,3}(t) \triangleq \tau \int_{-\tau}^{0} \int_{t+\theta}^{t} \dot{x}(\eta)^* S_1 \dot{x}(\eta) \mathrm{d}\eta \mathrm{d}\theta$$

with P_1, R_1 and S_1 being positive-definite Hermitian matrices of appropriate dimensions and $g_1(t)$ in (8.30). The proof is presented as follows:

Proof (Proof of Theorem 8.1) Using a similar technique as in the second part of the former proof of Theorem 8.1 and taking the derivative of $V_1(t)$ along the trajectory of system (8.4), we have

$$\Delta V_1(t) \leq g^*(t) F_2^* \Xi_1 F_2 g(t) \tag{8.42}$$

with Ξ_1 in (8.8). Define index

$$J \triangleq \int_0^\infty \left[e(t)^* e(t) - \gamma^2 w(t)^* w(t) \right] \mathrm{d}t. \tag{8.43}$$

Then in view of the fact that $V_1(t) \geq 0$ and the relations in (8.31) and (8.42), under the zero initial condition, it follows that

$$\begin{aligned} J &\leq \int_0^\infty \left[e(t)^* e(t) - \gamma^2 w(t)^* w(t) \right] \mathrm{d}t + V_1(\infty) - V_1(0) \\ &= \int_0^\infty \left[e(t)^* e(t) - \gamma^2 w(t)^* w(t) + \dot{V}_1(t) \right] \mathrm{d}t \\ &\leq \int_0^\infty g(t)^* \left[\left(M_1^* M_1 - \gamma^2 E_1^* E_1 \right) + F_2^* \Xi_1 F_2 \right] g(t) \mathrm{d}t \end{aligned} \tag{8.44}$$

Let $\Theta = M_1^* M_1 - \gamma^2 E_1^* E_1 + F_2^* \Xi_1 F_2$, and by Parseval's theorem [14], we get

$$\int_0^\infty g^*(t) \Theta g(t) \mathrm{d}t = \frac{1}{2\pi} \int_{-\infty}^{\infty} g_s^* \Theta g_s \mathrm{d}\omega, \tag{8.45}$$

where g_s is the Laplace transform of the time-domain vector-valued function $g(t)$ and $s = \mathrm{j}\omega$. Noting that the frequency of noise $w(t)$ resides in an FF range $\mathbf{S}$ and combining (8.44) and (8.45), we have that if the following FDI holds:

$$g_s^* \Theta g_s < 0, \forall s \in \mathbf{S}, \tag{8.46}$$

then $J < 0$ holds for all $\omega \in \mathbf{\Omega}$. This means that (8.46) is a sufficient condition guaranteeing the FF performance specification in (ii). Further noting the fact that g_s belongs to $\mathbf{N}_s$ defined in (8.23) due to $\Gamma_s F_1 g_s = 0$, we have

$$\mathbf{N}_s^* \Theta \mathbf{N}_s < 0, \ \forall s \in \mathbf{S} \Rightarrow g_s^* \Theta g_s < 0, \ \forall s \in \mathbf{S}. \tag{8.47}$$

By the GKYP Lemma, the LHS of "⇒" in (8.47) is equivalent to that the LMI

$$\Theta + F_1^* (\Phi \otimes P + \Psi \otimes Q) F_1 < 0 \tag{8.48}$$

holds for some $P, Q \in \mathbf{H}_{2n_x}$ with $Q > 0$. Noting the term $\Delta_2 \equiv 0$ in (8.18), we can remove positiveness and only reserve conjugate symmetry for R_1, and then the LMI in (8.48) is exactly the one in (8.25).

In the sequel, by using Lemma 1.3 and following the remaining lines of the first part of the previous proof, we can complete the proof in this section.

Remark 8.3 Compared with the former frequency-domain approach, the LKF-based one naturally facilitates to derive delay-dependent results from the time-domain perspective, since many existing delay-dependent results on EF H_∞ filtering [12, 13, 15, 16] are obtained by this method. Results presented in this chapter would be further improved by other time-domain techniques that have been proven to be effective when treating delayed systems, for instance, the introduction of higher order integral terms into the LKF as in [17]. However, results for this case are not easily explored by the former frequency-domain approach since it has been difficult to find such relations as in (8.19) directly from the frequency-domain perspective. This merit of the LKF-based method provides not only an easy-to-derive approach to the proof but a promising prospect for further improving the results.

Remark 8.4 If the filter state space realization is given, (8.26) or (8.48) is already an LMI-based condition ensuring the FF H_∞ filtering performance in (ii). However, if the filter parameters are to be determined, one cannot obtain the filter parameters straightly from (8.26) or (8.48) since the filter state space matrices are multiplied by two Hermitian matrix variables, i.e., P_1 and Q. Still for this reason, the routine change-of-variable method used in [12, 13, 15, 18] is not applicable for linearizing (8.26) for filter design. Although additional variables Y_1 in (8.8) and Y_2 in (8.9) do not contribute to reducing the conservatism inherent to results for time-delay systems, the feature that the products between the unknown matrices P_j, $j = 1, 2$, Q and the filter matrices are eliminated facilitates the derivation of filter synthesis results with no attention paid to P_j, $j = 1, 2$, Q anymore.

8.3.4 Delay-Independent FF BRL

Delay dependence of Theorem 8.1 is associated with the matrices S_j, $j = 1, 2$. Indeed, noting the LKF in (8.33) and (8.41), if we remove the terms $V_{j,2}(k)$, $j = 1, 2$, then $V_j(k)$, $j = 1, 2$ are the classic one introduced in [12] for deriving delay-independent results for the single delay case. It is noted that the delay-partitioning idea does not work any more in delay-independent results. Therefore, as a counterpart of the delay-independent EF H_∞ filtering results in [12], we give the following delay-independent FF BRL for analyzing the FF H_∞ performance of the filtering error system.

Corollary 8.1 *Consider the system in* (8.1) *and suppose that a scalar* $\gamma > 0$ *is given. For a given filter, the filtering error system in* (8.4) *is asymptotically stable with an FF* H_∞ *performance bound* γ *in the frequency range* $\mathbf{S}$*, if there exist matrices* $P_1 \in \mathbf{H}_{2n_x}$, $0 < P_2 \in \mathbf{H}_{2n_x}$, $0 < Q \in \mathbf{H}_{2n_x}$, $R_1 \in \mathbf{H}_{n_x}$, $0 < R_2 \in \mathbf{H}_{n_x}$ *and* $Y_j \in \mathbb{C}^{4n_x \times 2n_x}$, $j = 1, 2$ *such that*

$$\begin{bmatrix} -I_{n_z} & M \\ M^* & \tilde{\Sigma} + \text{sym}\left\{X_1^* Y_1 Z_1\right\} \end{bmatrix} < 0, \tag{8.49}$$

$$\tilde{\Xi}_2 + \text{sym}\left\{X_2^* Y_2 Z_2\right\} < 0, \tag{8.50}$$

where

$$\begin{aligned}
\tilde{\Sigma} &\triangleq \text{diag}\{\tilde{\Xi}_1 + \tilde{\Xi}_Q, -\gamma^2 I_{n_w}\}, \\
\tilde{\Xi}_j &\triangleq \begin{bmatrix} 0_{2n_x \times 2n_x} & P_j & 0_{2n_x \times n_x} \\ P_j & K^{\mathrm{T}} R_j K & 0_{2n_x \times n_x} \\ 0_{n_x \times 2n_x} & 0_{n_x \times 2n_x} & -R_j \end{bmatrix}, \ j = 1, 2, \\
\tilde{\Xi}_Q &\triangleq \begin{bmatrix} \Psi \otimes Q & 0_{4n_x \times n_x} \\ 0_{n_x \times 4n_x} & 0_{n_x \times n_x} \end{bmatrix},
\end{aligned}$$

and other notations are defined in Theorem 8.1 with $m = 1$.

8.4 Filter Design

By virtue of the obtained FF BRLs for H_∞ performance analysis, we focus on solving the FF H_∞ filter design problem in this section. Delay-dependent and delay-independent approaches will be investigated, respectively.

8.4.1 Delay-Dependent Approach

Based on Theorem 8.1, we have the following delay-dependent LMI condition for the existence of FF H_∞ filters for continuous time-delay systems in (8.1), which can be used to solve the problem of FF H_∞ filter design.

Theorem 8.2 *Consider the system in* (8.1) *and suppose that scalars* $\gamma > 0$, ε_j, $j = 1, 2, 3, 4$, *and an integer* $m > 0$ *and a scalar* $\tau > 0$ *satisfying* $d = m\tau$ *are given. A filter in the form* (8.3) *exists such that the filtering error system in* (8.4) *is asymptotically stable with an FF* H_∞ *performance bound* γ *in the frequency range* $\mathbf{S}$*, if there exist routine matrices* $\mathcal{Y}_{j,1} \in \mathbb{C}^{n_x \times n_x}$, $\mathcal{Y}_{j,2} \in \mathbb{C}^{n_x \times n_x}$, $\mathcal{Y}_{j,3} \in \mathbb{C}^{n_x \times n_x}$, $\mathcal{Y}_{j,4} \in \mathbb{C}^{n_x \times n_x}$, $\mathcal{Y}_5 \in \mathbb{C}^{n_x \times n_x}$, $\mathcal{A}_{\mathcal{F}} \in \mathbb{C}^{n_x \times n_x}$, $\mathcal{B}_{\mathcal{F}} \in \mathbb{C}^{n_x \times n_y}$, $\mathcal{C}_{\mathcal{F}} \in \mathbb{C}^{n_z \times n_x}$,

$\mathcal{D}_{\mathcal{F}} \in \mathbb{C}^{n_z \times n_y}$, *and Hermitian matrices* $P_1 = \begin{bmatrix} P_{1,1} & P_{1,2} \\ P_{1,2}^* & P_{1,3} \end{bmatrix} \in \mathbf{H}_{2n_x}$, $0 < P_2 = \begin{bmatrix} P_{2,1} & P_{2,2} \\ P_{2,2}^* & P_{2,3} \end{bmatrix} \in \mathbf{H}_{2n_x}$, $0 < Q = \begin{bmatrix} Q_{1,1} & Q_{1,2} \\ Q_{1,2}^* & Q_{1,3} \end{bmatrix} \in \mathbf{H}_{2n_x}$, $R_1 = \begin{bmatrix} R_{1,1} & R_{1,2} \\ R_{1,2}^* & R_{1,3} \end{bmatrix} \in \mathbf{H}_{mn_x}$, $0 < R_2 = \begin{bmatrix} R_{2,1} & R_{2,2} \\ R_{2,2}^* & R_{2,3} \end{bmatrix} \in \mathbf{H}_{mn_x}$, $0 < S_j \in \mathbf{H}_{n_x}$, $j = 1, 2$ *such that the following LMIs hold:*

$$\begin{bmatrix} -I_{n_z} & \mathcal{M} \\ \mathcal{M}^* & \Sigma + \mathrm{sym}\{\Upsilon_1\} \end{bmatrix} < 0, \tag{8.51}$$

$$\Xi_2 + \mathrm{sym}\{\Upsilon_2\} < 0, \tag{8.52}$$

where

$$\mathcal{M} \triangleq \begin{bmatrix} 0_{n_z \times n_x} & 0_{n_z \times n_x} & H - \mathcal{D}_{\mathcal{F}} C & -\mathcal{C}_{\mathcal{F}} & (H_d - \mathcal{D}_{\mathcal{F}} C_d) E_d & L - \mathcal{D}_{\mathcal{F}} D \end{bmatrix},$$

$$\Upsilon_1 \triangleq \begin{bmatrix} -\mathcal{Y}_{1,1} & -\mathcal{Y}_5 & \mathcal{Y}_{1,1} A + \mathcal{B}_{\mathcal{F}} C & \mathcal{A}_{\mathcal{F}} & (\mathcal{Y}_{1,1} A_d + \mathcal{B}_{\mathcal{F}} C_d) E_d & \mathcal{Y}_{1,1} B + \mathcal{B}_{\mathcal{F}} D \\ -\mathcal{Y}_{1,2} & -\mathcal{Y}_5 & \mathcal{Y}_{1,2} A + \mathcal{B}_{\mathcal{F}} C & \mathcal{A}_{\mathcal{F}} & (\mathcal{Y}_{1,2} A_d + \mathcal{B}_{\mathcal{F}} C_d) E_d & \mathcal{Y}_{1,2} B + \mathcal{B}_{\mathcal{F}} D \\ -\mathcal{Y}_{1,3} & -\varepsilon_1 \mathcal{Y}_5 & \mathcal{Y}_{1,3} A + \varepsilon_1 \mathcal{B}_{\mathcal{F}} C & \varepsilon_1 \mathcal{A}_{\mathcal{F}} & (\mathcal{Y}_{1,3} A_d + \varepsilon_1 \mathcal{B}_{\mathcal{F}} C_d) E_d & \mathcal{Y}_{1,3} B + \varepsilon_1 \mathcal{B}_{\mathcal{F}} D \\ -\mathcal{Y}_{1,4} & -\varepsilon_2 \mathcal{Y}_5 & \mathcal{Y}_{1,4} A + \varepsilon_2 \mathcal{B}_{\mathcal{F}} C & \varepsilon_2 \mathcal{A}_{\mathcal{F}} & (\mathcal{Y}_{1,4} A_d + \varepsilon_2 \mathcal{B}_{\mathcal{F}} C_d) E_d & \mathcal{Y}_{1,4} B + \varepsilon_2 \mathcal{B}_{\mathcal{F}} D \\ 0 & 0 & 0 & 0 & & \end{bmatrix},$$

$$\Upsilon_2 \triangleq \begin{bmatrix} -\mathcal{Y}_{2,1} & -\mathcal{Y}_5 & \mathcal{Y}_{2,1} A + \mathcal{B}_{\mathcal{F}} C & \mathcal{A}_{\mathcal{F}} & (\mathcal{Y}_{2,1} A_d + \mathcal{B}_{\mathcal{F}} C_d) E_d \\ -\mathcal{Y}_{2,2} & -\mathcal{Y}_5 & \mathcal{Y}_{2,2} A + \mathcal{B}_{\mathcal{F}} C & \mathcal{A}_{\mathcal{F}} & (\mathcal{Y}_{2,2} A_d + \mathcal{B}_{\mathcal{F}} C_d) E_d \\ -\mathcal{Y}_{2,3} & -\varepsilon_3 \mathcal{Y}_5 & \mathcal{Y}_{2,3} A + \varepsilon_3 \mathcal{B}_{\mathcal{F}} C & \varepsilon_3 \mathcal{A}_{\mathcal{F}} & (\mathcal{Y}_{2,3} A_d + \varepsilon_3 \mathcal{B}_{\mathcal{F}} C_d) E_d \\ -\mathcal{Y}_{2,4} & -\varepsilon_4 \mathcal{Y}_5 & \mathcal{Y}_{2,4} A + \varepsilon_4 \mathcal{B}_{\mathcal{F}} C & \varepsilon_4 \mathcal{A}_{\mathcal{F}} & (\mathcal{Y}_{2,4} A_d + \varepsilon_4 \mathcal{B}_{\mathcal{F}} C_d) E_d \\ 0 & 0 & 0 & 0 & 0 \end{bmatrix},$$

and Σ *and* E_d *are in* (8.8), Ξ_2 in (8.9). *Moreover, if the previous conditions are satisfied, an acceptable state space realization of the FF* H_∞ *filter is given by*

$$\begin{bmatrix} A_F & B_F \\ C_F & D_F \end{bmatrix} = \begin{bmatrix} \mathcal{Y}_5^{-1} & 0 \\ 0 & I \end{bmatrix} \begin{bmatrix} \mathcal{A}_{\mathcal{F}} & \mathcal{B}_{\mathcal{F}} \\ \mathcal{C}_{\mathcal{F}} & \mathcal{D}_{\mathcal{F}} \end{bmatrix}. \tag{8.53}$$

Proof Suppose that (8.51) and (8.52) hold for some $\mathcal{Y}_{j,i}$, $j = 1, 2$, $i = 1, 2, 3, 4$, $\mathcal{Y}_5$, $\mathcal{A}_{\mathcal{F}}$, $\mathcal{B}_{\mathcal{F}}$, $\mathcal{C}_{\mathcal{F}}$, $\mathcal{D}_{\mathcal{F}}$, $P_1 = P_1^*$, $P_2 = P_2^* > 0$, $Q = Q^* > 0$, $R_j = R_j^*$, $S_j = S_j^* > 0$, $j = 1, 2$. (8.52) implies $-\mathcal{Y}_5 - \mathcal{Y}_5^* < -P_{2,3} < 0$. Thus, $\mathcal{Y}_5$ is nonsingular, and by virtue of $\mathcal{Y}_5$, $\mathcal{A}_{\mathcal{F}}$, $\mathcal{B}_{\mathcal{F}}$, $\mathcal{C}_{\mathcal{F}}$ and $\mathcal{D}_{\mathcal{F}}$, the matrices A_F, B_F, C_F and D_F in (8.53) are well defined. Construct matrices Y_1 and Y_2 as follows:

$$Y_1 \triangleq \begin{bmatrix} \mathcal{Y}_{1,1} & \mathcal{Y}_5 \\ \mathcal{Y}_{1,2} & \mathcal{Y}_5 \\ \mathcal{Y}_{1,3} & \varepsilon_1\mathcal{Y}_5 \\ \mathcal{Y}_{1,4} & \varepsilon_2\mathcal{Y}_5 \end{bmatrix}, \quad Y_2 \triangleq \begin{bmatrix} \mathcal{Y}_{2,1} & \mathcal{Y}_5 \\ \mathcal{Y}_{2,2} & \mathcal{Y}_5 \\ \mathcal{Y}_{2,3} & \varepsilon_3\mathcal{Y}_5 \\ \mathcal{Y}_{2,4} & \varepsilon_4\mathcal{Y}_5 \end{bmatrix}, \tag{8.54}$$

and matrices A_F, B_F, C_F and D_F in (8.53). In view of the inequality in (8.4), matrices $X_1,\ Z_1,\ M$ in (8.8) and matrices $X_2,\ Z_2$ in (8.9), we have

$$\Upsilon_i = X_i^* Y_i Z_i,\ i = 1, 2,\ \mathcal{M} = M, \tag{8.55}$$

from which, one knows that (8.8) and (8.9) are satisfied for matrices $P_1 = P_1^*$, $P_2 = P_2^* > 0$, $Q = Q^* > 0$, $R_j = R_j^*$, $S_j = S_j^* > 0$, $j = 1, 2$ and matrices Y_j, $j = 1, 2$ defined in (8.54), A_F, B_F, C_F and D_F in (8.53). This means that (8.8) and (8.9) follow from (8.51) and (8.52), respectively. Thus, Theorem 8.2 can guarantee the asymptotic stability and an FF H_∞ performance level γ of the filtering error system. Moreover, the matrices A_F, B_F, C_F and D_F given by (8.53) form an acceptable state space realization of the filter. The proof is completed.

8.4.2 Delay-Independent Approach

Similarly, we have a delay-independent LMI condition for the existence of FF H_∞ filters. This delay-independent method for FF H_∞ filter design for continuous-time-delay systems can be obtained from Corollary 8.1, and also is a particular case of Theorem 8.2 by setting $S_j = 0$, $j = 1, 2$.

Corollary 8.2 *Consider the system in* (8.1) *and suppose that scalars* $\gamma > 0$ *and* ε_j, $j = 1, 2, 3, 4$ *are given. A filter in the form* (8.3) *exists such that the filtering error system in* (8.4) *is asymptotically stable with an FF* H_∞ *performance bound* γ *in the frequency range* **S**, *if there exist routine matrices* $\mathcal{Y}_{j,1} \in \mathbb{C}^{n_x\times n_x}$, $\mathcal{Y}_{j,2} \in \mathbb{C}^{n_x\times n_x}$, $\mathcal{Y}_{j,3} \in \mathbb{C}^{n_x\times n_x}$, $\mathcal{Y}_{j,4} \in \mathbb{C}^{n_x\times n_x}$, $j = 1, 2$, $\mathcal{Y}_5 \in \mathbb{C}^{n_x\times n_x}$, $\mathcal{A}_\mathcal{F} \in \mathbb{C}^{n_x\times n_x}$, $\mathcal{B}_\mathcal{F} \in \mathbb{C}^{n_x\times n_y}$, $\mathcal{C}_\mathcal{F} \in \mathbb{C}^{n_z\times n_x}$, $\mathcal{D}_\mathcal{F} \in \mathbb{C}^{n_z\times n_y}$ *and Hermitian matrices* $P_1 = \begin{bmatrix} P_{1,1} & P_{1,2} \\ P_{1,2}^* & P_{1,3} \end{bmatrix} \in \mathbf{H}_{2n_x}$, $0 < P_2 = \begin{bmatrix} P_{2,1} & P_{2,2} \\ P_{2,2}^* & P_{2,3} \end{bmatrix} \in \mathbf{H}_{2n_x}$, $0 < Q = \begin{bmatrix} Q_{1,1} & Q_{1,2} \\ Q_{1,2}^* & Q_{1,3} \end{bmatrix} \in \mathbf{H}_{2n_x}$, $R_1 \in \mathbf{H}_{n_x}$, $0 < R_2 \in \mathbf{H}_{n_x}$ *such that the following LMIs hold:*

$$\begin{bmatrix} -I_{n_z} & \mathcal{M} \\ \mathcal{M}^* & \tilde{\Sigma} + \mathrm{sym}\{\Upsilon_1\} \end{bmatrix} < 0, \tag{8.56}$$

$$\tilde{\Xi}_2 + \mathrm{sym}\{\Upsilon_2\} < 0, \tag{8.57}$$

where $\tilde{\Sigma}$ *and* $\tilde{\Xi}_2$ *are in* (8.49) *and* (8.50) *with* $m = 1$, *respectively. Moreover, if the previous conditions are satisfied, an acceptable state space realization of the FF* H_∞ *filter is given by* (8.53).

Remark 8.5 Note that there are several tuning parameters ε_j involved in the inequality conditions in (8.51), (8.52), (8.56) and (8.57). On one hand, if these tuning parameters are prescribed in advance, the inequality conditions in Theorem 8.2 and Corollary 8.2 are LMIs. On the other hand, smaller disturbance attenuation levels may be achieved by Theorem 8.2 and Corollary 8.2 through searching for some better combinations of these scalar parameters. For instance, the `fminsearch` function in MATLAB®[1] can be used for this purpose. The effectiveness of the proposed design methods will be illustrated by some numerical examples.

Remark 8.6 [11] shows that larger m generates larger allowable maximum time delay d. However, the number of decision variables would increase sharply, which means heavier computational burden. A reasonable trade-off between less conservatism and lighter computational burden is to let $m = 2$. As is shown in the example, $m = 2$ suffices to yield results with sufficiently low conservatism, outperforming the methods in the literature.

Based on Theorem 8.2 and Corollary 8.2, a delay-dependent approach, Algorithm 25, and a delay-independent approach, Algorithm 26, can be obtained, respectively, for the design of suboptimal FF H_∞ filters for continuous time-delay systems using the LMI technique.

Algorithm 25 Delay-Dependent Approach to FF H_∞ Filter Design

- Prescribe an integer $m \in \mathbb{N}^+$ and scalars ε_j, $j = 1, 2, 3, 4$;
- Solve the minimization problem:

$$\min_{\substack{P_1 = P_1^*,\, P_2 = P_2^* > 0,\, Q = Q^* > 0,\\ R_j = R_j^*,\, S_j = S_j^* > 0,\, j = 1, 2,\\ \mathcal{Y}_{j,i},\, j = 1, 2,\, i = 1, 2, 3, 4,\, \mathcal{Y}_5,\\ \mathcal{A}_\mathcal{F}, \mathcal{B}_\mathcal{F}, \mathcal{C}_\mathcal{F}, \mathcal{D}_\mathcal{F}, \mu}} \mu = \gamma^2 \quad \text{s.t. (8.51) and (8.52)}$$

- Compute the filter matrices from (8.53).

8.5 Examples

In this section, we give two illustrative examples to exhibit the effectiveness and advantages of the proposed FF H_∞ filter design methods.

Example 8.1 For the continuous-time state-delayed system model in (8.1), consider the following nominal matrix parameters borrowed from [16, 18] with delay d assumed to be time-invariant.

[1] MATLAB® is a registered trademark of The MathWorks, Inc.

Algorithm 26 Delay-Independent Approach to FF H_∞ Filter Design

- Prescribe scalars ε_j, $j = 1, 2, 3, 4$;
- Solve the minimization problem:

$$\min_{\substack{P_1 = P_1^*,\, P_2 = P_2^* > 0,\\ Q = Q^* > 0,\, R_j = R_j^*,\, j = 1,2,\\ \mathcal{Y}_{j,i},\, j = 1,2,\, i = 1,2,3,4,\, \mathcal{Y}_5,\\ \mathcal{A}_\mathcal{F},\, \mathcal{B}_\mathcal{F},\, \mathcal{C}_\mathcal{F},\, \mathcal{D}_\mathcal{F},\, \mu}} \mu = \gamma^2 \text{ s.t. (8.56) and (8.57)}$$

- Compute the filter matrices from (8.53).

$$\begin{aligned} A &= \begin{bmatrix} -2 & 0 \\ 0 & -0.7 \end{bmatrix}, \quad A_d = \begin{bmatrix} -1 & -1 \\ -1 & -1 \end{bmatrix}, \quad B = \begin{bmatrix} -0.5 \\ 2 \end{bmatrix}, \\ C &= \begin{bmatrix} 0 & 1 \end{bmatrix}, \quad C_d = \begin{bmatrix} 1 & 2 \end{bmatrix}, \quad D = 1, \\ H &= \begin{bmatrix} 2 & 1 \end{bmatrix}, \quad H_d = \begin{bmatrix} 0 & 0 \end{bmatrix}, \quad L = 0. \end{aligned} \tag{8.58}$$

Suppose that the FF ranges of the disturbance $w(t)$ are $\omega_l = 1.5$ rad/s, $\{\omega_1, \omega_2\} = \{2, 10\}$ rad/s and $\omega_h = 10$ rad/s, respectively. By Algorithm 25 in this chapter and the methods in [15, 16, 18], we design LTI full-order filters of the form in (8.3) and then compare the achieved minimum H_∞ performance level γ^* in the FF ranges. For brevity, we just specify the tuning parameters as $\varepsilon_1 = \varepsilon_2 = 5$, and $\varepsilon_3 = \varepsilon_4 = 1$ for the methods in this chapter and [16]. When $d = 0.06$ s, the guaranteed minimum H_∞ performance γ^* obtained by Algorithm 25 ($m = 2$) is 0.0460 for $\omega \le 1.5$ rad/s, 0.0681 for $2 \le \omega \le 10$ rad/s and 0.0387 for $\omega \ge 10$ rad/s, respectively, each of which is smaller than 0.0744 by Algorithm 25 ($Q = 0$, $m = 2$) and 0.0760 by the EF H_∞ filter design methods in [15, 16, 18].

The corresponding state space realizations of the filters generated by Algorithm 25 are

$$\text{LF}: \begin{bmatrix} A_F & B_F \\ C_F & D_F \end{bmatrix} = \left[\begin{array}{cc|c} -3.3045 & -0.0471 & 0.4323 \\ 0.5984 & -5.2810 & -1.7029 \\ \hline -1.9304 & -0.9527 & 0.0059 \end{array}\right], \tag{8.59}$$

$$\text{MF}: \begin{bmatrix} A_F & B_F \\ C_F & D_F \end{bmatrix} = \left[\begin{array}{cc|c} -2.5328 & 0.4793 & 0.5018 \\ -4.6936 & -8.5246 & -2.0672 \\ \hline -1.9288 & -0.9685 & 0.0016 \end{array}\right], \tag{8.60}$$

$$\text{HF}: \begin{bmatrix} A_F & B_F \\ C_F & D_F \end{bmatrix} = \left[\begin{array}{cc|c} -2.6561 & 0.4224 & 0.4961 \\ -0.8770 & -6.6515 & -2.0219 \\ \hline -1.3548 & -0.8131 & 0.0060 \end{array}\right], \tag{8.61}$$

$$\text{EF}: \begin{bmatrix} A_F & B_F \\ C_F & D_F \end{bmatrix} = \left[\begin{array}{cc|c} -2.4423 & 0.5103 & 0.5017 \\ -3.9456 & -8.1099 & -2.0596 \\ \hline -2.0000 & -1.0000 & 0 \end{array}\right]. \tag{8.62}$$

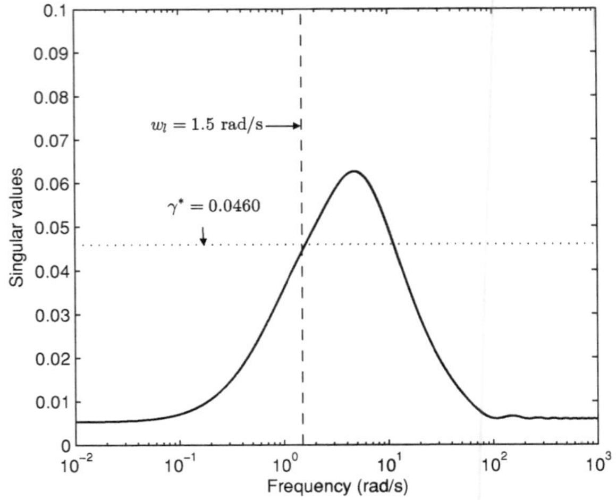

Fig. 8.1 Singular value curve with $w \leq 1.5$ rad/s and the filter in Eq. (8.59)

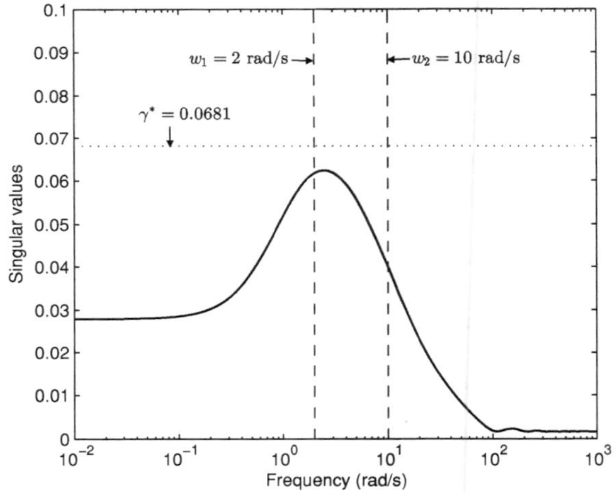

Fig. 8.2 Singular value curve with $2 \leq w \leq 10$ rad/s and the filter in Eq. (8.60)

To illustrate the effectiveness of the designed filters above, by connecting these filters to the original system in (8.58), we depict the singular value curves of the filtering error systems in Figs. 8.1–8.4. All the singular values in these figures are smaller than the achieved H_∞ filtering performance bound γ^* in the respectively considered frequency range, which demonstrates the effectiveness of the proposed delay-dependent FF H_∞ filter design method, Algorithm 25.

Table 8.1 includes detailed results for more FF ranges and different d. From the table, we have the following observations:

- The guaranteed minimum FF H_∞ performance level γ^* obviously depends on the delay d;

Fig. 8.3 Singular value curve with $w \geq 10$ rad/s and the filter in Eq. (8.61)

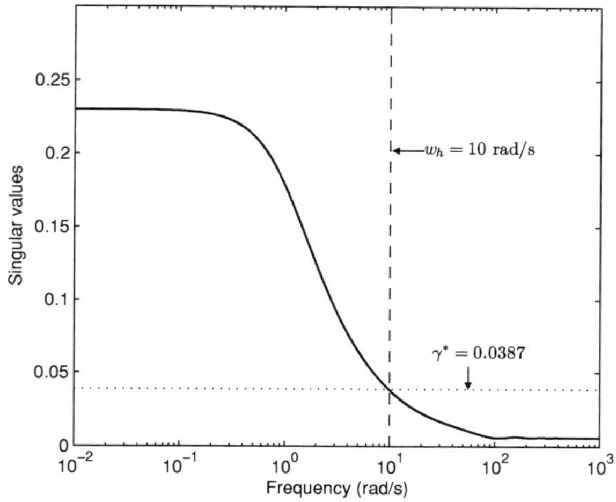

Fig. 8.4 Singular value curve for the EF range and the filter in Eq. (8.62)

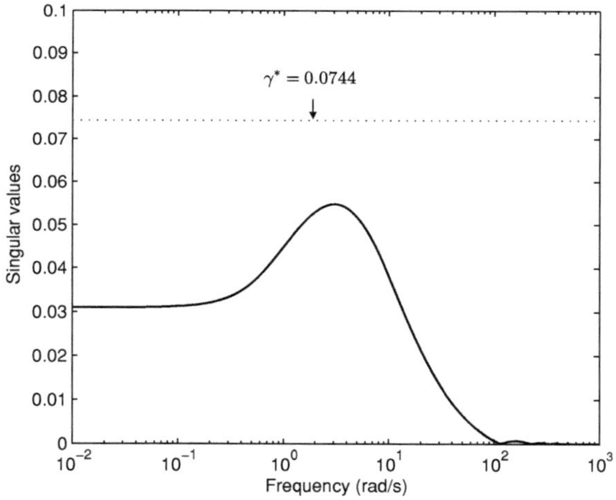

- If FF specifications are considered, the FF method in this chapter, Algorithm 25, yields much better results than that obtained by the EF methods in the literature;
- Even for the EF case, Algorithm 25 ($Q = 0$, $m = 2$) also outperforms the EF H_∞ filter design methods in [15, 16, 18]. Especially when d increases up to 1.2, Algorithm 25 is still able to find some feasible solutions while the methods in [15, 16, 18] fail.

To further illustrate the performance difference between the FF and EF H_∞ filters, we perform some time-domain simulations for the model in (8.58) with the LF filter in (8.59) and the EF filter in (8.62). Based on the FF specification $\omega_l = 1.5$ rad/s, we assume that system (8.58) is perturbed by a disturbance signal $w(t)$ of the sinusoidal function form as

Table 8.1 Achieved minimum γ^* for different delay d and FF ranges

Method	Frequency (rad/s)	d		
		0.06	0.6	1.2
[18]	$0 \leq \omega < \infty$	0.0760	1.7876	∞
[16]	$0 \leq \omega < \infty$	0.0760	1.7876	∞
Algorithm 25 ($Q = 0, m = 2$)	$0 \leq \omega < \infty$	0.0744	1.3240	7.8350
Algorithm 25 ($m = 2$)	$0 \leq \omega \leq 0.5$	0.0216	0.2162	0.4359
Algorithm 25 ($m = 2$)	$0 \leq \omega \leq 1.5$	0.0460	0.4875	2.2511
Algorithm 25 ($m = 2$)	$2 \leq \omega \leq 10$	0.0681	1.2965	2.4315
Algorithm 25 ($m = 2$)	$1 \leq \omega \leq 20$	0.0703	1.3184	7.6450
Algorithm 25 ($m = 2$)	$5 \leq \omega < \infty$	0.0571	0.4472	0.4472
Algorithm 25 ($m = 2$)	$10 \leq \omega < \infty$	0.0387	0.0872	0.0872

$$w(t) = \begin{cases} \sin 0.2t, & 0 \leq t \leq 100 \text{ s} \\ 0, & \text{otherwise} \end{cases} \tag{8.63}$$

The actual output $z(t)$ and its estimate $z_F(t)$ with these two filters are shown in Fig. 8.5 and the filtering error $z(t) - z_F(t)$ is depicted in Fig. 8.6. It is shown that the LF filter in (8.59) and the EF filter in (8.62) both can estimate the actual output well, and the filtering errors of the two filters converge to zero after $w(t)$ becomes 0. However, obviously, the LF filter has a better filtering performance than the EF one for the considered "FF" disturbance $w(t)$ in (8.63). Moreover, by calculation, we have $\|w\|_{L_2}^2 = 43.1226$ and

$$\text{LF:} \ \ \|e\|_{L_2}^2 = 0.0068, \ \frac{\|e\|_{L_2}}{\|w\|_{L_2}} = 0.0125 \leq 0.0460,$$

$$\text{EF:} \ \ \|e\|_{L_2}^2 = 0.0456, \ \frac{\|e\|_{L_2}}{\|w\|_{L_2}} = 0.0325 \leq 0.0760,$$

which further exhibits the effectiveness of the designed filters.

It should be pointed out that we attempt to design filters for the above example by the delay-independent method, Algorithm 26, but it doesn't work, which further shows the role of the delay-dependence of Algorithm 25. In the following, we provide another example to show that the delay-independent method is also useful.

Example 8.2 Consider another state-delayed system borrowed from [12] with a single delay as follows:

$$A = \begin{bmatrix} 0 & 3 \\ -4 & -5 \end{bmatrix}, \ A_d = \begin{bmatrix} -0.1 & 0 \\ 0.2 & -0.2 \end{bmatrix}, \ B = \begin{bmatrix} -0.5 \\ 1 \end{bmatrix},$$

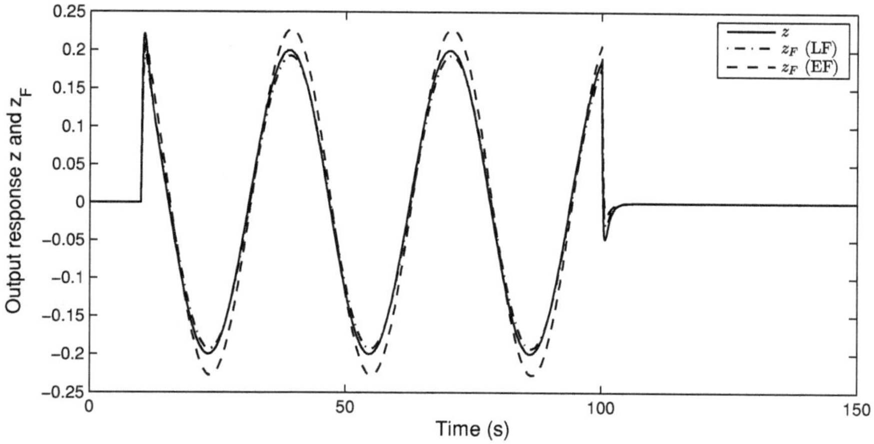

Fig. 8.5 Actual output and its estimation

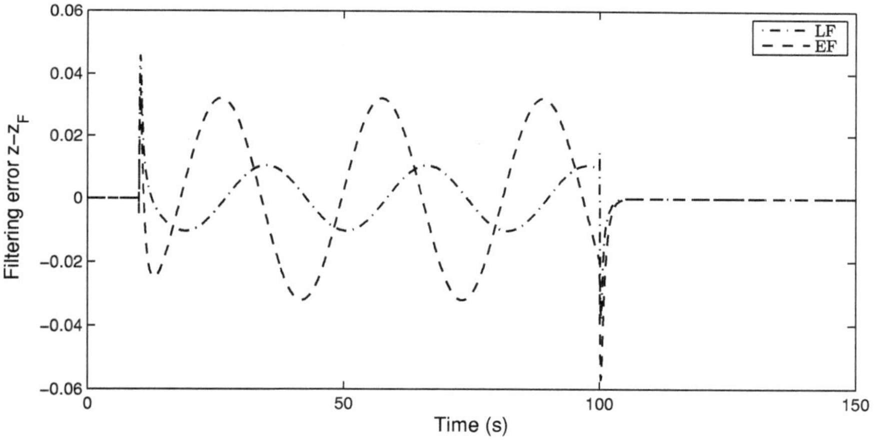

Fig. 8.6 Filtering error

$$
\begin{aligned}
C &= \begin{bmatrix} 0 & 100 \end{bmatrix}, \ C_d = \begin{bmatrix} 0 & 10 \end{bmatrix}, \ D = 1, \\
H &= \begin{bmatrix} 0 & 100 \end{bmatrix}, \ H_d = \begin{bmatrix} 0 & 0 \end{bmatrix}, \ L = 0,
\end{aligned} \tag{8.64}
$$

which is used to illustrate the potentials of the proposed delay-independent FF H_∞ filter design method, Algorithm 26.

For this example, we just specify the tuning parameters as $\varepsilon_1 = \varepsilon_2 = \varepsilon_3 = \varepsilon_4 = 1$. The achieved minimum H_∞ filtering performance level γ^* by Algorithm 26 is 1.9968 for $\omega \le 1.5$ rad/s, 2.2255 for $4 \le \omega \le 20$ rad/s and 1.9438 for $\omega \ge 5$ rad/s, respectively, and the corresponding state space realizations of the filters designed by Algorithm 26 are

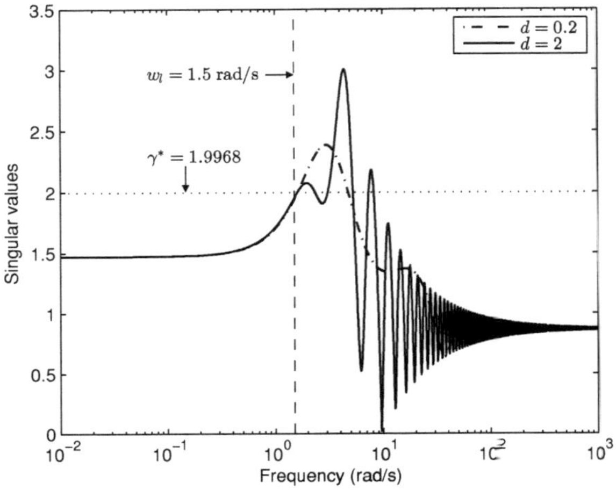

Fig. 8.7 Singular value curve with $0 \leq w \leq 1.5$ rad/s

$$\text{LF: } \begin{bmatrix} A_F & B_F \\ C_F & D_F \end{bmatrix} = \left[\begin{array}{cc|c} 0.1062 & 2.6492 & 0.0202 \\ -4.1805 & -5.6192 & -0.0531 \\ \hline -0.8012 & -8.8950 & 0.8691 \end{array}\right], \tag{8.65}$$

$$\text{MF: } \begin{bmatrix} A_F & B_F \\ C_F & D_F \end{bmatrix} = \left[\begin{array}{cc|c} 5.0289 & 6.0882 & 0.0300 \\ -13.4263 & -14.3758 & -0.0941 \\ \hline -8.1674 & -10.1091 & 0.8987 \end{array}\right], \tag{8.66}$$

$$\text{HF: } \begin{bmatrix} A_F & B_F \\ C_F & D_F \end{bmatrix} = \left[\begin{array}{cc|c} 8.4144 & 8.9996 & 0.0420 \\ -19.7437 & -19.6818 & -0.1155 \\ \hline -13.0214 & -14.4543 & 0.8782 \end{array}\right]. \tag{8.67}$$

Connect the filters in (8.65)–(8.67) to the original system in (8.64) and depict the corresponding singular value curves of the filtering error systems for delay $d = 0.2$ s and $d = 2$ s, which are shown in Figs. 8.7–8.9, respectively. All the singular values are smaller than the obtained minimum FF H_∞ filtering performance bound for both $d = 0.2$ s and $d = 2$ s, showing the effectiveness of the designed filters.

More detailed results for different frequency ranges are tabulated in Table 8.2, which includes the minimum $\gamma^* = 2.3758$ obtained by the delay-independent EF approach in [12]. From this table, one can clearly see that if FF specifications are taken into consideration, the delay-independent FF method in the chapter has an advantage over the EF method in [12]. □

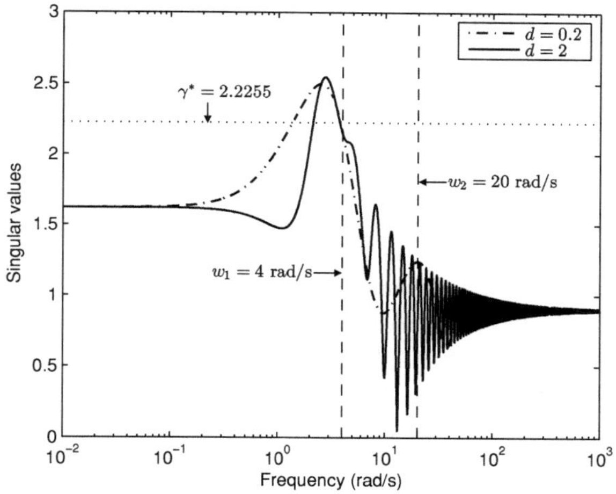

Fig. 8.8 Singular value curve with $4 \leq w \leq 20$ rad/s

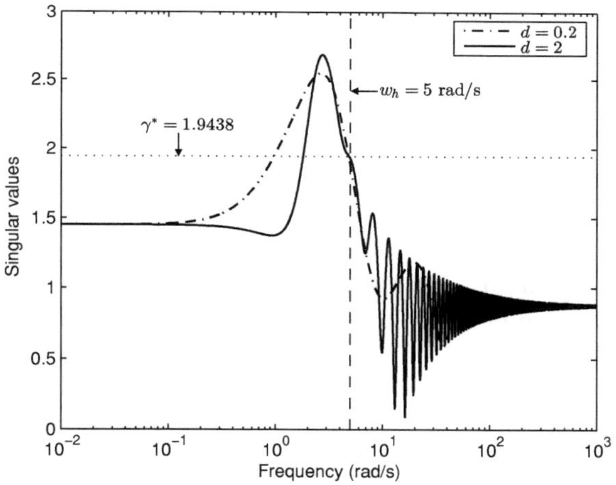

Fig. 8.9 Singular value curve with $5 \leq w < \infty$ rad/s

Table 8.2 Achieved minimum γ^* for different frequency ranges

Method	Frequency (rad/s)	γ^*
[12]	$0 \leq \omega < \infty$	2.3758
Algorithm 26	$0 \leq \omega \leq 0.5$	1.6398
Algorithm 26	$0 \leq \omega \leq 1.5$	1.9968
Algorithm 26	$0.2 \leq \omega \leq 2$	2.1961
Algorithm 26	$4 \leq \omega \leq 20$	2.2255
Algorithm 26	$5 \leq \omega < \infty$	1.9438
Algorithm 26	$50 \leq \omega < \infty$	0.2693

8.6 Summary and Notes

8.6.1 Summary

In the chapter, we have considered the problem of FF H_∞ filtering for continuous time-delay systems. The disturbance is assumed to reside in an LF/MF/HF range. By applying the GKYP Lemma and Projection Lemma, sufficient conditions have been proposed for analyzing and designing FF H_∞ filters for continuous-time state-delayed systems. A transfer function-based method and an LKF-based method are presented to prove the FF BRL from a frequency-domain point and a time-domain point of view, respectively. To reduce the conservatism of the obtained results, the delay-partitioning idea has been made use of. The proposed delay-dependent and -independent approaches cast the FF H_∞ design problem into solving a set of LMIs. Finally, the effectiveness and advantages of the proposed design methods have been demonstrated by two examples.

8.6.2 Notes

FF H_∞ filtering problems have been researched in [7–10, 19], where [7, 19] are concerned with linear systems using the standard GKYP Lemma [3], the authors in [8, 9, 20] investigate FF H_∞ filtering for 2-D systems in light of the 2-D GKYP Lemma [3], while the authors in [10] develop a delay-dependent FF BRL and apply it to the FF H_∞ filtering problem for discrete time-delay systems. Applications of the GKYP Lemma to solve other system and control problems with FF specifications can be found in [2, 22–24]. For all these results except [24], the Projection Lemma still plays an instrumental role in handling the strong coupling between Lyapunov matrices and system matrices in the GKYP Lemma for FF H_∞ filter design.

References

1. Richard, J.P.: Time-delay systems: an overview of some recent advances and open problems. Automatica **39**(10), 1667–1694 (2003)
2. Iwasaki, T., Hara, S.: Feedback control synthesis of multiple frequency domain specifications via generalized KYP lemma. Int. J. Robust Nonlinear Control **17**, 415–434 (2007)
3. Iwasaki, T., Hara, S.: Generalized KYP lemma: unified frequency domain inequalities with design applications. IEEE Trans. Autom. Control **50**, 41–59 (2005)
4. Iwasaki, T., Meinsma, G., Fu, M.: Generalized S-procedure and finite frequency KYP lemma. Math. Probl. Eng. **6**, 305–320 (2000)
5. Chen, Y., Zhang, W., Gao, H.: Finite frequency H_∞ control for building under earthquake excitation. Mechatronics **20**(1), 128–142 (2010)
6. Du, C., Xie, L., Guo, G., Teoh, J.N.: A generalized KYP lemma based approach for disturbance rejection in data storage systems. Automatica **43**, 2112–2118 (2007)

7. Wang, H., Yang, G.H.: A finite frequency approach to filter design for uncertain discrete-time systems. Int. J. Adaptive Control Signal Process. **22**, 533–550 (2008)
8. Li, X., Gao, H.: Robust finite frequency H_∞ filtering for uncertain 2-D Roesser systems. Automatica **48**(6), 1163–1170 (2012)
9. Li, X., Gao, H.: Robust finite frequency H_∞ filtering for uncertain 2-D systems: The FM model case. Automatica **49**(8), 2446–2452 (2013)
10. Gao, H., Li, X.: H_∞ filtering for discrete-time state-delayed systems with finite frequency specifications. IEEE Trans. Autom. Control **56**(12), 2935–2941 (2011)
11. Gouaisbaut, F., Peaucelle, D.: Delay-dependent stability analysis of linear time delay systems. In: IFAC Workshop on Time Delay Systems (TDS'06). Aquila, Italy (2006)
12. de Souza, C.E., Palhares, R.M., Peres, P.L.D.: Robust H_∞ filter design for uncertain linear systems with multiple time-varying state delays. IEEE Trans. Signal Process. **49**(3), 569–576 (2001)
13. Gao, H., Wang, C.: Robust L_2-L_∞ filtering for uncertain systems with multiple time-varying state delays. IEEE Trans. Circuits Syst. I, Fundam. Theory Appl. **50**(4), 594–599 (2003)
14. Zhou, K., Doyle, J.C., Glover, K.: Robust and Optimal Control. Prentice-Hall, NJ (1996)
15. Gao, H., Chen, T.: H_∞ estimation for uncertain systems with limited communiation capacity. IEEE Trans. Autom. Control **52**, 2070–2084 (2007)
16. Qiu, J., Feng, G., Yang, J.: A new design of delay-dependent robust H_∞ filtering for continuous-time polytopic systems with time-varying delay. Int. J. Robust Nonlinear Control **20**(3), 346–365 (2010)
17. Sun, J., Liu, G.P., Chen, J.: Delay-dependent stability and stabilization of neutral time-delay systems. Int. J. Robust Nonlinear Control **19**, 1364–1375 (2009)
18. Zhang, X.M., Han, Q.L.: Robust H_∞ filtering for a class of uncertain linear systems with time-varying delay. Automatica **44**(1), 157–166 (2008)
19. Ding, D., Yang, G.: Finite frequency H_∞ filtering for uncertain discrete-time switched linear systems. Prog. Nat. Sci. **19**, 1625–1633 (2009)
20. Li, X., Gao, H., Karimi, H.R.: Robust H_∞ filtering for 2-D FM systems: A finite frequency approach. In: 51st IEEE Conference on Decision and Control, pp. 3526–3530. Maui, Hawaii, USA (2012)
21. Yang, R., Xie, L., Zhang, C.: Generalized two-dimensional Kalman-Yakubovich-Popov lemma for discrete Roesser model. IEEE Trans. Circuits Syst. I, Regul. Pap. 55(10), 3223–3233 (2008)
22. Li, X., Gao, H., Wang, C.: Generalized Kalman-Yakubovich-Popov Lemma for 2-D FM LSS model. IEEE Trans. Autom. Control **57**(12), 3090–3103 (2012)
23. Li, X., Gao, H.: A heuristic approach to static output-feedback controller synthesis with restricted frequency-domain specifications. IEEE Trans. Autom. Control **59**(4), 1008–1014 (2014)
24. Li, X., Yu, C., Gao, H.: An iterative LMI approach to IIR noise transfer function optimization for Delta-Sigma modulators. In: 2013 Australian Control Conference, pp. 67–72. Perth, Australia (2013)

Chapter 9
Conclusion

This chapter will draw conclusion and summarize all the LMI algorithms for robust filter design presented in the book.

9.1 Conclusion

This book is concerned with robust filtering problems with polytopic uncertainty, and attention has been paid to the parameter-dependent approaches to robust filter design. Robust filtering for complicated systems (time-delay and 2-D systems) also have been investigated in the parameter-dependent framework, such that the conservatism of the presented design methods has been reduced. Moreover, some emerging problems have been addressed to reflect the new trends in the field of robust filtering. The technical contents of the book have been organized as three parts, seven chapters.

Part I has been concerned with quadratic approaches (Chap. 2) and parameter-dependent approaches (Chap. 3) to robust filter design for uncertain LTI systems with polytopic uncertainty entering the state space matrices. All the quadratic and parameter-dependent approaches in the two chapters have been formulated in terms of LMIs. Numerical examples have been presented to demonstrate the effectiveness of each design method.

- In Chap. 2, we have investigated the robust filtering problems by virtue of the quadratic approach. For each filtering scheme, the optimal filter design methods for nominal systems have been derived first, and then extended to deal with robust filtering problems based on the quadratic stability notion in robust control theory. On one hand, due to the convex combination nature of the LMI formulation, the quadratic approaches suffice to guarantee the existence of robust filters through placing the same robust filtering performance constraint on each vertices of the convex bounded uncertainty domain. On the other hand, since a single quadratic Lyapunov function has been employed to handle uncertainty, the quadratic approaches are too conservative for some specific problems and leave much room

H. Gao and X. Li, *Robust Filtering for Uncertain Systems*,
Communications and Control Engineering, DOI: 10.1007/978-3-319-05903-7_9,

for further improvement. Three filtering schemes have been addressed in this chapter, namely, (robust) H_2, H_∞ and energy-to-peak filtering. H_2 filters assume Gaussian white noises, and robust H_2 filters have improved the robustness against parametric uncertainty in systems; H_∞ and energy-to-peak filters assume energy-bounded noises, so that they can deal with both Gaussian and non-Gaussian noises, more robust than H_2 filters in dealing with uncertainty in signals.

- In Chap. 3, parameter-dependent approaches have been derived for reducing the conservatism of quadratic approaches in robust H_2 and H_∞ filtering. By virtue of the Projection Lemma, dilated conditions have been obtained, and slack matrices have been introduced to eliminate the product terms between Lyapunov matrices and system matrices. Linearly parameter-dependent approaches to robust filter design have been derived by imposing the Lyapunov matrices to be linearly parameter-dependent and fixing the slack matrices to be constant, respectively. Polynomially parameter-dependent approaches have been developed by rendering the Lyapunov matrices and slack matrices both to be homogeneous polynomials of the uncertainty parameters with the same degree. Numerical examples have clearly shown the improvement of parameter-dependent approaches in this chapter.

Part II has been focused on the robust filtering problems for time-delay systems (Chaps. 4 and 5) and 2-D systems (Chap. 6). Parameter-dependent sufficient conditions have been obtained in terms of LMIs for the existence of robust H_∞ filters for these two types of complicated systems. Numerical examples have been provided to illustrate the effectiveness of the presented robust filtering methods in each chapter of this part.

- In Chap. 4, we have studied the robust H_∞ filtering problem for uncertain continuous-time delayed systems with polytopic uncertainty and a single delay in the state. By constructing an LKF and applying the free weighting matrix technique, a parameter-dependent LMI condition first has been derived, under which the H_∞ filtering performance can be guaranteed. Then, by performing some change of variables, a parameter-dependent LMI condition for the existence of robust H_∞ filters has been obtained, which, incorporated with the polynomially parameter-dependent idea, finally gives rise to a polynomially parameter-dependent approach to designing robust H_∞ filters for uncertain continuous time-delay systems. The presented design method also depends on the size of the delay bound.
- In Chap. 5, we have addressed the discrete-time counterpart of the same problem of Chap. 4. Based on a discrete-time LKF, a sufficient parameter-dependent LMI condition has been established for the H_∞ filtering performance. To decouple the product of the Lyapunov matrices and system matrices, slack matrices then have been introduced to derive a new performance analysis condition, and further to obtain an existence condition of robust H_∞ filters. By applying the polynomially parameter-dependent idea, a parameter-dependent approach in terms of LMIs has been finally developed for designing robust H_∞ filters for uncertain discrete time-delay systems. Like the method in Chap. 4, the one in this chapter is also delay-dependent.

- In Chap. 6, the robust H_∞ filtering problems of polytopic uncertain 2-D systems have been addressed. By introducing slack matrices into the existing 2-D BRLs, new LMI conditions have been established for analyzing the H_∞ performance of 2-D systems. Then by performing some change of variables and using the polynomially parameter-dependent idea, parameter-dependent approaches have been derived for robust H_∞ filter design for uncertain 2-D systems. The commonly used FM model and Roesser model of discrete-time 2-D systems have been considered, respectively.

Part III has addressed two robust filtering problems with new specifications, namely, the NCS environment with limited communication capacity (Chap. 7) and time-delay systems with FF specifications (Chap. 8), respectively. These new specifications reflect some practical requirements in filtering, so that the filtering methods in this chapter can be applied to more realistic situations.

- In Chap. 7, the problem of H_∞ filtering for NCSs with limited communication capacity has been dealt with. First, a delay system model with norm-bounded uncertainty and two successive delay components has been established for describing the considered filtering problem over a network with limited communication capacity, where three typical phenomena in the NCS environment, that is, measurement quantization, data package dropout, and transmission delay, are successively handled in the same framework. Then, by utilizing the Lyapunov-Krasovskii method, delay-dependent sufficient conditions in terms of LMIs have been proposed for filtering performance analysis and filter design, respectively. Furthermore, the design method has been extended to cope with polytopic uncertainty by virtue of the quadratic approach and the linearly parameter-dependent approach, respectively. Finally, we have exploited two numerical examples to demonstrate the effectiveness of the proposed filter design method.
- In Chap. 8, the problem of FF H_∞ filtering for continuous-time systems with a constant state delay has been investigated. The FF specification is motivated by the fact that many practical signals have their energy within an FF range. By combining the GKYP Lemma and the Projection Lemma, delay-dependent and delay-independent FF BRLs have been developed for FF H_∞ filtering performance analysis. Two proofs for the FF BRLs have been provided from a frequency-domain perspective and a time-domain perspective, respectively. Correspondingly, delay-dependent and delay-independent approaches in terms of LMIs have been proposed for designing filters with a guaranteed FF H_∞ performance. For the delay-dependent approach, the delay partitioning idea has been applied to reduce conservatism. At last, we have employed two numerical examples to illustrate the effectiveness and advantages of the filter design methods developed in the chapter.

Table 9.1 Algorithms (optimal approaches, quadratic approaches and parameter-dependent approaches)

Algorithm #	CT/DT[a]	Index	Method	Scalar variables	Rows of matrix inequalities
Algorithm 1	CT	H_2	Optimal	$2n_x^2+0.5n_z^2+n_xn_y+n_xn_z+n_x+0.5n_z+1$[b]	$4n_x+2n_z+n_w$[b]
Algorithm 2/4	CT/DT	H_2	Optimal	$2n_x^2+0.5n_z^2+n_xn_y+n_x+0.5n_z+1$	$5n_x+2n_z+n_w$
Algorithm 3	DT	H_2	Optimal	$2n_x^2+0.5n_z^2+n_xn_y+n_xn_z+n_x+0.5n_z+1$	$6n_x+2n_z+n_w$
Algorithm 5	CT	H_2	Quadratic	$2n_x^2+0.5n_z^2+n_xn_y+n_xn_z+n_x+0.5n_z+1$	$4sn_x+(s+1)n_z+sn_w$[b]
Algorithm 5	DT	H_2	Quadratic	$2n_x^2+0.5n_z^2+n_xn_y+n_xn_z+n_x+0.5n_z+1$	$6sn_x+(s+1)n_z+sn_w$
Algorithm 6	CT	H_2	Quadratic	$2n_x^2+0.5n_z^2+n_xn_y+n_x+0.5n_z+1$	$(2s+3)n_x+2n_z+sn_w$
Algorithm 6	DT	H_2	Quadratic	$2n_x^2+0.5n_z^2+n_xn_y+n_x+0.5n_z+1$	$(4s+1)n_x+2n_z+sn_w$
Algorithm 7	CT/DT	H_∞	Optimal	$2n_x^2+n_xn_y+n_xn_z+n_yn_z+n_x+1$	$4n_x+n_z+n_w$
Algorithm 8	CT	H_∞	Quadratic	$2n_x^2+n_xn_y+n_xn_z+n_yn_z+n_x+1$	$(2s+2)n_x+sn_z+sn_w$
Algorithm 8	DT	H_∞	Quadratic	$2n_x^2+n_xn_y+n_xn_z+n_yn_z+n_x+1$	$4sn_x+sn_z+sn_w$
Algorithm 9	CT	EP[a]	Optimal	$2n_x^2+n_xn_y+n_xn_z+n_x+1$	$4n_x+n_z+n_w$
Algorithm 9	DT	EP	Optimal	$2n_x^2+n_xn_y+n_xn_z+n_x+1$	$6n_x+n_z+n_w$
Algorithm 10	CT/DT	EP	Optimal	$2n_x^2+n_xn_y+n_x+1$	$5n_x+n_z+n_w$
Algorithm 11	CT	EP	Quadratic	$2n_x^2+n_xn_y+n_xn_z+n_x+1$	$4sn_x+sn_z+sn_w$
Algorithm 11	DT	EP	Quadratic	$2n_x^2+n_xn_y+n_xn_z+n_x+1$	$6sn_x+sn_z+sn_w$
Algorithm 12	CT	EP	Quadratic	$2n_x^2+n_xn_y+n_x+1$	$(2s+3)n_x+n_z+sn_w$
Algorithm 12	DT	EP	Quadratic	$2n_x^2+n_xn_y+n_x+1$	$(4s+1)n_x+n_z+sn_w$
Algorithm 13	CT/DT	H_2	LPD[a]	$(2s+6)n_x^2+0.5sn_z^2+n_xn_y+n_xn_z+n_xn_w$ $+sn_x+0.5sn_z+1$	$s(6n_x+2n_z+n_w)$
Algorithm 14	CT/DT	H_∞	LPD	$(2s+6)n_x^2+n_xn_y+n_xn_z+n_xn_w+n_yn_z+sn_x+1$	$s(6n_x+n_z+n_w)$
Algorithm 15	CT/DT	H_2	PPD[a]	$(6J(s,g)+2)n_x^2+0.5J(s,g)n_z^2+n_xn_y+$ $n_xn_z+J(s,g)(n_xn_w+n_x+0.5n_z)+1$[c]	$J(s,g+1)(6n_x+2n_z+n_w)$
Algorithm 16	CT/DT	H_∞	PPD	$(6J(s,g)+2)n_x^2+n_xn_y+n_xn_z+n_yn_z$ $+J(s,g)(n_xn_w+n_x)+1$	$J(s,g+1)(6n_x+n_z+n_w)$

[a] *CT* continuous-time, *DT* discrete-time, *EP* energy-to-peak, *LPD* linearly parameter-dependent, *PPD* polynomially parameter-dependent

[b] n_x system order, n_z estimated signal dimension, n_y measured signal dimension, n_w disturbance dimension, s vertex number of uncertainty domain

[c] g is the degree of the polynomial, and $J(s,g)=\frac{(s+g-1)!}{g!(s-1)!}$

Table 9.2 Algorithms (time-delay systems, 2-D systems, NCSs and FF)

Algorithm #	System	Index	Method	Scalar variables	Rows of matrix inequalities
Algorithm 17	CT TDS[a]	H_∞	PPD+DD[b]	$(4.5J(s,g)+1.5)n_x^2+n_xn_y+n_xn_z+n_yn_z+(1.5J(s,g)+0.5)n_x+1$[c]	$(7J(s,g+1)+2J(s,g))n_x+J(s,g+1)(n_z+n_w)$[c]
Algorithm 18	DT TDS[a]	H_∞	PPD+DD	$(8J(s,g)+2)n_x^2+n_xn_y+n_xn_z+n_yn_z+2J(s,g)n_x+1$	$(7J(s,g+1)+4J(s,g))n_x+J(s,g+1)(n_z+n_w)$
Algorithm 19	DT TDS	H_∞	PPD+DI[b]	$(5.5J(s,g)+2)n_x^2+n_xn_y+n_xn_z+n_yn_z+1.5J(s,g)n_x+1$	$(5J(s,g+1)+3J(s,g))n_x+J(s,g+1)(n_z+n_w)$
Algorithm 20	FM	H_∞	PPD	$(6J(s,g)+3)n_x^2+2n_xn_y+n_xn_z+2J(s,g)n_x+1$	$J(s,g+1)(6n_x+n_z+n_w)$
Algorithm 21	Roesser	H_∞	PPD	$2J(s,g)n_1^2+2J(s,g)n_2^2+(2J(s,g)+2)n^2+J(s,g)n_1+J(s,g)n_2+nn_y+nn_z+1$[d]	$J(s,g+1)(4n+n_z+n_w)$
Algorithm 22	NCSs	H_∞	Nominal	$11.5n_x^2+n_x(n_y+n_z+2n_w+2n_v)+n_yn_z+2.5n_x+2$[e]	$10n_x+n_y+n_z+n_w+n_v$
Algorithm 23	NCSs	H_∞	Quadratic	$11.5n_x^2+n_x(n_y+n_z+2n_w+2n_v)+n_yn_z+2.5n_x+2$	$(8s+2)n_x+s(n_y+n_z+n_w+n_v)$
Algorithm 24	NCSs	H_∞	LPD[b]	$(10s+1.5)n_x^2+n_yn_z+(2s+0.5)n_x+2+n_x(n_y+n_z+2sn_w+2sn_v)$	$0.5s(s+1)(8n_x+n_y+n_z+n_w+n_v)+(s+1)n_x$
Algorithm 25	CT TDS	FF H_∞	DD	$(17+m^2)n_x^2+n_xn_y+n_xn_z+n_yn_z+(4+m)n_x+1$[f]	$(14+3m)n_x+n_z+n_w$
Algorithm 26	CT TDS	FF H_∞	DI	$17n_x^2+n_xn_y+n_xn_z+n_yn_z+4n_x+1$	$15n_x+n_z+n_w$

[a] *CT* continuous-time, *DT* discrete-time, *TDS* time-delay systems

[b] *PPD* polynomially parameter-dependent, *LPD* linearly parameter-dependent, *DD* delay-dependent, *DI* delay-independent

[c] n_x system order, n_z estimated signal dimension, n_y signal dimension, n_w disturbance dimension, s vertex number of uncertainty domain, g degree of the polynomial, and $J(s,g)=\frac{(s+g-1)!}{g!(s-1)!}$

[d] n_1 horizontal state dimension, n_2 vertical state dimension, and $n=n_1+n_2$

[e] n_v measurement noise dimension

[f] m delay partitioning number

9.2 Algorithms

For reader's convenience, the algorithms presented in this book are summarized in Tables 9.1 and 9.2, where the information includes the system class each algorithm applies to, the employed index, the underlying key method, the number of scalar variables, and the number of rows of the matrix inequalities for each algorithm. All the listed algorithms recast the (robust) H_2, H_∞ or energy-to-peak filtering problems into convex programming problems subject to LMI constraints.

Index

H. Gao and X. Li, *Robust Filtering for Uncertain Systems*, Communications and Control Engineering, DOI: 10.1007/978-3-319-05903-7,

MIX
Papier aus verantwortungsvollen Quellen
Paper from responsible sources
FSC® C105338

If you have any concerns about our products,
you can contact us on
ProductSafety@springernature.com

In case Publisher is established outside the EU,
the EU authorized representative is:
Springer Nature Customer Service Center GmbH
Europaplatz 3, 69115 Heidelberg, Germany

Printed by Libri Plureos GmbH
in Hamburg, Germany